Motion and Time Study

for Lean Manufacturing

Third Edition

Fred E. Meyers

James R. Stewart
NORTHERN ILLINOIS UNIVERSITY

Prentice Hall

Upper Saddle River, New Jersey
Columbus, Ohio

Library of Congress Cataloging-in-Publication Data

Meyers, Fred E.

    Motion and time study for lean manufacturing / Fred E. Meyers, James R. Stewart.—3rd ed.

       p. cm.

    ISBN 0-13-031670-9

    1. Motion study.   2. Time study.   3. Methods engineering. I. Stewart, James R. (James Robert), II. Title

T60.7 .M48 2002

658.5′42—dc21   00-068468

Vice President and Editor in Chief: Stephen Helba
Executive Editor: Debbie Yarnell
Associate Editor: Michelle Churma
Production Editor: Louise N. Sette
Production Supervision: Clarinda Publication Services
Design Coordinator: Robin G. Chukes
Cover Designer: Tom Mack
Cover Art: © Visual Edge Imaging
Production Manager: Brian Fox
Marketing Manager: Jimmy Stephens

This book was set in Times Roman by The Clarinda Company and was printed and bound by R. R. Donnelley & Sons Company. The cover was printed by Phoenix Color Corp.

Prentice-Hall International (UK) Limited, *London*
Prentice-Hall of Australia Pty. Limited, *Sydney*
Prentice-Hall Canada, Inc., *Toronto*
Prentice-Hall Hispanoamericana, S.A., *Mexico*
Prentice-Hall of India Private Limited, *New Delhi*
Prentice-Hall of Japan, Inc., *Tokyo*
Prentice-Hall Singapore Pte. Ltd
Editora Prentice-Hall do Brasil, Ltda., *Rio de Janeiro*

10 9 8 7 6 5

ISBN: 0-13-031670-9

Dedicated to my wife, Mary,
our children, and grandchildren

Fred E. Meyers

Dedicated to those who set the crafts to work
And to those who labor performing their usual vocations
And finding time for deity and the distressed
Using the tools of the craftsman

James R. Stewart

Preface

With the publication of the third edition, a new author, James R. Stewart, has been added to the team. Certainly this does not represent any change in the unique, down-to-earth teaching method of senior author and scholar Fred Meyers. Having been a user and supporter of this book since it was noted at a NAIT seminar a decade ago, the junior author intends to keep it as a text for the teaching of basic courses in using the tools of motion study and time study. Although the book has undergone a number of changes, it provides a practical education in the basic principles of motion and time study.

Over the past decade, lean manufacturing has become the philosophy for manufacturers who want to use the tools for improving operations. We see the concept as so important that a new Chapter 2 has been devoted to it. We describe it as a lean manufacturing environment. As an environment, it nurtures and supports many types of improvement systems and methods. And, it is open to the concepts of various cultures and methodologies. Because it sets the tone for all that follows, the chapter has been placed after the overview chapter and before the history chapter.

A number of persons important to work measurement have been added to the history chapter, Chapter 3. These include pioneers Henry Gantt and Harrington Emerson, early text authors, consultants, and educators Ralph Barnes and Marven Mundel, as well as Shigeo Shingo, the early Toyota pioneer of what is becoming the lean manufacturing environment. Although we considered many others, the lessons of the lives of these pioneers provide the guidance to install the lean manufacturing environment and to continue the innovation and improvement of our manufacturing system.

Chapter 4 has pulled together the introductory material scattered over several chapters. It now provides a concise but complete summary of what is to follow. Chapter 5 is directed toward teaching process charting and process improvement. Chapter 6 adds the SIMO chart to the other tools of operations analysis. Chapter 7 includes a new section on ergonomics and on the environmental aspects of the workplace. Predetermined Time Standards (PTS) Systems, Chapter 8, includes new descriptions of two commercial systems to the usable system that Fred Meyers designed. Time study has been left as developed in prior editions. Chapter 10 still includes the description of standard data but also has the calculation of line balancing and the concepts of lean manufacturing environment plant balancing. All of the material previously spread through three chapters is consolidated here. In Chapter 11, the principles of work sampling have been augmented by a new auditing procedure and form. A process for scheduling and measuring work performance supplements the indirect labor types described in Chapter 12. The last four chapters have been kept as written, as have the appendices.

References, sample problems, worked examples, tests, and other supplemental materials are published separately in the teacher's supplement. We hope students and teachers will find that the many changes in this edition add to their ability to learn and use the tools of motion and time study in the lean environment.

Acknowledgments

For the third edition, we wish to thank Rodha Balamuralikrishna for his figures. We also thank reviewers Kenneth Currie of Tennessee Technological University and Donna C. S. Summers, Ph. D., of the University of Dayton.

Fred E. Meyers
James R. Stewart

Preface to the Second Edition

The purpose of this how-to motion and time study book is to provide students and practitioners with a resource that describes the techniques and procedures of motion and time study. This book has appropriately been called a "cookbook." Practical, detailed advice is given on all aspects of motion and time study, including work station design, job analysis, and the techniques of setting time standards.

The mathematics requirement of this textbook is high school algebra. A few simple formulas are included in the standard data chapter. These formulas require the insertion of a variable to calculate the time requirement. Two more complicated formulas are used to show how tables are developed. The practitioner can use the tables to save time.

Motion study is accomplished before time standards are set. When a company decides to introduce a new product, a technician is asked to provide a plan to produce, for example, 1,500 units per day. The technician must design work stations for every fabrication, assembly, and packout operation. From the work station drawing, a left-hand/right-hand analysis of the work content is made. A predetermined time standard has been set for every body motion, so the times for every motion required to do the job are added together. This will be the time standard, and it was set before the company had the first part, machine, or operator.

Modern management requires constant vigilance of its industrial engineers and technicians to reduce costs, reduce effort, and improve the working environment. Lean manufacturing (the Toyota production system) has a word, *muda*, which means waste. More specifically, any activity that uses resources but does not add value is *muda*. Lean thinking is one solution to *muda*. Lean thinking promotes using less effort by

1. Eliminating useless motions,
2. Combining motions,
3. Changing the sequence of motions to make flow smoother, and
4. Simplifying motions.

Lean thinking results in the elimination or reduction of waste.

Motion and time study has finally found a home in the modern plant by helping employees to

1. Understand the nature and true costs of work,
2. Assist management in reducing unnecessary costs, and
3. Balance work cells to make work flow more smoothly.

Motion and time study has also contributed the concept of time standards, so that important management decisions can be made intelligently. Motion and time study can

1. Reduce and control costs,
2. Improve working conditions and environment, and
3. Motivate employees.

Manufacturing plant management needs time standards. Many major decisions would be only a guess without time standards. How would we determine how many machines to buy, how many people to hire, how much to sell the product for; how would we schedule the plant, how would we justify new methods or equipment, how would we ensure a balanced work load on assembly lines, and how would we evaluate employees or pay for increased effort? Chapter 4 answers these questions and inspires an appreciation of the importance of motion and time study.

This book will equip engineers and managers with the purposes, attitudes, methods, and techniques of motion and time study to make their plants leaner. Chapters 5 through 8 discuss methods analysis techniques.

Stopwatch time study can be accomplished only after the machines have been installed and the operators fully trained. In a proposed new plant, no machines or employees are available to time study, but an operating plant can use stopwatch time study very effectively. The stopwatch technique is the oldest technique of setting time standards and it is entrenched in many companies. Chapter 9 examines this technique.

Standard data is another technique of setting time standards before production begins, but it is developed from in-plant experience. Standard data is very personal to a specific company, and companies cannot normally use another's standard data. This is the most accurate, least costly method of setting time standards, and every motion and time study department should be developing its own. Chapter 10 examines this technique.

Work sampling is based on the laws of probability and it is a scientific technique of setting quality time standards. Office work, engineering departments, maintenance craft, and even equipment can be work sampled. Everyone who has worked with others has work sampled. Chapter 11 discusses this common practice in a scientific light. Consultants often use work sampling first to establish the beginning efficiency of the operation. Potential savings forecasts will be based on current efficiency.

Labor is a significant portion of manufacturing cost and must be controlled. Performance control systems based on time standards give management the control they need. History and research have shown that operations working without a performance control system average 60% of normal. When a performance control system is established, 85% performance results. Industrial plants on incentive average 120% performance. The size of these cost reductions is spectacular, and no industrial engineer, technologist, or manager will go unnoticed when such improvements are made. Chapter 13 discusses performance control systems.

Chapter 12 discusses uses of time standards that could be an important part of a technologist's career. Wage payment (Chapter 14) includes incentive systems, which is a fun area in which technologists can work. Assembly line balancing (Chapter 10) includes instructions on setting up assembly lines. This is a big area in many plants.

Most of the textbook deals with direct factory labor, but every area of business can be positively influenced by motion and time study. Chapter 12 discusses 10 of the largest indirect labor categories.

Chapter 15, the time management techniques chapter, is aimed at making the motion and time study technologist more productive.

Human relationships are an important part of motion and time study. The successful attitudes and goals of motion and time study technologists are discussed in Chapter 16.

A step-by-step procedure, real-life examples, sample problems, and blank forms are included for every technique. This book will remain a good reference years after a course or seminar on motion and time study.

Your feedback will be valued and considered for future editions. You may write to me in care of Prentice Hall, One Lake Street, Upper Saddle River, New Jersey 07458. Our objective is to provide a practical, usable, how-to text for motion and time study.

Acknowledgments

I am a student of Ralph Barnes, Peter August, and Mitchell Fein. All three have influenced me greatly, and their attitudes are part of me. I must thank Dr. Matthew P. Stevens, of Purdue University, for his statistical expertise and help, and Dr. Richard Edwards, of the University of Kentucky, for his continuing support and encouragement.

I also thank the reviewers of the second edition for their helpful comments: S. Deivanayagam, Tennessee Technical University; Alfred R. Hamelin, Walters State Community College; Matthew R. Meyer, Asheville-Buncombe Technical College; and Ross Udey, Peru State College.

Fred E. Meyers

About the Authors

Fred E. Meyers is president of Fred Meyers and Associates, an industrial engineering management consulting company. He designs and implements production improvement and motivation systems. Mr. Meyers is a registered professional industrial engineer and a senior member of the Institute of Industrial Engineers. He has 35 years of industrial engineering experience. He has worked for Caterpillar Tractor Co., Boeing's aerospace division, Mattel Toy Co., Times Mirror Corp., Ingersoll-Rand's proto tool division, Spaulding's golf club division, and Southern Illinois University–Carbondale, College of Engineering, where he taught for 20 years while starting and operating his consulting business. He was director of applied research and an associate professor.

Mr. Meyers has worked for over 100 companies as a consultant responsible for installing incentive systems, performance control systems, plant layouts, new product startup, and cost estimating systems. He has worked in heavy equipment manufacturing, aerospace, consumer products, appliance manufacturing, lumber, plywood, paper, oil blending and packaging, furniture, tooling, fiberglass, and many other areas. The variety of his assignments has given him the ability to see the wide-ranging uses of motion and time study.

Fred E. Meyers has taught motion and time study to over 130 classes and 5,000 people, including professional engineers and managers, union stewards, and college students. He has presented seminars to the National Association of Industrial Technology, industrial plants, the U.S. Air Force and Navy, and labor unions.

James R. Stewart is Associate Professor of Technology at Northern Illinois University. For the past decade, he has taught plant layout, engineering economy, manufacturing philosophy, production and inventory systems, industrial quality control, ergonomics, and work measurement and improvement. He is a Fellow in the World Academy of Productivity Science. He is a senior member of the Institute of Industrial Engineers and is a founding member and is on the board of directors of the Society for Work Science. He is also on the board of The International MODAPTS Association. He is an active member of a number of other societies, including NAIT, ASQ, and Human Factors and Ergonomics Society. He has 30 years of experience in work measurement in education, government, and industry.

Dr. Stewart has served on the faculty of several universities; worked in city, county and state productivity programs; and has managed engineering programs in electronics assembly, electronics component fabrication, pulp and paper fabrication, fiberglass processing, industrial tape manufacturing, and engineering consulting. He has published many articles about unique applications of work measurement. James R. Stewart has taught motion and time study in credit and noncredit courses for over 25 years. He has been certified and taught a number of predetermined time systems, including MOST, Work Factor, MTM-1, and MODAPTS.

Contents

Introduction to Motion and Time Study

A new vocabulary has developed in the past decade that stems from the Toyota production system and a book titled *Lean Thinking* by James Womack and Daniel Jones. Lean manufacturing is a concept whereby all production people work together to eliminate waste. Industrial engineering, industrial technologists, and other groups within management have been attempting this by themselves since the beginning of the Industrial Revolution, but now that we have a well-educated, motivated production work force, modern manufacturing management has discovered the advantage of seeking the work force's help in eliminating waste. The Japanese have a word for waste, *muda,* which is the focus of much attention all over the world. Who knows better than the production employee, who spends eight hours a day on a job, how to reduce waste? The goal is to tap this resource by giving production employees the best tools available, and the techniques within a motion and time study course are some of the tools they need to do their new job.

Applications of lean manufacturing have begun to appear in American industry. These have been documented in a book called *Becoming Lean: Inside Stories of U.S. Manufacturers,* edited by Jeffrey Liker, which describes the process and problems of installing and maintaining lean manufacturing. However, much of the force behind the lean manufacturing environment is attributable to Shigeo Shingo and his landmark work in industrial engineering at Toyota. Firmly based on the works of the pioneers of motion study and methods improvement, he has provided a challenge to improve our methods, our operations, and our systems. Indeed, systems do not have to be confined to the walls of a plant nor of a company—the entire response to the needs of an individual can be improved: cross-cutural, cross-national, cross-company, one efficient system for the manufacture of goods and services, all designed, manufactured, and delivered in the lean environment. All utilizing—no, all requiring—the tools and approaches presented in this book.

Motion and time study helps employees understand the nature and true costs of work, and it helps them assist management in reducing unnecessary costs and balancing work cells to make work flow smoother. In addition, time standards help managers make important management decisions intelligently. For example, manufacturing plant management needs time standards, even before production starts, to determine how many people to hire, how many machines to buy, how fast to move conveyors, how to divide work among employees, and how much the product will cost; and, after production starts, to determine how much cost reductions will return, who works the hardest, and perhaps who should earn more money. Motion and time study can reduce and control costs, improve working conditions and environment, and motivate people. This book will equip engineers, technologists, and managers with the purposes, attitudes, methods, and techniques of motion and time study to make plants leaner and, in turn, to train the work force in the techniques of motion and time study.

Upon completion of a course in motion and time study, you will have an appreciation of the techniques that measure and control costs, the confidence to apply them to your organization, and a desire to involve all employees in your company's cost reduction and cost control efforts.

Motion and time study is purely the study of techniques. There are about 25 techniques that assist in the study and measurement of work, and these techniques are the main subject of this text. These techniques are constantly improving, but their basic purpose is to improve the world of work and to reduce *muda* (waste). This book examines the techniques in the following general categories:

1. Motion analysis techniques
2. Time study techniques
3. Uses of time standards.

Manufacturing management and engineering students are being prepared to design work stations, develop efficient and effective work methods, establish time standards, balance assembly lines, estimate labor costs, develop effective tooling, select proper equipment, and lay out manufacturing facilities. However, the most important thing to learn is how to train production workers in these skills and techniques so they can become motion and time conscious.

A person working with motion and time study will study an individual job or series of jobs to learn the details of that work and make changes. Changes may be small, but improvements must be made continually to keep the company competitive. Without change, no growth occurs and failure is imminent. A company must never stop looking for improvements or it will become obsolete. A company that can involve all its employees in this effort toward improvement will have a competitive advantage that will lead to a larger market share.

Very few industries have new technology that is exclusively theirs. However, industries have something that is more important than exclusive technology: employees who understand that improvement comes only by hard work and attention to detail. There is no easy way.

Breaking down a job into its smallest components and putting it back together again using motion study techniques will result in an improvement. A motion and time study person will have the following attitudes:

"We can reduce the cost of any job."

"Cost is our measuring rod."

"Cost reduction is our job."

American industry must continue to deliver quality products at a reasonable price. Quality and price are the most important considerations for staying competitive. Motion and time study people concentrate on reducing costs but must never lose sight of quality. The following attitudes are critical:

"We never propose a method that will reduce quality."

"We never set standards for producing scrap."

"Lower cost and high quality are our competitive edge. One without the other leads to failure."

"Work smarter, not harder" has been the motto of every industrial engineer, manager, and technologist, but the new motto should be "Work smarter *and* harder."

Motion study offers a great potential for savings in any area of human effort. We can save the total cost of an element of work by *eliminating* it. We can greatly reduce the cost by *combining* elements of one task with elements of another task. We can *rearrange* the elements of a task to make it easier. We can also *simplify* the task by moving parts and tools close to the point of use, or pre-positioning parts and tools, or providing mechanical assists, or downgrading elements of work to less time-consuming elements; and we can even have the part redesigned to make it easier to produce. Simplification is the most time-consuming way of reducing costs, and its savings are small compared to eliminating or combining elements, but we can always simplify. These subjects and techniques are called work simplification or the cost reduction formula, and we will discuss them in great detail (see Chapters 5 through 8).

Motion study uses the principles of motion economy to develop work stations that are friendly to the human body and efficient in their operation. The field of ergonomics studies the effect of motions on the human body and has become an extremely important part of developing work methods. Ergonomics is a complex subject and should be a course or even a field of study of its own. This textbook cannot do it justice and still cover all the other areas of motion and time study, but people who want to make motion and time study, job design, or any other area of manufacturing management or engineering their chosen career field should take as many courses in ergonomics as are offered. This text will study the principles of motion economy, which just touches the surface of the larger field of ergonomics. People who design work stations must be aware of the impact their designs will have on people's lives. The designers can make work harder than it needs to be if they do not pay attention to the principles of motion economy and the field of ergonomics.

Motion study has improved the quality of work life beyond belief. If we could go back 50 to 100 years and look at what work used to be like, we would see no resemblance to work today. Material-handling devices have taken the drudgery out of work, and other machines have taken away the physical effort required of our ancestors and replaced it with what the human does best: think, solve problems, take corrective action, and use the senses to monitor operations. One of my favorite terms for modern manufacturing workers is *machine or work center managers*. Motion study has made work safer and easier than ever before, and we are just getting started. You can help by applying the principles of motion study and learning as much as possible about ergonomics. Today, we are pulling machines together into cells to produce one or a few complicated parts in mass. The machine or work center manager operates and is responsible for five to ten machines. This is what a supervisor or work leader used to do.

Motion study must consider the operator's safety above all else. No one wants to be responsible for getting someone hurt or producing injury due to long-term exposure to a condition or environment. The only way you can minimize the chance of designing bad work stations is to learn all you can about safe and effective design. Work center designers need to be the company's expert on safety, ergonomics, and motion economy principles.

Motion and time study must also be aware of product quality. We do not want to recommend any changes that will negatively affect quality. Quality control efforts can also be the subject of motion study. We can make any operation more effective, and quality control is no different.

Time study can reduce cost significantly as well. Time standards are goals to strive for. In organizations that operate without time standards, 60% performance is typical. This statistic can be proven by work sampling that operation. When time standards are set, performance improves to an average of 85%. This is a 42% increase in performance:

$$\frac{85\% - 60\%}{60\%} = 42\% \text{ productivity increase.}$$

Incentive systems can improve performance even further. Incentive system performances average 120%, another 42% increase in productivity:

$$\frac{120\% - 85\%}{85\%} = 42\% \text{ productivity increase.}$$

1. Manufacturing plants with no standards average 60% performance.
2. Manufacturing plants with time standards average 85% performance.
3. Manufacturing plants with incentive systems average 120% performance.

If additional production output is required, don't buy more machinery, don't add a second shift, don't build a new plant: Just establish a motion and time study program. A description of the uses of time study follows in Chapter 4.

Motion and time study is a lot of work and creates some labor/management conflict, but by bringing labor into the process of motion and time study, conflict can be replaced with cooperation and a feeling of being part of something important.

It is said that successful people do what other people do not want to do:

1. Work long
2. Work hard
3. Criticize
4. Accept criticism
5. Become involved

Motion and time study techniques require all of these, and the person who adopts these habits is likely to reap great responsibility and rewards. As stated previously, our motto is "Work smarter *and* harder." Motion and time study requires both.

Motion and time study is considered to be the backbone of industrial engineering, industrial technology, and industrial management programs because the information that time studies generate affects so many other areas, including the following:

1. Cost estimating
2. Production and inventory control
3. Plant layout
4. Materials and processes
5. Quality
6. Safety.

Motion and time study creates a cost consciousness that is desired of every manufacturing employee, and cost conscious employees have a competitive advantage. It is said that an engineer or a manager who does not know the economic ramifications of his or her decisions is worthless to industry.

Motion study precedes the setting of time standards. An industrial engineer's time would be wasted setting time standards for poorly designed jobs. Cost reduction via motion study is automatic and can be significant. Motion study is a detailed analysis of the work method in an effort to improve it. Motion studies are used to

1. Develop the best work method.
2. Develop motion consciousness on the part of all employees.
3. Develop economical and efficient tools, fixtures, and production aids.
4. Assist in the selection of new machines and equipment.
5. Train new employees in the preferred method.
6. Reduce effort and cost.

Motion study is for cost reduction, and time study is for cost control. Motion study is the creative activity of motion and time study. Motion study is design, while time study is measurement.

This book focuses on techniques. Once the importance of motion and time study is understood and accepted, the techniques of motion and time study are introduced. Technique is the *how* of motion and time study. The following techniques of motion study are covered in this book:

1. Process charts
2. Flow diagrams
3. Multiactivity charts
4. Operation charts
5. Flow process charts
6. Operations analysis chart
7. Work station design
8. Motion economy
9. Flow patterns
10. Predetermined time standards system (PTSS).

The techniques of time study start with the last motion study technique, which shows the close relationship between motion study and time study. The techniques of time study are as follows:

1. Predetermined time standards system (PTSS)
2. Stopwatch time study
3. Standard data formula time standards
4. Work sampling time standards
5. Expert opinion and historical data time standards.

Each of these techniques is covered in Chapter 4 and again in great detail in a later chapter. *How-to* is the core of this book; this is a techniques book. Each technique includes a step-by-step procedure for using this industrial technology tool. Each technique also includes a completed example and a problem to solve. At the back of the book are blank forms to be used and reproduced as needed. You have the author's permission to copy these forms for your or your company's use. The heading of the forms can be easily changed to your company name.

QUESTIONS

1. What is lean thinking?
2. What is the difference between process and method?

3. What are the two most important considerations for a company to stay competitive?

4. What are three good attitudes of a motion and time study analysis?

5. What should be the motto of industry?

6. What are four ways motion study can reduce cost? What is the most important way?

7. At what percentage of performance do manufacturing plants operate in the following situations?

   a. No time standards _____

   b. With time standards _____

   c. With incentive standards _____

8. What are the five things successful people do that unsuccessful people do not want to do?

Motion and Time Study for the Lean Environment

This chapter discusses the nature of the lean production environment and its basic concepts and describes some of the techniques utilized in attaining it. The philosophy of Shigeo Shingo and its application to motion and time study tools is the basis for the lean environment approach. We include examples of the tools of motion and time study as they apply to achieving this environment. Specifically, the chapter covers

1. The nature of the lean environment,
2. The implementation of the environment,
3. The use of motion study tools to improve the environment,
4. The use of time study tools to maintain the environment,
5. Measures of performance within the lean environment.

THE NATURE OF THE LEAN ENVIRONMENT

The term *lean production environment* was coined by James Womack to differentiate the practices he observed in Japan at Toyota from the practices of mass production. Focusing on eliminating all forms of waste in processes, the lean production environment has several objectives that are in direct opposition to mass production orthodoxy, to the concepts of line and staff organizational structure that supported the large-order-quantity system, and, indeed, to the very nature of leadership and management philosophy.

The nature of the lean environment demands a leadership approach that implements a search for leanness throughout the organization. Vision, culture, and strategy are integrated, says Womack, into customer service with high quality, low cost, and short delivery times. The concept of a lean environment requires taking an aggressive management approach to finding ways of improving performance. It involves the entire work force and utilizes concepts from methods and time study, quality and process control, and many other areas that belong to separate management functions in the mass production system. Although many of these tools have been available for decades, the application at appropriate times and for appropriate reasons is what establishes the lean environment.

The Toyota Experience

Toyota has always been a Japanese company that kept close tabs on American practices. Toyota executives visited Ford when the River Rouge Plant was the envy of the world. Taylor and Gilbreth both made trips to Japan. The concepts of mass production worked well in Japan until the end of World War II, but the war necessarily left one nation, the United States, with a rapidly growing economy and the other with a loss of facilities and market.

As in the American mass production system, the Japanese manufacturing system paid lip service to quality but, when given the choice, would side with keeping the line going. Quality experts Joe Juran and W. Edwards Deming made extensive postwar trips to Japan. But the welcome mat was also out for the motion and time experts, most notably Marvin Mundel. The Japanese listened and adapted. People like Yasuhiro Monden, Taiichi Ohno, Kaoru Ishikawa, and Shigeo Shingo learned from them a combination that would work for Toyota.

There are numerous books about the Toyota production system and how it developed written from many points of view and showing the unique experiences of successful corporate adaptation to a new environment. In the 1960s, Toyota was struggling. Auto lots had unsold and undriven cars as much as three years old. Inventory was choking the company. It is small wonder that the company discovered the muda of excess inventory. It was stacked around them in raw materials, work in process, and finished goods. Quality improvement was necessary to convince the U.S. and world markets that the oil-miser cars, ignored by the American companies, could be reliable when manufactured in Japan. But quality improvement could not solve many of Toyota's problems.

Toyota needed to make in one factory what American automobile companies made in many different factories. They could not set the line for one model and run all the orders, then change over to another model. They needed to schedule the line to make any combination of cars and trucks. Size, color, interiors—all needed to be in the right place at the right time and, importantly, at the right cost. Thus, the problem was apparent, but the solution was clearly not mass production in the traditional way. The Second World War had changed that.

American industry, which obviously supplied the tools and philosophy for the Toyota production system, took a decade to realize what was occurring in Japan even as it became successful in NUMMI (New United Motor Manufacturing), an early and very successful joint venture by Toyota that worked by rewriting labor relations history in a unionized plant known for mistrust of management. Later, Toyota successfully made the system work in Georgetown, Kentucky, a plant of nonunionized employees with little prior mass manufacturing experience. The proof was there that leadership and effort could succeed.

America Responds

Throughout the eighties, American automakers revisited and restructured the concepts of customer satisfaction, quality, productivity, and scheduling to find the Japanese secret. Jeffrey Liker states the situation succinctly: "For some reason, which I do not claim to understand, it took us at least a decade from the early 1980s to the early 1990s to figure out that there was more to Japanese manufacturing than just techniques. . . ."* In 1997, the Big Three found their new jargon. After a number of years of testing model lines and studying the concepts, Chrysler, Ford, and General Motors declared that Toyota had the finest production system and that they were transforming their production systems to versions of the Toyota production system. Executive after executive made stirring speeches about the new lean manufacturing at the University of Michigan Lean Manufacturing conference in Dearborn. To which Jim Womack stated, "So we are all Toyota guys now. What does that mean?"**

The Implementation of the Environment

The lean environment requires an aggressive attack on waste, called *muda* in Japanese. There are muda everywhere: The material that is waiting to be processed, the labor waiting for tools, the raw materials that were not inspected and may be defective, the wrong paint pigments for the unit, even the form that requires extra clerical efforts are muda. And the customer getting less or more than he is happy with is a form of muda. In addition, the negative impact that a product may have on society in the form of preplanned replacement, pollution, and unforeseen disposal costs or risks are forms of muda. There are many other forms of waste that can be found at work, play, or school.

Lean thinking is the antidotal approach promoted by Womack and the exponents of lean manufacturing systems. It includes

1. Specifying value as action steps,
2. Sequencing value-created actions,
3. Interruption-proof sequences,
4. Demand rather than supply sequences,
5. Ever more effective performance through learning.

The result of this approach has been phenomenal in a number of disciplined, world-class organizations. Each company had to find its own mix of tools, but in analyzing such organizations, Liker found that there were four major success factors apparent in companies that became lean. These are

1. *Preparing and motivating people.* It is important that people who will be expected to step out in new roles have both the tools necessary to succeed, the motivation to complete

---

* Jeffrey K. Liker, ed. *Becoming Lean: Inside Stories of US Manufacturers* (Cambridge, Mass: Productivity Press, 1998.)

** James P. Womack and Daniel T. Jones. *Lean Thinking* (New York: Simon & Schuster, 1996.)

the work, and the leadership to maintain focus. While the tools require knowledge and understanding of application, the motivational experience requires a deeper understanding and commitment. Figure 2-1 shows the process of gaining such commitment. Remember, if this looks too easy, at each level the student undergoes a progressively deeper fundamental transformation in beliefs about production systems. This chapter and text only touch on the first level shown in the figure.

a. Measurement and feedback are powerful motivators for change.

b. Crisis will motivate, but it is better to act before it hits.

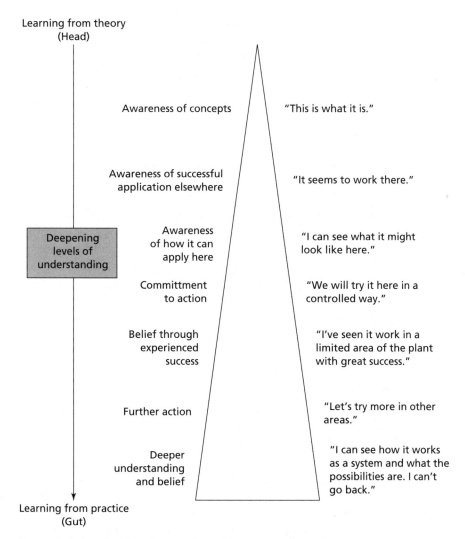

FIGURE 2-1 Levels of understanding of lean manufacturing.

    c. Focused training may be necessary in traditionally hostile workplaces.

    d. Learning by doing is the practical approach. Success leads to bigger successes.

2. *Roles in the change process*

    a. Lean requires creative engineers. It is not enough to motivate novices to lean concepts and expect them to struggle. Competent resources must be provided.

    b. Hourly employees need to influence the shop floor. Outside experts may have preconceived notions of what is needed, and their ideas may not be best for the process and the employees who live with it.

    c. External coaches in lean practices are needed. Usually the plant has almost no expertise in coaching, change, and lean vision. Someone, either hard or soft, depending on the workplace, must provide that day-to-day assistance.

    d. Leadership. A high level of commitment, understanding, and participation is required in all successful lean production facilities.

3. *Methodologies for change*

    a. Flow: The most important parameter of the system. Nearly every improvement is directly reflected in the flow through the plant, which then indirectly affects the plant costs.

    b. System vision: The view of the lean organization must begin with the supplier and continue through to the customer's successful use of the product. Although implemented incrementally, each change must contribute to overall customer needs.

    c. Model line: It is often possible to set up one operation and gain rapid successes there. These achievements reward workers and provide incentive to expand the implementation.

    d. Blitz Kaizens: Although they focus on individual accomplishments rather than system-level changes, this tool is useful at the beginning to provide resolution of a crisis and rewards for creative ideas and rapid implementation.

4. *Environment for change*

    a. Trust: It is earned through actions not words. Problems of trust have usually occurred for many years before the introduction of lean production, and trust will be regained only through fair treatment of employees.

    b. Guiding principles: Written guidelines are required for places where work is done according to the rulebook. They are a useful way of committing the organization's support to the participating employees and may be necessary where trust is low.

    c. Job security: One major commitment, formally or informally, given by most companies is that no employees will lose jobs due to lean production systems. According to

Liker, this commitment has been kept with hourly employees. Managers who actively opposed the system were a different matter.

The Toyota system evolved out of the practical need of a company to survive to become a model for how any company could survive under difficult circumstances. Each of the tools developed to improve Toyota manufacturing could be implemented and installed, in any company. But although groups of tools, or even all of them, could be used by another firm, the effect would not be the same as at Toyota. Although some observers would attribute Toyota's massive improvements to some sort of symbiosis, Shingo attributes them to the application of traditional motion and time study problem-solving techniques. And, it must be noted, these techniques were applied as needed to obvious bottlenecks.

If the unique application of tools to particular problems is the real secret to Toyota production, or even lean manufacturing, as Shingo claimed, then history is repeating itself. Shingo was certainly familiar with the pioneers of Scientific Management, the Gilbreths and Taylor. Certainly, Taylor made just such claims about his systems. He called people who adopted his ideas piecemeal "jackals" and "wolves," because they were tearing off small techniques and tools and leaving the main, analytical part unutilized.

Analysis and Improvement of Systems

Shigeo Shingo, the Japanese industrial engineer of whom Norman Bodek, president of Productivity Inc., says, "I am certain that the name of Shigeo Shingo will rank with those of Henry Ford, Fredrick Taylor, Eli Whitney . . . and others." We agree. His legacy is that a change-responsive manufacturing system can be designed and that the traditional tools of industrial engineering can be used to evaluate and aggressively improve the lean environment.

In addition to recognizing the differences between the outcomes of a methods-focused system and a time recording system in motion study, Shingo recognized and emphasized a difference in level. That is, that process analysis is an important candidate for improvement and learning, but that operations analysis (called by the authors *activity analysis*) also has potential for labor and cost savings. Note that the Japanese definition of *process* is a continuous flow by which raw materials are converted into finished goods. *Operation* is any action performed by man, machine, or equipment on raw materials or intermediate or finished products. Shingo saw production as a network of operations and processes. We reserve the use of the word *operation* for one of the phases of process traditionally called Operation, Transportation, Inspection, and Delay. Substitution of the word *activity* fits the definition used by such classic work measurement authors as Barnes and Mundel.

Although our wording is slightly different, we have no quarrel with the philosophy, approach, and case study method of Shingo and recommend his books. Further, we present one of Dr. Shingo's final thoughts for your consideration:

Why is it that Gilbreth's ideas about making work more rational attracted so little attention in his own country? Even in Japan (as of 1990), I have the impression that the idea of efficient thinking is thought to be out of date, a not particularly highly regarded relic of classical industrial engineering.

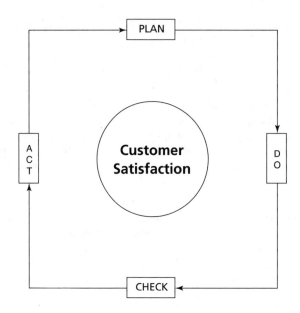

FIGURE 2-2 The PDCA chart.

Yet I believe this particular element of Gilbreth's thinking to be the mother of improvement and progress. Why do people think so little of it? Efficient thinking is the true parent of a number of creative production systems such as:

Single minute exchange of die (SMED).

Source inspection and mistake proofing poka-yoke devices.

Drastic lead time reductions.

Non-stock production systems.*

Many of Shingo's ideas represented a major change in thinking. Often, this type of change is called a paradigm shift. It opens entirely new challenges and opportunities to the creative marketing, management, and engineering staff (particularly product and process staff). Tools for brainstorming and the categorization of information are useful for gathering and sorting solutions to company problems.

Continuous improvement (PDCA) is one way to improve systems. (See Figure 2-2.) It has many names in the quality field and has been called the Ishikawa, Deming, Juran, or even Shewhart cycle. As described by Deming and adopted as a methodology by Juran, it consists of action by steps—first planning, then doing, then checking or studying the results, and finally acting to modify the plan for the next round. As a result of successful improvement—what has been known as Juran's trilogy (see Figure 2-3)—movement from a controlled state to an out-of-control state to an improved state will occur. Continuous improvement is impor-

---

* Shigeo Shingo. *The Shingo Production Management System: Improving Process Functions* (Cambridge, Mass: Productivity Press, 1992, p. 188.)

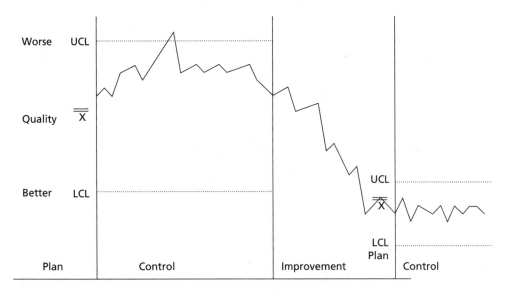

FIGURE 2-3 Juran's trilogy.

tant for achieving growth toward goals even when the specific approach has yet to be developed and as a resource for tapping the intelligence and creativity of the work force.

Some improvement ideas require coordination and scheduling. Project management tools are often used to plan such activities. The Gantt chart (Figure 2-4), dating back to the Scientific Management pioneers, is the primary tool. In addition, many network tools are available. Mundel developed a system for assigning a level of difficulty to individual projects based on the number of different specialists one would need to team with. They include

1. Change work instruction	Industrial Engineer
2. Change method	Industrial, Manufacturing Engineer
3. Change process	Industrial, Manufacturing, Process Engineer
4. Change raw material	Industrial, Manufacturing, Process Engineer, Buyer
5. Change product	Industrial, Manufacturing, Process Engineer, Buyer, Marketing*

Projects will, of course, require more resources and have greater risk if more organizational units are involved. Rewards must be evaluated also.

---

*Adapted from Marvin E. Mundel and David L. Danner. *Motion and Time Study: Improving Productivity* (Englewood Cliffs: Prentice Hall, 1994, Chapter 3.)

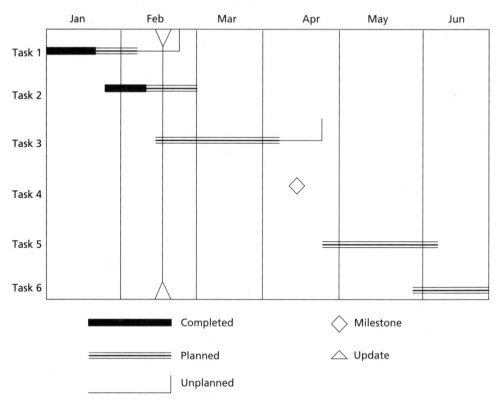

FIGURE 2-4 Gantt chart.

APPROACHES TO QUALITY MEASUREMENT AND IMPROVEMENT

Quality in Lean Systems

The changes in the manufacturing approach to quality involve philosophy, tools, and people. In the context of lean manufacturing, a stable (controlled) quality environment is required before other improvements can be implemented. Very low nonconformities provide the stable manufacturing environment required for lean manufacturing lot sizes and scheduling practices. The resultant environment will then permit improvements in operations and activities.

Total Quality Management

A necessary component of lean manufacturing is a functioning quality system. The pull production management process cannot exist without a superior level of product quality, and one of the major benefits of lean manufacturing is the lack of traditional quality costs. However, quality improvement efforts come in many forms. Total Quality Management (TQM), a term originally coined to represent organizational efforts to improve quality performance, as opposed to departmental efforts to reduce shipment of defects through inspection or efforts to measure process stability, has become an umbrella phrase for all types of corporate initiatives. Fundamental to the concept of total quality in the lean environment is the development of the continuous quality improvement company goal. Many of the ideas developed at Toyota and included in the Japanese manufacturing system are commonly thought of as a part of TQM. These are employee involvement, quality circles, suggestion systems, 100% screening, and team concepts.

Quality Systems Auditing

The concepts of lean manufacturing are closely connected to the concept of supply chain management. Toyota began the relationships with suppliers that permitted the just-in-time delivery of materials from supplier to production line. An immense amount of trust on the part of both parties is necessary for the success of this system. The supplier must be completely candid about its manufacturing process, and the purchaser must be comfortable with a direct-to-line materials flow. It is possible, as many Japanese engineers such as Shingo believed, that no one will ever duplicate the "small village" atmosphere of Toyota in the 1970s. However, over a decade of successes in the automobile and electronics industries shows that such supplier relationships can be maintained on a global scope.

As an outgrowth of these two-party relationships, the need for a third party, the uninvolved quality assessor, has become apparent. During the nineties, major spending on quality measurement was in the area of certification to International Standards Organization (ISO 9000) quality systems standards. Most suppliers today have certified quality systems. The concept of auditing is an important new feature of the lean manufacturing environment. Internal auditing is required for ISO certification and is useful for continuous improvement—extremely so for all phases of waste reduction.

Six-Sigma Innovation

Over the past decade, a concept called six-sigma, which refers to the difference between specification tolerance and process variation in a parts-per-million environment, has been developed at Motorola and General Electric. It is a proactive approach that involves utilizing the tools of measurement and improvement on targets of opportunity. Perhaps unique to this concept is the view of quality as a result of the contribution of each process, allowing concentration on particular process quality problems and the identification of bottlenecks to quality. It is becoming a major quality tool for the lean factory.

PLANNING AND SCHEDULING THE PRODUCTION SYSTEM

Cycle Time in Lean Systems

If quality is the necessary prerequisite, and process improvement the never-ending quest, then the ultimate goal of the lean manufacturing company is the customer response provided by the planning and scheduling system. This system must deliver the proper number of good units at the proper time. It is the combination of small-lot manufacturing (indeed, an ultimate objective of single-lot manufacturing may be expected) and of short cycle times that define the lean system. While the small lot is achieved through traditional methods described throughout later chapters in this book, the reduction in cycle time has been achieved through changes in the way production planning and scheduling are conceived and executed. Many attribute the results achieved at Toyota to its unique scheduling system, but once again, it is important to keep in mind that Toyota's system provided what was needed at the time rather than the ultimate superiority of the system. Good design requires that tools appropriate to the process need to be installed; there is as yet no optimal system for all occasions.

Establishing Goals and Priorities

Although much has been written about the planning, scheduling, and dispatching of work through a manufacturing facility, the typical approach to the problem is fitting each order into the sequence of existing orders. When applying this approach, the solution obtained varies depending on the goals that are selected to schedule the work and the priorities given to these goals. Before choosing a scheduling system, the requirements must be determined. Otherwise, a system could be selected that will not effectively meet the requirements.

For a goal of the least expensive manufacturing costs, each order would be set at the economic order quantity (EOQ) and the largest order would be scheduled first. No consideration would be given to customer needs. This would cause frequent dispatching of orders to customers; even so, delivery performance could be poor. Setup costs would be uncontrolled.

For a goal of least expensive costs while producing the most orders, reversing the schedule of largest order first—that is, scheduling the smallest order first—will ship the most orders. Setup costs and customer delivery considerations are again ignored.

With orders set at economic order quantities, changing the manufacturing sequence should not affect manufacturing costs. Product runs could, therefore, be scheduled in groups to achieve the lowest setup cost. Note that this would have effects on customer delivery similar to the other EOQ models. Scheduling that combines orders prioritized first for setup costs and then for size is possible. There are two ways to measure customer delivery lateness and tardiness. In measuring for lateness, orders that are received early reduce the effect of orders that arrive late. The assumption is that the early orders offset the late ones. If that is not acceptable, only those orders that are overdue are counted. Tardiness is the indicator of the amount of overdue orders.

Even after the selection of a combination of such indicators, the goal-setting problem is not concluded. Is the company willing to be very tardy on a few orders or a little tardy on many orders? Is the company willing to disregard the schedule in the face of a customer demand? Is the company willing to manufacture lots at less than EOQ quantities and absorb the loss of profit? These are all day-to-day problems for the production control professional and for the company policymakers.

It is when the company has marketing problems requiring it to manufacturer outside the traditional EOQ model, such as the problem at Toyota in the 1970s, that innovative scheduling systems are devised. Shingo describes not only the just-in-time and kanban systems but also flow and clustered systems of process, block assembly, and mixed-model production. The approach is not one of selecting a better system, but one of applying the rules of planning and scheduling to continuously deliver the correct product on time.

Push vs. Pull Scheduling

The traditional manufacturing system, which lasted over a century, involves calculating the amount of finished product that will be needed in the future. Quality losses and unallocated materials are calculated, and the number of units that must be started are scheduled. How many parts to order is calculated on the basis of the economic order quantity. Storage space for the unused work in process (WIP) and salvageable waste must, of course, be provided, but when machines are run at nearly full capacity, such storage areas can become large and costly and material handling becomes costly as well. Scheduling by starting a lot and nursing it through production is called *push* production. Some of the traditional problems of this type of production scheduling can be reduced through computerized systems such as MRP-II, but the system will still be wasteful.

The Toyota production system, with just-in-time as a core philosophy, reduces finished goods inventory. Products are individually produced as they are needed. Because the schedule is driven by the output, the system is a *pull* scheduling system—pulled by the need for a final product. After final assembly, parts from the supplying operations replace the parts that are used. Since parts are made as needed, there are no work in process materials except those required to meet anticipated needs. The quality of the parts must be excellent for the assembler cannot choose from a stock of parts. Setups may be more frequent under pull scheduling and require single minute exchange of die (SMED) analysis. Planning must be done from the work unit back to the supplier, so that the unit has all the correct parts available in time to supply a customer order. To achieve this, total and individual process cycle time must be reduced.

Product Cycle Time Improvement

Perhaps the major difference between the traditional push systems and the Toyota pull system is the philosophy of improvement that Toyota infused into its implementation. To be successful, Toyota had to reduce the length of time it took to manufacture a particular unit. The company had seen cars sit for over a year as finished goods on dealer car lots around the world in the late 1960s. There were finished goods and raw materials on skids and racks throughout its manufacturing plants. The cost of all this was obvious. While machines kept running, units that were needed could not be made because some parts were not available. Schedules for particular units were in terms of months, but drivers wanted their cars in days. Whatever system could be devised to correct these problems had to be profitable. Shingo, realizing that process time controlled the cycle time and that long setup time increased the EOQ and severely limited the options for flexible manufacturing, applied many traditional motion improvement and time study techniques to reducing cycle time. Thus, the tools discussed throughout this book are directly applicable to lean manufacturing.

Balancing of Line Stations (Takt Time)

It is relatively easy to balance a line station if there is only one sequence of activities and thus only one router and product, but this is not the case in many manufacturing facilities. Where there are product variations, a concept called *Takt* (from a German word for musical meter) links customer requirements with the production flow. If blue hatchbacks are selling at four per day, four blue ones should be made. Further, if three of them have air condition-ing, three air conditioners should be installed. The time required at each work station must be calculated for the types of models being produced and the schedule must be updated fre-quently according to changes in customer needs. The time allotted to many stations may remain constant, while at a few it will vary with the difference in accessories. Takt times are established by customer needs, but alternative methods of production, such as cells, progres-sive lines, or individual stations, can also be considered in terms of costs and cycle times.

STEPS TOWARD ESTABLISHING A LEAN MANUFACTURING ENVIRONMENT

Womack and Jones provide a five-year time frame for developing the lean organization. It requires very concentrated efforts and fundamental organizational change, the ability to ana-lyze all of the activities in the organization and eliminate the muda, and the leadership to support and encourage appropriate change.

First Six Months: Getting Started

Step 1: Find a change agent. Internally or externally, there is someone who is right to lead the change to a lean environment. If no one is qualified, perhaps the company is not ready.

Step 2: Get lean knowledge. A teacher (or sensei) may be useful. In any case, the change agent and the senior management must be prepared to think in lean management terms.

Step 3: Find a lever for change. Crisis helps force change. Applying lean management to a business unit in crisis and achieving dramatic results will impress others in the orga-nization. Care must be taken, however, not to lose the company while trying to build the lean organization.

Step 4: Map the value streams. Investigate each product family and map the sequence step by step. Be sure to identify issues with suppliers and customers.

Step 5: Select an important and visible activity. Start with something that is performing poorly but that is necessary. Success is highly likely from applying lean techniques and will increase interest. Demand immediate results without investment.

Step 6: Expand the scope. Once momentum occurs, take the opportunity to attack other problems. Processes in design, development, and marketing as well as production will be found to contain muda.

To Year Two: Creating the New Organization

Step 1: Reorganize by product families. The value stream exists for the customer and is unique for that market. Most companies have multiple value streams and should be organized to supply the needs of each value stream family separately.

Step 2: Create a lean promotion function. A permanent staff is needed to develop and implement the ideas generated by the process maps. This may be combined with quality assurance in many organizations.

Step 3: Devise a policy for excess people. The better it works, the more excess. Is the organization ready to expand production? Is downsizing possible? Is the organization "mean"?

Step 4: Devise a growth strategy. As margins increase, the opportunity exists to absorb excess people into new areas. In many companies, sales have tripled while the work force remained constant.

Step 5: Remove the anchor-draggers. Most people in every organization will learn and adapt to new ideas. Those that actively oppose must be removed.

Step 6: Instill perfection. Once reorganization has been completed, it should be reexamined to find better ways to improve. Often, one improvement will lead to the discovery of a bigger improvement. The organization should be ready to back up from the great leap to a lesser step.

To Year Four: Installing Business Systems

Step 1: Introduce lean accounting. At this point, allocated overheads are reduced and application of actual costs to the product families can begin. Thus, persons responsible for the product can control much of the product cost. With traditional accounting methods, the overhead percent increases as the labor, machines, and floor space are reduced. Savings disappear into the accounting overhead account.

Step 2: Pay for performance. Because lean manufacturing provides more profits but must be cared for continuously by management, profit sharing is an equitable reward.

Step 3: Make everything transparent. Goals should be scoreboarded and processes should be benchmarked. Simple, graphical information about performance successes and bottlenecks should be provided.

Step 4: Teach lean thinking. Policy deployment and lean learning should be synchronized so that all learn what they need to know at the proper time and support is available to implement the new ideas and approaches.

Step 5: Right-size the tools. The simple should be chosen over the complex. What permits flow in the product family should be sought out.

Year Five: Completing the Transformation

Step 1: Convince suppliers and customers to become lean. Firms rarely actively record more than a third of their cost and lead time. Convincing suppliers to follow a lean system can reduce finished goods cost and build a more stable supplier. Customers who follow a lean system can increase their market share—and increase their orders.

Step 2: Develop a global strategy. Lean manufacturing provides the way to cut costs while shortening production lead times and time from concept to market, improving quality, and providing the customer with the right item when it is wanted. What is it that the company needs? A strategy must be developed.

Step 3: Convert from top-down to bottom-up initiatives. Managers must become coaches and employees must become proactive. The role of change agent becomes one of collegial leader. Is the person who has been the change agent up to this point ready for major behavior change? Often, the change agent moves on.*

THE USE OF MOTION STUDY TOOLS TO IMPROVE THE ENVIRONMENT

Improving Processes and Cycle Time (Process Charts)

Lean manufacturing requires maps of the sequence of the process. This is done with the traditional tool of the Gilbreths, the process chart. The steps in the process need not be a part or even a physical object. Process charts can be made for the sequence a person performs, a flow of material through a factory, the steps of an automated piece of equipment, the flow of information, or even the flow of money through an operating unit.

Shingo saw the relationship between processes to be fundamental to major improvements and cited the work of the Gilbreths. The process chart provides the means for analyzing and improving processes. It is the primary tool for lean manufacturing analysis. While specifying the steps required in any existing process, it also provides the questions for gaining insight into that process (who, what, where, why, when, and how) and the approach required for improving the process (eliminate, combine, change sequence, simplify). Process analysis is important to anyone wanting to understand processes and particularly important to the lean manufacturing analyst. Chapter 5 describes how the process chart is constructed.

Improving Operations and Task Performance (Operations Charts)

A number of techniques aid in the detailed analysis of a specific operation in the process chart. Motion sequences, for example, of individuals, machines, or combinations of both are used. It is important that motions occur in rhythm with each other and that each motion occurs at its proper time. Often, delays can be eliminated through such analysis. The measures of improvement in operations charts have a common root, the reduction of cycle time, which is important to the concepts of lean manufacturing and is also the basis for SMED. The person-machine operations chart graphically show the work patterns before and after changes are made based on the principles of internal work, external work, machine time, and idle time for person and machine. Chapter 6 discusses the design and analysis of various types of operations charts.

Improving Worker Quality of Life (Psychology, Ergonomics, and Safety)

Each work environment is a very complex interaction of persons, tools, machines, and other aspects of the environment. What one employee considers concern from a supervisor may

---

* Adapted from James P. Womack and Daniel T. Jones. *Lean Thinking* (New York: Simon and Schuster, 1996, p. 270.)

be considered meddling by a different employee. Opening the window may make it too hot or cold; closing a window may make it boring or still or noisy. Environments may be dirty or clean; others may be hazardous. The problems of designing a quality environment are complex. A number of principles, or guidelines, can be used to increase employees' ability to consistently produce while reducing the work effort. Although these guidelines do not solve the problems of providing a happy, safe, comfortable, and effective workplace, they are a start. The chapter on workplace design, Chapter 7, includes a discussion of these guidelines.

Chapters 5–7 discuss the tools that are useful in improving processes and operations. These are also the key topics in the study of the science of work advocated by the Gilbreths.

THE USE OF TIME STUDY TOOLS TO MAINTAIN THE LEAN ENVIRONMENT

Measuring Individual and Group Performance (Time Study)

Sometime during the development of an improved lean environment, as the process chart was perhaps completed, some worker is bound to have said, "But I can't do all of that." The Takt time had been developed and operations had been eliminated, combined, and rearranged. But the work station could not operate and the Toyota line-stopping lever was pulled. Some workers could do the job, but others couldn't.

During a time study, jobs are divided into timeable elements. Since non-value-added elements are eliminated in the methods study, no elements are discarded in the time study. Time values for different workers are recorded during the study. It is true that the study reveals that some people can't do the job and a few can. Indeed, the range from slowest to fastest employee may be about 2.6 to one (one person can make 50, another 130). The question of what a skilled operator working at a consistent pace with the proper method can do is what the entire complicated procedure called time study attempts to determine. Shingo felt that American analysts spent too much time on standards systems at the expense of methods improvements. However, time study will determine performable cycle times better than any other measurement approach. Chapter 9 describes how to conduct a time study.

Determining Appropriate Allowances (Work Sampling)

Allowances are an important part of time study. Determining personal time and fatigue recovery time for the worker is necessary, as is the time for the delays that occur during the task. Delays, a controllable item, can be measured with a technique called *work sampling* (ratio delay sampling), which is a necessary technique for the computation of task times under Shingo's approach. A discussion of work sampling can be found in Chapter 11.

Predetermined Time Systems and Prediction of Standards

Although stopwatch time study is the general tool for determining standards, predetermined time standards systems, described in Chapter 8, provide a way to quickly use the accumulated results of prior time studies. It is likely, with the current emphasis on lean manufacturing, that predetermined time analyses will be used to provide data consistently and rapidly for new scheduling tools.

Combining Time Standards and Operations Charts to Implement SMED

A complex stamping press SMED analysis can require huge amounts of data. Details about all of the operations must be gathered. Activities that can be combined or eliminated are identified, and finally, ideas about how to make internal work (the machine is stopped) into external work (the machine is running) are tested. Throughout, accurate times for all of the activities are needed. Both the present, long-downtime process and the new, SMED process must be documented. The study and analysis use the tools of both motion and time study.

MEASURES OF PERFORMANCE WITHIN THE LEAN ENVIRONMENT

Performance Reporting as a Team Improvement Tool

Scoreboarding is important for any team improvement project. As the processes become complex, the performance of the group or of individuals is difficult to measure. Performance reporting, traditionally used to measure a department or a group operation, is useful in giving individuals and teams the necessary measurement of their performance. Chapter 13 describes several performance control systems.

Progressive Line Balancing and Scheduling

The idle time available in a many-person line can be extremely high because the line runs at the pace of the slowest worker. Only when lines are well balanced do they provide the most economic efficiency. Often, two lines with fewer employees may provide better labor utilization, flexibility, and reduced cycle time. The techniques for balancing a line for least delay are shown in Chapter 10.

Work Sampling, Auditing, and Visual Inspection

A body of literature exists that shows how to recognize visual signs of the performance of the employees in a manufacturing plant. These signs are a measure of output since the plant condition is the result of the combined efforts of the employees and management. Another measure of performance is the internal audit for quality conducted under ISO 9000. Such audits ensure that procedures are understood and followed. Work sampling can also be a very useful tool for determining the actual type and pace of activities being completed in the plant. The sampling does not depend on any secondary report, but, like the other measures, is a result of direct observation.

QUESTIONS

1. Who coined the term *lean environment* and why?
2. How does "leanness" differ from mass production?
3. What motion study tools help improve the environment?
4. Describe the concept of muda.
5. What are the characteristics of the lean environment?
6. How does the customer affect lean design?
7. What were the differences between Toyota and the U.S. automakers?
8. How is time study used to maintain the environment?
9. Who was the industrial engineer at Toyota during the 1960s and what did he provide?
10. Can lean manufacturing work in American plants?

History of Motion and Time Study

The history of motion and time study is not long but has been full of controversy. Time study was developed in about 1880. Frederick W. Taylor is said to be the first user of the stopwatch to measure work content. His purpose was to define "a fair day's work." In about 1900, Frank and Lillian Gilbreth started working with methods study. Their goal was to find the one best method. During 1928, Elton Mayo started what is known as the human relations movement. By accident, he discovered that people work better when their attitude is better. These four pioneers of motion and time study are discussed in this chapter. But first, we provide a little earlier background.

Labor has always been a major factor in the cost of a product. As labor productivity improves, costs go down, wages go up, and profits go up. From the earliest industrial history, management has looked for labor-saving techniques. Industrial technology's objective and purpose for being is to increase productivity and quality. Output per labor hour is the most commonly used measurement of productivity. Motion and time study techniques give management the tools to measure and improve productivity.

Concern for productivity has always been a major motivation of production managers. Productivity is a concern of anyone in business. Look at the farmer, for example. How much more work can be accomplished with a tractor over horses? How many more acres can be plowed, planted, and reaped by a single person? How many more bushels per manhour can be harvested? Bushels per manhour is a good measure of farm productivity. Take this concept into manufacturing, and we have number of units produced per hour worked. Coal mining is another good example. Everyone agrees that mining machines are more productive than pick and shovel. Would anyone disagree that the mining job of today is better than 100 years ago? Of course not! The tons of coal per laborer per day has continued to improve.

Historical examples of how new technology increased productivity in every area of business and industry exist by the thousands. The steam engine replaced horsepower, making the Industrial Revolution possible, and interchangeable parts replaced one-of-a-kind parts, making mass production and the assembly line possible. During the nineteenth century, early manufacturers were vying for competitive advantages, and technology was expanding at a fever pitch. Secrecy was of utmost importance. Sharing of information and ideas was minimal.

In 1832, Charles Babbage wrote his book *On the Economy of Machinery and Manufacturers* and included ideas on division of labor, organization charts, and labor relations. It would be 50 to 100 years before they were used extensively. Higher education and professional societies are given credit for opening up the information on manufacturing management techniques, but the sharing of information was slow because of the secrecy mentality of most managers. Only in Frederick W. Taylor's later life did he write about what he had done.

FREDERICK W. TAYLOR (1856–1915)

Frederick W. Taylor is known as the father of scientific management and industrial engineering. He is the first person to use a stopwatch to study work content and, as such, is called the father of time study.

He was born in Philadelphia, Pennsylvania, to wealthy parents. He passed the Harvard University entrance exams with honors, but eye problems stopped him from attending. Taking his doctor's advice, Taylor entered the labor force as an apprentice machinist. Four years later, at the age of 22 (1878), Taylor went to work for Midvale Steel Works as a laborer. He worked his way up to time keeper, journeyman, lathe operator, gang boss, and foreman of the machine shop. At age 31, he was chief engineer of Midvale Steel Works. In 1883, after years of night school, he earned a B.S. in mechanical engineering from Stevens Institute.

Many years later, Taylor was able to explain what he accomplished through his four Principles of Scientific Management:

1. Develop a science for each element of a person's work, thereby replacing the old rule-of-thumb methods.
2. Select the best worker for each task and train that worker in the prescribed method developed in Principle 1.
3. Develop a spirit of cooperation between management and labor in carrying out the prescribed methods.
4. Divide the work into almost equal shares between management and labor, each doing what they do best.

Before Taylor, the work force developed its own methods through trial and error. The work force was responsible for seeing that everything was available to do the job, such as laborers bringing their own tools to work. Frederick Taylor wanted management to reject opinion for a more exact science. Taylor

1. Specified the work method,
2. Instructed the operator in that method,
3. Maintained standard conditions for performing that work,
4. Set time standard goals,
5. Paid premiums for doing the task as specified.

Frederick Taylor is responsible for the following innovations:

1. Stopwatch time study
2. High-speed steel tools
3. Tool grinders
4. Slide rules
5. Functional-type organization.

Taylor's Shoveling Experiment

Between 400 and 600 men moved mountains of coal, coke, and iron ore around the two-mile-long yards of Midvale Steel Works. Each man brought his own shovel from home and was assigned to a gang for moving materials. Taylor noticed the different sizes of shovels and wondered which shovel was the most efficient. Taylor talked management into a formal study of this operation. He asked a laborer, known today only as John, if he would be willing to help him study the coal, coke, and iron ore shoveling job. Taylor told John he would double his salary, and I'll bet John answered positively within 10 seconds. Taylor watched John with a stopwatch and measured everything he did. He varied the shovel size, duration, number of breaks, and work hours. The results were fantastic. He purchased quantities of three types of shovels—one for coal, one for coke, and one for iron ore. The results of Taylor's shoveling experiment are summarized in Table 3-1.

Table 3-1

	BEFORE STUDY	AFTER STUDY
No. people	400–600	140
Pounds/shovel	3½–38	21½
Bonus	No	Yes
Work unit	Teams	Individual
Cost/ton	7¢ to 8¢	3¢ to 4¢

A savings of $78,000/year

FRANK (1868–1924) AND LILLIAN (1878–1972) GILBRETH

Frank and Lillian Gilbreth are known as the parents of motion study. In their lifetime search for the one best method of doing a specific job, they developed many new techniques for studying work. That they are the parents of motion study is universally accepted.

Frank Gilbreth started his work life as a bricklayer's apprentice. He was immediately aware of motions. He noticed that when the instructor showed him how to lay bricks the instructor used one set of motions, another set of motions when he was working by himself, and a third set of motions when in a hurry. Frank questioned this practice and went about searching for the one best method. He set up his own construction business with the competitive advantages of

1. Adjustable scaffolding—previously bricklayers built the wall from their toes to their highest reach, then built some more scaffolding and started over.
2. Bricklayer's helpers—at about one half the cost of a bricklayer. The helper would sort, carry, and stack bricks for the bricklayer.
3. Constant mortar mix.
4. Improved motion pattern.
5. Three hundred and fifty bricks per hour instead of the previous 120.

Lillian Gilbreth (shown in Figure 3-1) was a trained psychologist and a people-oriented person. In addition to her work in methods engineering, she raised 12 children and authored books. I know that Lillian's response to Frank when he came home from work

FIGURE 3-1 An outstanding pioneer of motion study.

saying, "You should have seen how I designed that job today," was "Frank, you can't do that to people." Lillian kept Frank from dehumanizing work and made him conscious of the human element.

Frank and Lillian's continued success in motion study led them away from the construction business into consulting. Frank's training in engineering and Lillian's training in psychology made them a powerful team. Not only did they develop methods study techniques like the cyclograph (shown in Figure 3-2), chronocyclographs, movie cameras, etc., they also studied fatigue, monotony, and transfer of skills and assisted the handicapped in becoming more mobile. Their work has become a tradition of industrial engineering.

The knack that Frank and Lillian possessed for analyzing work motions enhanced their ability to substitute shorter and/or less fatiguing motions to improve the work environment. Their research has called attention to the fact that great gains may be realized, even in the simpler trades where one would not suppose such possibilities exist. Their systematic study of motions reduced costs greatly and founded a new profession of methods analysis.

The elimination of all useless motions and the reduction of the remaining motions were the foundation of the Gilbreths' work. The elimination of this unwanted waste has become known as *work simplification.* The Gilbreths used flow diagrams to show movement of product around an entire plant because they gave an accurate geographical picture of the entire process. The Gilbreths developed process charts to show diagrammatically the sequence and relationship of elements of a process. The operations chart showed details of the individual operations. These charts were able to show the interrelationship of laborer and machine, a gang of people working together, and left hand versus right hand. Some motions were so fast that the Gilbreths incorporated a high-speed motion picture camera and a special clock

FIGURE 3-2 Gilbreths' cyclograph: The light path the hands make in the process of producing one part or cycle.

called a microchronometer to study this work. Increments of 1/2000 of a minute were used. From this work came micromotion study.

Among the generally accepted theories of efficient motions developed by the Gilbreths was the terminology defining the entire range of manual motions. To refer to these 17 elementary subdivisions of motion, later engineers coined a short word, *therblig,* which is Gilbreth spelled backward (except for the *th*). Therbligs are the basic components of the motion pattern. Essentially, they name the different activities of the hand. Each therblig has a symbol, color designation, and letter symbol for the purpose of charting. Search, select, grasp, transport, hold, position, inspect, and assemble are a few of the therbligs.

The predetermined time standards system that is studied in this book is a newer, easier-to-learn, easier-to-apply technique that is built on the Gilbreths' work.

HENRY LAURENCE GANTT (1861–1919)

Henry Laurence Gantt, a close associate of Taylor's, was among the major pioneers of motion and time study. He worked for Taylor at Midvale and Bethlehem Steel, graduated from Stevens Hoboken Institute of Technology the year after Taylor, and served in various positions at ASME (American Society of Mechanical Engineers) with the rest of the Taylorites. He invented the task and bonus system or earned-hour plan. Rather than penalizing the less proficient worker as Taylor did with his multiple piecework plan, Gantt advocated a livable wage with a sizable bonus for performance over 100%. While Taylor emphasized the analytical and organizational aspects of work, Gantt was more interested in operator selection, training, and motivation.

Gantt visualized industrial management problems as national productivity problems. He worked to educate management because he felt that the new ideas required a complete overhaul of the industrial system. Although many of his ideas were extremely radical and have no context in today's workplace, his basic notion of the importance of a leadership function is exactly in line with today's thinking.

During the First World War, Gantt developed a technique for scheduling work. It was used, along with his ideas for motivating people, to rebuild the merchant marine fleet. He also designed the antisubmarine tactic known as convoy zigzagging that permitted escort ships to protect the slow freighters for which he received a congressional citation. (Unfortunately, Mr. Gantt suffered an untimely death. After dinner at a meeting at the New York Engineer's Club he excused himself before hearing the speaker. He retired to bed with severe pains and died a few days later.) The performance control system described in Chapter 8 is a result of the Gantt efforts.

HARRINGTON EMERSON (1853–1931)

Harrington Emerson was a successful academic; as chair of the Modern Language Department at University of Nebraska at age 23, he had already studied in England, France, Germany, and Greece. But he wanted to make his fortune, so he went into business. He eventually found himself improving operations on the Burlington Railroad. For the rest of his life, he would be consulting about how to improve such operations. An expert in the maintenance of the engines, rolling stock, right of way, and the scheduling of trains, he saved the Santa Fe Railroad from near bankruptcy and formed it into an industrial giant.

Emerson was the expert that was needed to make Scientific Management, the Taylor system, a household name. As a principal witness in the Eastern Rate case, his experiences proved that the use of efficient methods would lead to tremendous savings. A very polished individual, his advice was sought after in the board rooms of America. For some reason, unlike Taylor, Gantt, and Gilbreth, accounts of his work were never extensively published and no comprehensive biography exists. Perhaps his work is best remembered as an example of how the creative engineer can find the tools to improve any operation.

RALPH M. BARNES (1900–1984)

Dr. Barnes was one of the first and best-known professors of engineering in the field of work measurement. During his 21 years at the University of Iowa, he attained the rank of full professor and was influential in the promotion of industrial engineering. After his retirement, he taught for 34 years, until his death, as professor emeritus at the University of California at Los Angeles. Prior to beginning his teaching career, he worked at U.S. Window Glass, Bausch and Lomb Optical Co., and Eastman Kodak.

Barnes's international consultancies included England, Uruguay, Norway, Spain, Mexico, Costa Rica, Japan, Peru, Norway, and Sweden. A Fellow in the International Academy of Management, he was a recipient of the Society of Management Gilbreth Medal (1941), Industrial Incentive Award of the ASME (1951), and the Frank and Lillian Gilbreth Award from the Institute for Industrial Engineers (1969). He was a member of numerous honorary fraternities and societies. His achievements included writing the longest published text on work measurement, a thorough description of the Gilbreths' micro-motion study, time study, and the procedure for work sampling. He conducted numerous methods studies of activities with motion picture cameras and developed rating films for training time study technologists.

MARVIN E. MUNDEL (1916–1996)

A student of Dr. Barnes in the late thirties, Dr. Mundel taught at Purdue from 1942 to 1952. He went on to direct the Army's Management Training Center at Rock Island Arsenal. This led to a successful government and consulting career, which lasted until his death at 90.

Mundel was a major contributor to the reconstruction of Japan after World War II and made many trips there. Several of his books were written for the Asian Productivity Council. He also worked in Australia, Taiwan, Hong Kong, the Philippines, and Singapore. During a 40-year consulting career, he also provided productivity systems to the Department of Defense, Department of Agriculture, Department of Housing, and the Department of Interior. He also assisted in landmark local government projects in the state of Washington, the city of Milwaukee WI, and in Prince George County, MD. Mundel's text simplified the description of the various tools of work measurement and provided numerous examples of the ways the tools could be used. He developed many manuals for government agencies, companies, and productivity agencies and taught seminars about these techniques.

SHIGEO SHINGO (1909–1990)

Dr. Shingo received an honorary doctoral degree at Utah State University in 1988 for his contributions to improving quality and productivity. This capped a 60-year career that began with a degree in mechanical engineering at Yamanashi Technical College.

Deeply impressed by a translation of Taylor's principles, at age 22 he read *Principles of Scientific Management* and decided to make the study of scientific management his life's work. He attended the first training session of the Industrial Engineering Training Session, and he was thoroughly instructed in the "motion mind" concept of Kenichi Horikome. By applying his learning, he made plant production double. In the decade following the World War II, Shingo worked in many industries for the Japan Management Association, where he learned the details of manufacturing that fill his books.

Shingo founded the Institute for Management in order to gain an opportunity to test his concept of systems. The key contributor to the concept of lean manufacturing, Shingo helped his clients Matsushita and Toyota make tremendous gains during the 1960s. For the rest of his life, he traveled the world teaching and consulting about his concepts.

PROFESSOR ELTON MAYO

Known as the father of the human relations movement, Professor Elton Mayo became involved in the Western Electric Company's Hawthorne plant productivity studies after the National Research Council of the National Academy of Science withdrew. The Hawthorne plant, near Chicago, undertook a research project to study what factors affected productivity. These studies took place between 1924 and 1933.

Phase I (1924–1927): Illumination Study

The basic premise of this study was that increased illumination of the work area would increase productivity. The results of this first three-year study were called inconclusive because too many other factors entered the picture. The National Research Council withdrew from the study, but Professor Elton Mayo of Harvard University became interested in Western Electric's efforts and joined forces for the next phase.

Phase II (1927–1929): Relay Assembly Study

A group of five women were selected, briefed, and relocated to an experimental assembly room where other factors could be controlled (other factors were blamed for the failure of the illumination study). The basic premise of the relay assembly experiments was that "a change in working conditions would result in a change in productivity." To isolate the factors being studied, the experimenters worked hard to keep the operators' attitudes positive toward the study, management, and their work. The experiment observers spent much time talking to and listening to the operators.

The factors being studied were

1. Incentive system
2. Rest periods
3. Paid lunch breaks
4. Elimination of Saturday work
5. Reduction in work hours
6. Free lunch and drinks.

The relay assembly experiments consisted of 13 phases. Table 3-2 shows the phases, factors, duration, and results.

Phase III (1929–1930): Interviewing Program

Twenty-one thousand employees of Western Electric Corporation were interviewed. The company wanted their opinions on what people want from their jobs. Learning how to ask questions and learning how to listen were the greatest contributions of this phase.

Table 3-2 Hawthorne Plant: Relay Assembly Experiment 13 Phases.

PHASE	NO. WEEKS DURATION	CUMULATIVE NO. WEEKS	FACTOR	HOURS WORKED PER PERSON	AVERAGE HOURLY OUTPUT	AVERAGE WEEKLY OUTPUT
1	2	2	In regular dept.	39.7	49.51	1,973
2	5	7	Experiment room	45.6	49.4	2,254
3	8	15	Incentive system	44.7	51.25	2,289
4	5	20	2–5 min. rest	42.4	53.11	2,251
5	4	24	2–10 min. rest	44.2	55.9	2,470
6	4	28	6–8 min. rest	44.2	55.9	2,470
7	11	39	11 min A.M.– Lunch—10 min. rest	41.4	56.1	2,305
8	7	46	Stop ½ hr. early	40.7	62.5	2,542
9	4	50	Stop 1 hr. early	39.0	64.5	2,516
10	1	62	Same as 7	43.6	61.7	2,691
11	9	71	No. 7 + eliminate Sat. A.M.	39.6	63.6	2,517
12	12	83	No. 3 no lunch/rests	45.9	61.0	2,802
13	31	114	Same as no. 7 but brought own lunch— given beverages	43.1	66.7	2,873

Phase IV (1931–1932): Bank Wire Observation Room

Informal organization and its influence on productivity were studied in this phase, for the results of the Hawthorne studies did not go as intended. The factors thought to improve performance did not lead to an automatic improvement. The studies did show that change affects the employee's attitude, which in turn affects results. The experimenters treated the operators differently than was typical of the period. As a result, the operators were made to feel important, and they were involved with something they thought important. When the other factors were changed negatively, production still improved, because the employees' attitudes continued to be positive. Even the early studies of illumination became proof of this new hypothesis, "Improve employee attitude and you will improve productivity." In modern times, company management is involving the work force in all phases of product development and implementation. Modern companies are finding the whole person to be of great help in all areas that affect them. No more do we consider people hired hands. We get the whole thing—hands, mind, mouth. It may have been easier in the old days, but bringing all minds to bear on a problem gives us better solutions.

Many other people have been involved with the development of motion and time study. Industrial engineers and technologists continue to improve techniques every day, but space prevents us from mentioning all the pioneers.

CONTROVERSY

We mentioned controversy over motion and time study in Chapter 1. We also mentioned that successful people do what others do not like to do, among which are to criticize and be criticized.

Frederick W. Taylor was criticized as being a management speed-up artist. Unscrupulous managers used Taylor's techniques, and when workers met the goals, management raised the standard. Taylor would have hated this process; we must never change a standard without due cause.

Lillian and Frank Gilbreth were charged with dehumanizing work. Because of the reduction of motions to the absolute best set of procedures possible, unions depicted the Gilbreths as antiworker and as wanting to make machines out of everyone. The Gilbreths are not properly given credit for removing the drudgery from work.

Elton Mayo was charged with being unscientific. His improvements were said to be caused by other factors.

Any time anyone does something new and different, people criticize. Criticism is easier than developing something new, so many would-be scientists try to show that some past study is in error.

Frederick Taylor, Frank and Lillian Gilbreth, and Elton Mayo were hardworking, ethical human beings who improved our world of work and had the courage to write about it.

QUESTIONS

1. How did Taylor become interested in time study?
2. What materials were shoveled? What similar jobs exist today?
3. Why is Lillian Gilbreth important?
4. What studies did Lillian Gilbreth do in the 1930s?
5. What studies was Frank Gilbreth noted for?
6. Describe the Hawthorne studies.
7. Why was Gantt a war hero?
8. What railroads did Emerson work for?
9. Who were Barnes and Mundel?
10. What is Shigeo Shingo noted for?

The Importance and Uses of Motion and Time Study

This chapter discusses the importance of motion and time study in manufacturing. It shows how motion and time study techniques are used to develop answers to many important manufacturing questions, to solve some of the largest manufacturing problems, and to reduce and control cost. As the title of this book and chapter indicate, motion and time study are two different subjects, each with their own set of techniques (tools). In this chapter, we discuss five areas:

1. Definition of a motion study
2. The importance and uses of motion study
3. The definition of a time standard
4. The importance and uses of time standards
5. Techniques of time study.

WHAT IS A MOTION STUDY?

Motion studies are performed to eliminate waste. The efforts and movements of the human employee are all valuable. If a person spends most of the day doing physical or mental tasks that do not contribute to the manufacture of product, the person will be just as tired as the person who only did things that contribute to product manufacture. It is the responsibility of management to ensure that the employee has appropriate and timely work, that the appropriate tools and materials are available, and that the employee knows what is expected.

Without any harm, and often to the employee's benefit, motion studies provide the proper way to consistently make good product. The technologist will use the tools of methods study to develop new methods, to correct ineffective work habits, to analyze processes and operations, and to develop appropriate person and machine combinations for production.

Before any improvement in quality or quantity of output, any study of operations time, any scheduling of work or balancing of workload, or any calculation of standard time, a study of the current and proposed methods is required. Often, studies of overall factory flow or process, called *macromotion studies,* are made and then additional studies of detail or operations, called *micromotion studies,* are completed for a project.

Motion studies were conducted by Frank and Lillian Gilbreth about a century ago in a search for the "one best way." It is important to note that such studies seek to minimize and simplify manual efforts. They are unconcerned with cycle times, because the proper method will have a minimal cycle time. The addition of times is for the completeness of the motion and time analysis as a tool for subsequent scheduling and measurement of work.

Macromotion Study

Any process can be studied by dividing it into process activity. Although each activity is different, depending on the product, there are five classes of activities that are included in all processes. Savings, often considerable, may be found in the process by reorganizing activities. These activities found in every sequence of processes are

Operations	Changes in the properties of the product
Transportations	Changes in the location of the product
Inspection	Confirmation that change conforms to specification
Delay	Wait for start of operation, transportation, or inspection
Storage	Wait until needed

When the process is first studied, each activity is recorded and sorted into one of the five classes. All observed activities are recorded, and activities not done are not recorded. The purpose of each activity is studied. Typically, the questions who? what? where? when? why? and how? must be answered. Other methods simply ask why? five times. In any case, one must understand what is occurring in order to propose any change. Next, each event is observed for potential improvement in the following sequence:

Can the activity be eliminated? If not,

Can the activity be combined and done with another activity? If not,

Can the activity be rearranged to occur in the sequence at an easier time? If not,

Can the activity be simplified with shorter distances, mechanical assist, or reduced complexity?

Once these questions are asked and the improvement sequence defined, it is necessary to draw a chart or diagram that shows the motion improvements. Such process charts include

Process Flow Plan	A plan-view plant layout with activities overlaid
Process Operations Chart	The sequence of serial and parallel operations only
Process Chart	All serial activities on a preprinted form
Flow Process Chart	All serial and parallel activities on a single page
Work Cell Load Chart	A plan view with repetitive operations
Route Sheet	A planning tool for scheduling operations

Macromotion studies are not limited to charting the activity flow of product. They may be used for tracing objects, persons, information, forms, money, energy, or other things through time (chart format) or space (plan view format). Macromotion study procedure and examples are given in Chapter 5.

Micromotion Study

Considerable wasted motion and idle time can occur within an operation. This time can't be found with macromotion studies because it is usually within one process operation. The improvement is gained from reducing the operation cycle time. Because the operation is often a small unit or a complex, infrequent event, micromotion studies are effective on highly repetitive operations or infrequent setup or maintenance procedures.

All of the charts provide columns for the tasks or delays incurred and a scale for the time involved. This time scale will vary considerably depending on the operation involved. The different types of operations charts and their characteristics are

Operations Analysis	Single person doing serial activities	Single column
Operator/Machine	Person and machine relationship	Dual column
Gang Chart	Team of workers	Multicolumn
Multimachine Chart	Person and several machines	Multicolumn
Left/Right Hand Chart	Person showing each hand separately	Dual column
Simo Chart	Person showing each arm separately	Dual column

All of the combinations of cooperative work, from individuals performing coordinated assemblies to sequences of automated machines, can be analyzed and improved with operations charts. Concepts as small as the time to grasp a penny or as broad as efficiently changing over models on an automotive assembly line can, and have been, improved with methods study.

The same basic concepts exist for improving operations as for improving processes: Divide the work into small groups, eliminate all unnecessary work, combine work to make the individual tasks easier, rearrange the sequence, and, finally, simplify each task whenever possible. Such improvements will provide an operation with the least amount of effort. To reduce cycle time, divide the work into tasks that can be done while the machine is running (internal work) and work that must be done when it is stopped (external work). Review all of the work that must be done when stopped and eliminate or recombine so that the manual work is either while the machine is running or is eliminated. Review all of the work remaining while the machine is stopped. Design simplified ways to reduce that time. Minimize external work time. Review the internal work. If the machine time is more than or equal to the internal work, the cycle is minimized. Without moving work from internal to external, eliminate, combine, change sequence, and simplify. Balance for the efficient use of persons and machines.

The application and procedures for applying the tools of micromotion are included in Chapter 6. More detailed approaches for using the knowledge gained from motion study include:

Work station design (Chapter 7)	An introduction to the factors involved in workplace design for motion and time study
Motion economy principles (Chapter 7)	The historical principles involved in designing and improving operations
Motion pattern analysis (Chapter 7)	A technique to plot the movements of the fingers and evaluate work station effectiveness
Predetermined Time Standards Systems (PTSS) (Chapter 8)	A procedure that uses coded motion study data to estimate time standards

THE IMPORTANCE AND USES OF MOTION STUDY

Motion study can save a larger percentage of manufacturing costs than anything else we can do in a manufacturing plant. By changing to a more automatic machine, we can eliminate or automate many steps in a process. For example, a progressive die can be used in a punch press to make bottoms and lids for tool boxes. The old method would call for

1. Two shear operations (to cut the steel to width and length)
2. Two punch press operations (to notch corners and punch holes)
3. Two press break operations (to form the four side walls).

The old method would require a total of 13 hours to make 1,000 parts. The progressive die will do all these operations at one time and will produce 1,000 parts per hour or one hour per thousand. If we take 1 hour away from the original 13 hours, we save 12 hours per 1,000 parts, or 24 hours per 1,000 tool boxes (because we save 12 hours each for the top and bottom). At the labor rate of $15.00 per hour, we will save $360.00 per 1,000 boxes, but, more impressively, we will remove 92% of the labor over the previous method, as well as eliminate the moving of parts between stations. Furthermore, we will eliminate the delays between operations, for even greater savings. If we use this new machine 8 hours per day, we will save $1,440.00 per day or $360,000.00 per year (1,000 parts per hour × 12 hours per 1,000 × $15.00 per hour × 8 hours per day × 250 days per year). We can buy some very impressive equipment for this kind of money.

A second example of improved methods (and a technique that is getting more attention today than any other) is work cells. There are many different kinds and purposes of work cells, and here we describe a cell to produce thousands of a single complicated part. A valve body used in hydraulics requires many long-cycle machine jobs. The old way was to run the part around the plant in big tubs, running each part through machines one at a time, with the operator waiting on the machine to do its thing. The work cell method would collect all the machines into one area cell and then have one or two operators unload, reload, start the machine, and move on to the next machine with the partially completed part until the part has been run through each machine in the cell. The machines are placed in a circle around the operator so the operator will walk as short a distance as possible. If there is only one operator, he or she will pick up a part from the in tub, take it to the first machine, remove the finished part from that machine, place the new part into the first machine, activate the machine, take the finished part to the second machine, unload the finished part from that machine, place the next part in that machine, activate the machine, and move the part on around the machines until a finished part is inspected and placed in the finished tub, which is next to the parts coming into the cell. The operator then picks up another part and starts all over again, making his or her way around the cell.

A more common, everyday manufacturing situation involves people looking at individual operations and asking themselves, "How can I make this job easier?" Maybe the answer is to move the parts or tools closer together, which will save time (about 0.001 minutes per inch), or to suspend a tool over the point of use, which will eliminate reaching for it and moving it so far. These savings are not as impressive as combining operations or eliminating operations, but they can be done on every job. Thus, we can reduce the cost of any job.

Motion study is design, and a job needs to be designed before a work station can be built, an operator trained, or a time study made. Motion study is a rewarding field in which to work. The techniques of motion study lead to many good solutions for cost reduction programs, which reflect favorably on the person doing the study.

WHAT IS A TIME STANDARD?

Before we can understand the importance and uses of time study, we must understand what we mean when we use the term *time standard*. The definition of a time standard is "the time required to produce a product at a work station with the following three conditions: (1) a qualified, well-trained operator, (2) working at a normal pace, and (3) doing a specific task." These three conditions are essential to the understanding of time study, so further discussion is needed.

A Qualified, Well-Trained Operator is required. Experience is usually what makes a qualified, well-trained operator, and time on the job is our best indication of experience. The time required to become qualified varies with the job and the person. For example, sewing machine operators, welders, upholsterers, machinists, and many other high-technology jobs require long learning periods. The greatest mistake made by new time study personnel is time-studying someone too soon. A good rule of thumb is to start with a qualified, fully trained person and to give that person two weeks on the job prior to the time study. On new jobs or tasks, predetermined time study systems are used. These standards seem tight (hard to achieve) at first because the times are set for qualified, well-trained operators.

Normal Pace is a concept we spend much time with in Chapter 9. Only one time standard can be used for each job, even though individual differences of operators cause different results. A normal pace is comfortable for most people. In the development of the normal pace concept, 100% will be the normal pace. Time standards of normal pace commonly used are

1. Walking 264 feet in 1.000 minutes (3 miles per hour),
2. Dealing 52 cards into four equal stacks in 0.500 minutes (at a bridge table),
3. Filling a 30-pin pinboard in 0.435 minutes (using two hands).

Training films for rating are also used to develop this concept.

A Specific Task is a detailed description of what must be accomplished. The description of the task must include

1. The prescribed work method,
2. The material specification,
3. The tools and equipment being used,
4. The positions of incoming and outgoing material,
5. Additional requirements like safety, quality, housekeeping, and maintenance tasks.

The time standard is only good for this one set of specific conditions. If anything changes, the time standard must change.

The written description of a time standard is important, but the mathematics is even more important. If a job takes 1.000 standard minutes to produce (Table 4-1), we can produce 60

Table 4-1 Time Conversion for Developing Time Standards.

MINUTES PER PIECE	PIECES PER HOUR[a]	HOURS PER PIECE[b]	HOURS PER 1,000 PIECES[c]
1.000	60	0.01667	16.67
.500	120	0.00833	8.33
.167	359	0.00279	2.79
2.500	24	0.04167	41.67
.050	1,200	0.000833	0.833

[a]Pieces per hour is calculated by dividing the minutes per piece into 60 minutes per hour.
[b]Hours per piece is calculated by dividing the pieces per hour into one hour (1/x).
[c]Hours per 1,000 pieces is calculated by multiplying the hours per piece by 1,000 pieces.

pieces per hour, and it will take 0.01667 hours to make one unit or 16.67 hours to make 1,000 units. The time in decimal minutes is always used in time study because the math is easier. The following three numbers are required to communicate a time standard:

1. The decimal minute (always in three decimal places, e.g., 0.001)
2. Pieces per hour (rounded off to whole numbers, unless less than 10 per hour)
3. Hours per piece (always in five decimal places, e.g., 0.00001). Many companies use hours per 1,000 pieces because the numbers are more understandable or meaningful.

Table 4-2 is a time standard conversion table that may be useful for a quick reference when needed. It can be used when either the minutes per unit, hours per unit, units per hour, or units per eight hours is known and you need to find the other three numbers pertaining to that standard. It can also be used to set goals for assembly lines or work cells. An interesting additional use is when jobs are added together and a new standard for the combined jobs is needed. Play with this table to understand the relationship between the different numbers that make up the term *standard time*. For example, if two jobs that need to be combined used to have the standards of 0.72 minutes per piece, or 83 pieces per hour, and 0.28 minutes per piece, or 214 pieces per hour, what is the new standard? Add 0.72 and 0.28 to get 1.00 minutes, or 60 pieces per hour combined.

Now that we understand what a time standard is, let's look at why time standards are considered to be one of the most important pieces of information produced in the manufacturing department.

THE IMPORTANCE AND USES OF TIME STUDY

The importance of time standards can be shown by the three statistics given in Chapter 1: 60%, 85%, and 120% performance. An operation that is not working toward standards typically works 60% of the time, whereas those operations working on standards work at 85% performance. This increase in productivity is equal to about 42%. In a small plant of 100 people, this improvement is equal to 42 fewer people, or about a million dollars per year in savings. Not only is the time standard very important, but it is extremely cost-effective.

The time standard is one of the most important pieces of information produced in the manufacturing department. It is used to develop answers for the following problems:

1. Determining the number of machine tools to buy
2. Determining the number of production people to employ
3. Determining manufacturing costs and selling prices
4. Scheduling the machines, operations, and people to do the job and deliver on time with less inventory
5. Determining the assembly line balance, determining the conveyor belt speed, loading the work cells with the correct amount of work, and balancing the work cells

Table 4-2 Time Standard Conversion Table: Minutes, Hours, Pieces per Hour, Pieces per Eight Hours.

STANDARD MINUTES	STANDARD HOURS	UNITS PER HOUR	UNITS PER 8 HOURS	STANDARD MINUTES	STANDARD HOURS	UNITS PER HOUR	UNITS PER 8 HOURS
480	8.000	0.1	1.0	0.98	0.01633	61.22	489.80
240	4.000	0.2	2.0	0.96	0.01600	62.50	500.00
160	2.667	0.4	3.0	0.94	0.01567	63.83	510.64
120	2.000	0.5	4.0	0.92	0.01533	65.22	521.74
96	1.600	0.6	5.0	0.9	0.01500	66.67	533.33
80	1.333	0.8	6.0	0.88	0.01467	68.18	545.45
70	1.167	0.9	6.9	0.86	0.01433	69.77	558.14
60	1.000	1.0	8.0	0.84	0.01400	71.43	571.43
50	0.833	1.2	9.6	0.82	0.01367	73.17	585.37
48	0.800	1.2	10.0	0.8	0.01333	75.00	600.00
45	0.750	1.3	10.7	0.78	0.01300	76.92	615.38
40	0.667	1.5	12.0	0.76	0.01267	78.95	631.58
38	0.633	1.6	12.6	0.74	0.01233	81.08	648.65
35	0.583	1.7	13.7	0.72	0.01200	83.33	666.67
32	0.533	1.9	15.0	0.7	0.01167	85.71	685.71
30	0.500	2.0	16.0	0.68	0.01133	88.24	705.88
28	0.467	2.1	17.1	0.66	0.01100	90.91	727.27
26	0.433	2.3	18.5	0.64	0.01067	93.75	750.00
25	0.417	2.4	19.2	0.62	0.01033	96.77	774.19
24	0.400	2.5	20.0	0.6	0.01000	100.00	800.00
23	0.383	2.6	20.9	0.58	0.00967	103.45	827.59
22	0.367	2.7	21.8	0.56	0.00933	107.14	857.14
21	0.350	2.9	22.9	0.54	0.00900	111.11	888.89
20	0.333	3.0	24.0	0.52	0.00867	115.38	923.08
19	0.317	3.2	25.3	0.5	0.00833	120.00	960.00
18	0.300	3.3	26.7	0.48	0.00800	125.00	1000.00
17	0.283	3.5	28.2	0.46	0.00767	130.43	1043.48
16	0.267	3.7	30.0	0.44	0.00733	136.36	1090.91
15	0.250	4.0	32.0	0.42	0.00700	142.86	1142.86
14	0.233	4.3	34.3	0.4	0.00667	150.00	1200.00
13	0.217	4.6	36.9	0.38	0.00633	157.89	1263.16
12	0.200	5.0	40.0	0.36	0.00600	166.67	1333.33
11	0.183	5.5	43.6	0.34	0.00567	176.47	1411.76
10	0.167	6.0	48.0	0.32	0.00533	187.50	1500.00
9	0.150	6.7	53.3	0.3	0.00500	200.00	1600.00
8	0.133	7.5	60.0	0.28	0.00467	214.29	1714.29
7	0.117	8.6	68.6	0.26	0.00433	230.77	1846.15
6	0.100	10.0	80.0	0.24	0.00400	250.00	2000.00
5	0.083	12.0	96.0	0.22	0.00367	272.73	2181.82
4	0.067	15.0	120.0	0.2	0.00333	300.00	2400.00
3	0.050	20.0	160.0	0.18	0.00300	333.33	2666.67
2	0.033	30.0	240.0	0.16	0.00267	375.00	3000.00
1	0.017	60.0	480.0	0.14	0.00233	428.57	3428.57
				0.12	0.00200	500.00	4000.00
				0.1	0.00167	600.00	4800.00
				0.08	0.00133	750.00	6000.00
				0.06	0.00100	1000.00	8000.00
				0.04	0.00067	1500.00	12000.00
				0.02	0.00033	3000.00	24000.00

6. Determining individual worker performance and identifying operations that are having problems so the problems can be corrected

7. Paying incentive wages for outstanding team or individual performance

8. Evaluating cost reduction ideas and picking the most economical method based on cost analysis, not opinion

9. Evaluating new equipment purchases to justify their expense

10. Developing operation personnel budgets to measure management performance.

A discussion of each of these uses of time study follows.

How would you answer the following questions without time standards?

1. How Many Machines Do We Need?

One of the first questions raised when setting up a new operation or starting production on a new product is "How many machines do we need?" The answer depends on two pieces of information:

a. How many pieces do we need to manufacture per shift?

b. How much time does it take to make one part? (This is the time standard.)

EXAMPLE:

1. The marketing department wants us to make 2,000 wagons per 8-hour shift.

2. It takes us 0.400 minutes to form the wagon body on a press.

3. There are 480 minutes per shift (8 hours/shift $\times$ 60 minutes/hr).

4. −50 minutes downtime per shift (breaks, clean-up, etc.)

5. There are 430 minutes per shift available @ 100%.

6. @ 75% performance (based on history or expectation) (0.75 $\times$ 430 = 322.5).

7. There are 322.5 effective minutes left to produce 2,000 units.

8. $\dfrac{322.5}{2,000 \text{ units}}$ = 0.161 minutes per unit, or 6.21 parts per minute.

The 0.161 minutes per unit is called the *takt* time[1] or plant rate. Every operation in the plant must produce a part every 0.161 minutes; therefore, how many machines do we need for this operation?

$$\frac{\text{Time standard} = 0.400 \text{ minutes/unit}}{\text{Plant rate} = 0.161 \text{ minutes/unit}} = 2.48 \text{ machines}$$

This operation requires 2.48 machines. If other operations are required for this kind of machine, we would add all the machine requirements together and round up to the next

---

[1]*Takt time* is a German word for plant rate. It is the available minutes divided by desired output.

whole number. In this example, we would buy three machines. (Never round down on your own. You will be building a bottleneck in your plant.)

2. How Many People Should We Hire?

Look at the operations chart shown in Figure 4-1. From a study of this chart, we find the time standard for every operation required to fabricate each part of the product and each assembly operation required to assemble and pack the finished product.

In the operation shown here (casting the handle), the 05 indicates the operation number. Usually, 05 is the first operation of each part. The 500 is the pieces per hour standard. This operator should produce 500 pieces per hour. The 2.0 is the hours required to produce 1,000 pieces. At 500 pieces per hour, it would take us 2 hours to make 1,000. How many people would be required to cast 2,000 handles per shift?

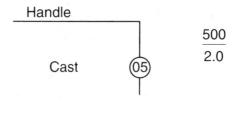

$$\begin{array}{r} 2{,}000 \text{ units} \\ \times\ 2.0 \text{ hours}/1{,}000 \\ \hline 4.0 \text{ hours at standard.} \end{array}$$

Not many people, departments, or plants work at 100% performance. How many hours would be required if we work at the rate of 60%, 85%, or 120%?

$$\frac{4 \text{ hours}}{60\%} = 6.66 \text{ hours}; \quad \frac{4 \text{ hours}}{85\%} = 4.7 \text{ hours}; \quad \frac{4 \text{ hours}}{120\%} = 3.33 \text{ hours}.$$

Therefore, depending on anticipated performance, we will be budgeted for a specific number of hours. Either performance history or national averages would be used to factor the 100% hours to make them practical and realistic.

Look again at the operations chart shown in Figure 4-1. Note the total 138.94 hours at the bottom right side. The operations chart includes every operation required to fabricate, paint, inspect, assemble, and pack out a product. The total hours is the total time required to make 1,000 finished products. In our water valve factory, we need 138.94 hours at 100% to produce 1,000 water valves. If this is a new product, we could expect 75% performance during the first year of production. Therefore,

$$\frac{138.94 \text{ hours per } 1{,}000}{75\% \text{ performance}} = 185 \text{ hours}/1{,}000,$$

where 75% = 0.75.

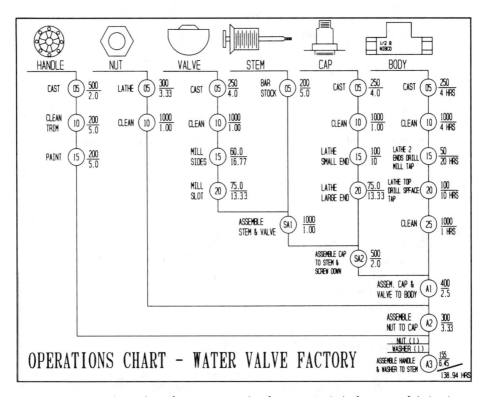

FIGURE 4-1 Operations chart for a water valve factory: A circle for every fabrication, assembly, and packout operation.

The marketing department has forecasted sales of 2,500 water valves per day. How many people are needed to make water valves?

$$185 \text{ hours}/1,000 \times 2.5 (1,000) = 463 \text{ hours/day needed.}$$

Divide this by 8 hours per employee per day, which is equal to 58 people. Management will be judged by how well it performs to this goal. If less than 2,500 units are produced per day with the 58 people, management will be over budget, and that is unforgivable. If it produces more than 2,500 units per day, management is judged as being good at managing, and the managers are promotable.

Most companies produce more than one product. The problem of how many people to hire is the same. For example, how many direct labor employees do we need for a

PRODUCT	HOURS/1,000	NO. OF UNITS NEEDED/DAY	HOURS AT 100%	ACTUAL %	ACTUAL HOURS NEEDED
A	150	1,000	150.0	70	214
B	95	1,500	142.5	85	168
C	450	2,000	900.0	120	750
					Total 1,132 hours

multiproduct plant? Per day, 1,132 hours of direct labor are needed. Each employee will work 8 hours; therefore,

$$\frac{1,132 \text{ hours}}{8 \text{ hours/employee}} = 141.5 \text{ employees.}$$

We will hire 142 employees, and management will be evaluated on the performance of these 142 employees. Without time standards, how many employees would you hire? Any other method would be a guess. The need for high-quality time standards is illustrated by this example. Management doesn't want to be judged and compared to unattainable time standards or production goals.

3. How Much Will Our Product Cost?

At the earliest point in a new product development project, the anticipated cost must be determined. A feasibility study will show top management the profitability of a new venture. Without proper, accurate costs, the profitability calculations would be nothing but a guess. Product costs consist of

		TYPICAL %
Manufacturing costs 50%	Direct labor	8
	Direct materials	25
	Overhead costs	17
	Plus	
Front-end costs 50%	Sales & distribution costs	15
	Advertising	5
	Administrative overhead	20
	Engineering	3
	Profit	7
		100%

Direct labor cost is the most difficult component of product cost to estimate. Time standards must be set prior to any equipment purchase or material availability. Time standards are set using predetermined time standards or standard data from blueprints and work station sketches. The time standards are collected on something like the operations chart shown in

Figure 4-1. On the bottom right side of the water valve operation sheet, we found the hours required to produce 1,000 units:

$$\frac{138.94 \text{ hours per } 1,000 \text{ units}}{85\% \text{ anticipated performance}} = 163.46 \text{ hours}/1,000$$

$$\begin{array}{l} 163.46 \quad \text{hours per } 1,000 \text{ water valves} \\ \underline{\times \$15.00 \quad \text{per hour labor rate}} \\ \$2,451.90/1,000, \text{ or } \$2.45 \text{ each.} \end{array}$$

Direct material is the material that makes up the finished product and is estimated by calling vendors for a bid price. Direct material costs are typically about 50% of the manufacturing cost (direct labor + direct materials + factory overhead). For our example, we will use 50%. On the operating chart, raw materials are introduced at the top of each line. Buyout parts are introduced at the assembly and packout station.

Factory overhead costs are all the expenses of running a factory except for the previously discussed direct labor and direct material. Factory overhead is calculated as a percent of direct labor. This percent is calculated using last year's actual costs. All manufacturing costs for last year are divided into three classifications:

Direct labor	$1,000,000
Direct material	3,000,000
Overhead	2,000,000
Total factory costs	$6,000,000.

The factory overhead rate for last year is

$$\frac{\$2,000,000 \text{ overhead cost}}{\$1,000,000 \text{ direct labor costs}} = 200\% \text{ overhead rate/labor dollar.}$$

Thus, each dollar in direct labor cost has a factory overhead cost of $2.00.

Example:	Water valve	
	Labor	$ 2.45 (from time standards)
	Overhead	$ 4.90 (200% overhead rate)
	Material	$ 7.35 (from our suppliers)
Total factory cost		$14.70
All other costs		14.70 (from ratios)
Selling price		$29.40.

Labor cost is the most difficult cost to calculate of all the costs that make up selling price. How could you calculate selling price without time standards? Anything else is a guess.

Cost estimating is an important part of any industrial management program and should be a complete course covering operations, product, and project costing. Motion and time study would, of course, be a prerequisite.

4. How Do We Schedule and Load Machines, Work Centers, Departments, and Plants?

Even the simplest manufacturing plant must know when to start an operation for the parts to be available on the assembly line. The more operations, the more complicated the scheduling.

EXAMPLE: One machine plant operates at 90%, 16 hours/day.

BACKLOG JOB	HOURS/1,000	UNITS REQUIRED	HOURS REQUIRED	BACKLOG—CUMULATIVE HOURS	BACKLOG—DAYS
A	5	5,000	27.8	27.8	1.74
B	2	10,000	22.2	50.0	3.12
C	4	25,000	111.1	161.1	10.07
D	3	40,000	133.3	294.4	18.40

The Gantt Chart in Figure 4-2 shows the same information as the preceding data. This plant operates a single machine 16 hours per day, five days per week. There are 294.4 hours of backlog at 16 hours per day = 18.4 days of work in the backlog. What if a customer comes in with a job and wants it in 10 days? The job is estimated to take only 48 hours of machine time. Can you deliver? What about the other four jobs? When have you promised them?

One scheduling philosophy is that operating departments are compared to buckets of time. The size of the bucket is the number of hours that each department can produce in a 24-hour day. For example,

DEPT.	NO. OF MACHINES	TWO SHIFTS HOURS PER DAY AVAILABLE	HISTORICAL DEPARTMENT PERFORMANCE	HOURS CAPACITY
Shears	2	32	85	27.2
Presses	6	96	90	86.4
Press breaks	4	64	80	51.2
Welding	4	64	75	48.0
Paint	3	48	95	45.6
Assembly line	1	80	90	72.0

Therefore, the scheduler can keep adding work to any department for a specific day until the hour capacity is reached; then it spills over to the next day.

Without good time standards, manufacturing management would have to carry great quantities of inventory to avoid running out of parts. Inventory is a great cost in manufacturing; therefore, knowledge of time standards will reduce inventory requirements, which will reduce cost. Production inventory control is an area of major importance in manufacturing, and time study should be a prerequisite.

As we've said before, how could you schedule a plant without time standards? Anything else is a guess.

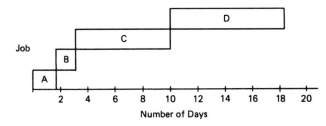

FIGURE 4-2 A picture of a machine's or department's time schedule of work.

5. How Do We Determine the Assembly Line and Work Cell Balance?

The objective of assembly line balancing is to give each operator as close to the same amount of work as possible. Balancing work cells has the same objective. It is of no value that one person or one cell has the ability to outrun the rest of the plant by 25%, because one person cannot produce more than comes to him or her or more than the subsequent operations can use. If the person has extra time, he or she could be given some of the work from a busier work station.

Assembly line balancing or work center loading can only be accomplished by breaking the job down into tasks that need to be performed and reassembling them into jobs or cells of near the same length of time. There will always be a work station or cell that has more work than any other. This station is defined as the 100% loaded station, or bottleneck station, and will limit the output of the whole plant. If we want to improve the assembly line (reduce the unit cost), we will concentrate on improving the 100% station. If we reduce the 100% station in the following example by 1%, we save an additional 1% for each person on the line because now they can all go 1% faster. We can keep reducing the 100% station until another work station becomes the 100% station (busiest station). Then our attention turns to this new 100% station for cost reduction. If we have 200 people on an assembly line and only one 100% station, we can save the equivalent of two people by reducing the 100% station by just 1%. We can use this multiplier to help us justify spending great sums of money to make small changes. (Assembly line balancing is discussed in detail in Chapter 10.)

EXAMPLE:

Assembly Line Balance.

OPERATOR NO.	OPERATION DESCRIPTION	STANDARD TIME IN MINUTES	LOAD %	PIECES PER HOUR	HOURS PER PIECE
1	Asemble 1, 2, 3	.250	84	200	.005
2	Assemble 4, 5, 6	.300[a]	100[a]	200	.005
3	Assemble 7, 8, 9	.275	92	200	.005
4	Assemble 10, 11, 12	.225	75	200	.005
				Total hours	.020[b]

[a]The busiest station on this line is work station 2, with .300 minutes of work. As soon as we identify the busiest station, we identify it as the 100% station, and by highlighting this standard and the 100% load, we communicate that only this standard is important. Even though they could work faster, every work station is limited to 200 pieces per hour, because station 2 is limiting the output of the whole assembly line.

[b].020 is the total hours required to assemble one finished unit. If we multiply this total hours by the average assembly wage rate, we have the total assembly labor cost. A better line balance is lower total hours.

See Figure 4-3 for an example of assembly line layout based on assembly line balance.

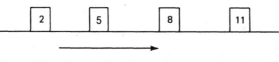

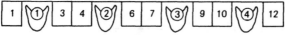

FIGURE 4-3 Assembly line layout based on assembly line balance.

Assembly line balancing usually includes a conveyor of some kind. The question that needs to be answered is, How fast should this conveyor run? Conveyor speed is measured in feet per minute, so if we know the distance between units on the assembly line, the size of the assembly, and the number of units required per minute, we can determine the conveyor speed to attain 100% performance.

For example, a charcoal gas grill assembly line has a *takt* time[2] of 0.225 minutes per assembled unit, and the space between box centers on the conveyor is 30 inches (2½ feet).

$$\frac{1 \text{ minute}}{0.225 \text{ minutes per grill}} = 4.444 \text{ grills per minute.}$$

The 4.444 grills per minute × 2.5 feet per grill = 11.11 feet per minute + allowance of 10% for personal fatigue and delay time = 12.22 feet per minute. Without adding the allowance back into the speed, we would never achieve 100% performance. (A much more detailed discussion of allowances is provided in Chapter 9.)

How else could we divide the work load equally and fairly without time standards? Any other way would be a guess. Line balancing is discussed in detail in Chapter 10.

6. How Do We Measure Productivity?

Productivity is a measure of output divided by input. If we are talking about labor productivity, then we are developing a number of units of production per hour worked.

EXAMPLE:

$$\text{Present} = \frac{\text{Output} = 1,000 \text{ units/day}}{\text{Input} = 50 \text{ people @ 8 hours/day}} = \frac{1,000}{400} = 2.5 \text{ units per hour worked.}$$

$$\text{Improved} = \frac{\text{Output} = 2,000 \text{ units/day}}{\text{Input} = 50 \text{ people @ 8 hours/day}} = \frac{2,000}{400} = 5.0 \text{ units per hour worked,}$$

or a 100% increase in productivity (a doubling of production).

---

[2]As noted previously, *takt* (a German word for plant rate and also referred to as the *R* value) time is the available minutes of production time divided by the rate of the line in pieces per day. For example, if our marketing department said we could sell 2,000 units per day, and we have 480 minutes available per shift less 30 minutes for breaks and clean-up, then we have a *takt* time of 450/2,000 = 0.225 minutes. This means that every machine or operation in the plant needs to produce a part every 0.225 minutes. If they don't, we will not produce 2,000 units. If one machine can't do it, we will need two machines, and so on.

We could also increase productivity by maintaining the output constant or reducing the number of people.

$$\text{Improved output} = \frac{1{,}000 \text{ units/day}}{40 \text{ people @ } 8 \text{ hours/day}} = \frac{2{,}000}{320} = 3.125 \text{ units per hour worked.}$$

These examples are good for plants or whole industries, but for individuals we use

$$\frac{\text{Earned hours}}{\text{Actual hours}} = \% \text{ performance.}$$

Earned hours are the hours of work earned by the operator based on the work standard and the number of pieces produced by the operator. For example, if an operator worked 8 hours and produced 1,000 units on a job with a time standard of 100 pieces per hour, we would have the following:

A. $\text{Earned hours} = \dfrac{1{,}000 \text{ pieces produced}}{100 \text{ pieces per hour}} = 10 \text{ hours.}$

B. Actual hours = 8 hours. Actual hours are the real time the operator works on the job (also called the time clock hours).

C. $\% \text{ performance} = \dfrac{\text{Earned hours}}{\text{Actual hours}} = \dfrac{10}{8} = 125\%.$

Industrial engineers will improve productivity by reporting performances of every operation, operator, supervisor, and production manager every day, week, month, and year. Performance reports are based on daily time cards filled out by operators and extended within the computer's performance control system. The performance control system is the subject of Chapter 13, and includes

1. Goal setting (setting time standards),
2. Comparison of actual performances with the goals,
3. Tracking results (graphing),
4. Reporting variances larger than acceptable limits,
5. Taking corrective action to eliminate causes of poor performances.

All five of these functions must be in place to have a functioning performance control system. A performance control system will improve performance by an average of 42% over performance with no control system. Companies without performance control systems typically operate at 60% of standard. Those companies with performance control systems will average 85% performance. This is accomplished by

1. Identifying nonproductive time and eliminating it,
2. Identifying poorly maintained equipment and fixing it,

3. Identifying causes for downtime and eliminating them,

4. Planning ahead for the next job.

Performance control systems hold problems up to the light of day and fix the problems. In plants that do not have standards, the employees know that no one cares how much they produce. Management's reaction to problems speaks louder than its words. How can supervisors know who is producing and who is not if they don't have standards? How would management know the magnitude of problems such as downtime for lack of maintenance, material, instruction, tooling, services, etc., if downtime is not reported?

We discuss the performance control system in Chapter 13.

7. How Can We Pay Our People for Outstanding Performance?

Every manufacturing manager would like to be able to reward outstanding employees. Every supervisor knows who he or she can count on to get the job done. Yet only about 25% of production employees have an opportunity to earn increased pay for increased output. A 1980, 400-plant study by an industrial engineering consultant, Mitchell Fein, found that when employees are paid via incentive systems, their performance improves by 41% over measured work plans and 65% over systems with no standards or no performance controls.

Stage I   Plants with no standards operate at 60% performance.

Stage II   Plants with standards and a performance control system operate at 85% performance.

Stage III   Plants with incentive systems operate at 120% performance.

A small company with 100 employees could save about $820,000 per year ($20,000/yr × 100 × 41%) on labor costs going from standards to incentive systems, and the employees will take home 20% more pay.

A National Science Foundation study found that when workers' pay was tied to their efforts,

1. Productivity improved,

2. Cost was reduced,

3. Workers' pay increased, and

4. Workers' morale improved.

Everyone can win with incentive systems. What is management waiting for? Chapter 14 discusses group and individual incentive systems.

8. How Can We Select the Best Method by Evaluating Cost Reduction?

A basic rule of production management is that all expenditures must be cost justified. Second, a basic rule of life is that everything changes. We must keep improving or become obsolete. To justify all expenditures, the savings must be calculated. This is called the *return*. The cost of making the change is also calculated. This is called the *investment*. When the return is divided by the investment, the resulting ratio indicates the desirability of the project. This ratio is called the ROI, or return on investment. To provide a uniform method of evaluating ROI, annual savings are used; therefore, all percents are per year.

EXAMPLE: We have been producing product A for several years and look forward to several more years of sales at 500,000 units per year or 2,000 units per day. The present method requires a standard time of 2.0 minutes per unit or 30 pieces per hour. At this rate, it takes 33.33 hours to make 1,000 units. All production will run on the day shift.

A. Present Method and Costs With a labor rate of $10.00/hour, the labor cost will be $333.30 to produce 1,000 units. The cost of 500,000 units per year would be $166,650.00 in direct labor.

$$\frac{1,000 \text{ pieces}}{30 \text{ pieces per hour}} = 33.33 \text{ hours/1,000 units.}$$

B. New Method and Costs We have a cost reduction idea. If we buy this new machine attachment for $1,000, the new time standard would be lowered to 1.5 minutes per unit. Will this investment be good for us?

First, how many attachments will we have to buy to produce 500,000 units per year?

$$\frac{500,000 \text{ units/year}}{250 \text{ days/year}} = 2,000 \text{ units/day}$$

 480 minutes/shift
$\underline{-50 \text{ minutes/shift downtime}}$
 430 minutes/shift @ 100%
$\underline{@80\% \text{ expected efficiency}}$
 344 effective minutes available to produce 2,000 units/shift

$$\frac{344 \text{ minutes}}{2,000 \text{ units}} = 0.172 \text{ minutes/unit.}$$

To produce 2,000 units per shift, we need a part every 0.172 minutes.

$$\text{Number of machines} = \frac{1.50 \text{ minutes/cycle}}{0.172 \text{ minutes/unit}} = 8.7 \text{ machines.}$$

We will purchase nine attachments at $1,000 each. Our investment will be $9,000 (9 × 1,000).

Second, what is our labor cost?

$$\text{Pieces per hour} = \frac{60 \text{ minutes/hour}}{1.5 \text{ minutes/part}} = 40 \text{ parts/hour or } 25 \text{ hours/1,000}$$

25 hours/1,000 $\times$ \$10.00/hour wage rate = \$250/1,000
500,000 units will cost 500 $\times$ \$250 = \$125,000.

New labor costs will be \$125,000 per year.

C. Savings: Direct Labor Dollars

Old method $\quad$ \$166,650 per year
New method $\underline{\$125,000}$ per year
Savings $\qquad$ \$ 41,650 per year

$$\frac{\text{Return (savings) \$41,650/year}}{\text{Investment (cost) 9,000}} = 463\%$$

ROI = 463%
463% = 0.216 years to pay off, or 2.59 months.

D. Return on Investment This investment will pay for itself in less than three months. If you were the manager, would you approve this investment? Of course you would, and so would anyone.

Cost reduction programs are important to the well-being of a company and the peace of mind of the industrial engineering department. A department that shows a savings of \$100,000 per employee per year doesn't have to worry about layoffs or elimination. A properly documented cost reduction program will update all time standards as soon as methods are changed. Every standard affected must be changed immediately.

Cost reduction calculations can be a little more complicated than our example, which did not account for

1. Taxes,
2. Depreciation,
3. Time value of money,
4. Surplus machinery—trade-in,
5. Scrap value.

9. How Do We Evaluate New Equipment Purchases to Justify Their Expense?

The answer to this question is the same as the answer to Question 8. Every new machine is a cost reduction. All other reasons can be converted to costs or benefits.

10. How Do We Develop a Personnel Budget?

This question was answered in Question 2 when we determined the number of people to hire. Budgeting is one of the most important management tools, and one must understand it fully to become an effective manager. It is said that you become a manager when you are responsible for a budget. You are a promotable manager when you come in under budget at the end of the year. Budgeting is a part of the cost estimating process. Labor is only one part of the budget, but it is one of the most difficult to estimate and control. How would we do it without time standards? It would be a very expensive guess.

How can managers guess at such important decisions as those just listed? Much of manufacturing management has received no formal training in making these decisions. It will be your job to show them the scientific way to manage their operations.

TECHNIQUES OF TIME STUDY

Time study covers a wide variety of situations. At one time a job must be designed, work stations and machines built, and a time standard set before the plant is built. In this situation, a PTSS (predetermined time standards system) or standard data would be the techniques used to set the time standard. Once a machine or work station has been operated for a while, the stopwatch technique is used. Some jobs occur once or twice a week, while others repeat thousands of times per day. Some jobs are very fast, while others take hours. Which technique do we use? The job of an industrial engineer and technologist is to choose the correct technique for each situation and correctly apply that technique.

Five techniques of time study are presented in this text:

1. Predetermined time standards systems
2. Stopwatch time study
3. Work sampling
4. Standard data
5. Expert opinion time standards and historical data.

A brief description of these five techniques is presented next. Each technique will be developed fully in its own chapter.

Predetermined Time Standards Systems

When a time standard is needed during the planning phase of a new product development program, the PTSS technique is used. At this stage of new product development, only sketchy information is available, and the technologist must visualize what is needed in the way of tools, equipment, and work methods. The technologist would design a work station for each step of the new product manufacturing plan. Each work station would be designed, a motion pattern would be developed, each motion would be measured, a time value would be assigned, and the total of these time values would be the time standard. This time standard would be used to determine the equipment, space, and people needs of the new product and its selling price.

Frank and Lillian Gilbreth developed the basic philosophy of predetermined time motion systems. They divided work into 17 work elements:

1. Transport empty
2. Search
3. Select
4. Grasp
5. Transport loaded
6. Pre-position
7. Position
8. Assemble
9. Disassemble
10. Release load
11. Use
12. Hold
13. Inspect
14. Avoidable delay
15. Unavoidable delay
16. Plan
17. Rest to overcome fatigue.

These 17 work elements, as mentioned in Chapter 3, are known as *therbligs*. Each therblig was reduced to a time table, and when totaled, a time standard for that set of motions was determined.

Methods time measurement (MTM)[3], MODAPTS (MODular Assignment at Predetermined TimeS), and Work Factors are popular predetermined time systems inspired by the Gilbreths' work. PTSS was developed from MTM and other predetermined time systems for the expressed reason to teach a system within a few hours. PTSS is a simplified system. It is a good system, but additional training is desirable. If you understand the concepts of PTSS, you will have a head start in learning any other system.

Figure 4-4 shows an example of a PTSS. PTSS is the first technique of time study covered in this text, and it is the last technique in the methods chapters (Chapter 8) because it is both a methods and a time study technique.

Stopwatch Time Study

Stopwatch time study is the method most manufacturing employees think of when talking about time standards. Frederick W. Taylor started using the stopwatch around 1880 for studying work. This technique is a part of many union contracts with manufacturing companies.

---

[3]MTM refers to all MTM systems—MTM-1, MTM-2, MTM-3, and MTM-4.

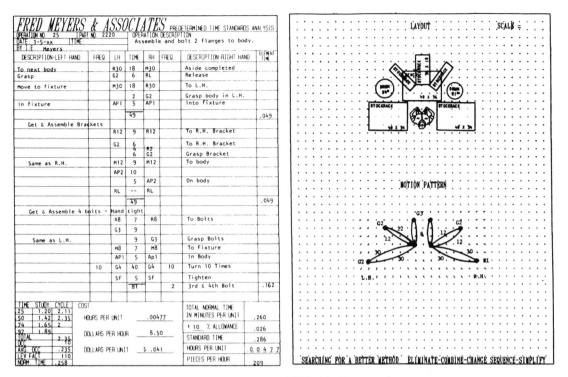

FIGURE 4-4 PTSS example.   PTSS side B example.

Time study is defined as the process of determining the time required by a skilled, well-trained operator working at a normal pace doing a specific task. Several types of stopwatches could be used:

1. Snapback: in one hundredths of a minute

2. Continuous: in one hundredths of a minute

3. Three watch: continuous watches

4. Digital: in one thousandths of a minute

5. TMU (time-measured unit): in one hundred thousandths of an hour

6. Computer: in one thousandths of a minute.

All but the TMU watch read in decimal minutes. The TMU watch reads in decimal hours. Digital watches and the computer are much more accurate, and many have memory functions that improve accuracy. Nearly 25% of this motion and time study book is dedicated to the stopwatch time study technique because of its traditional importance.

Two different time study procedures are covered in this book:

1. Continuous time study (Figure 4-5 is used on short-duration jobs)

2. Long-cycle time study (see Figure 4-6).

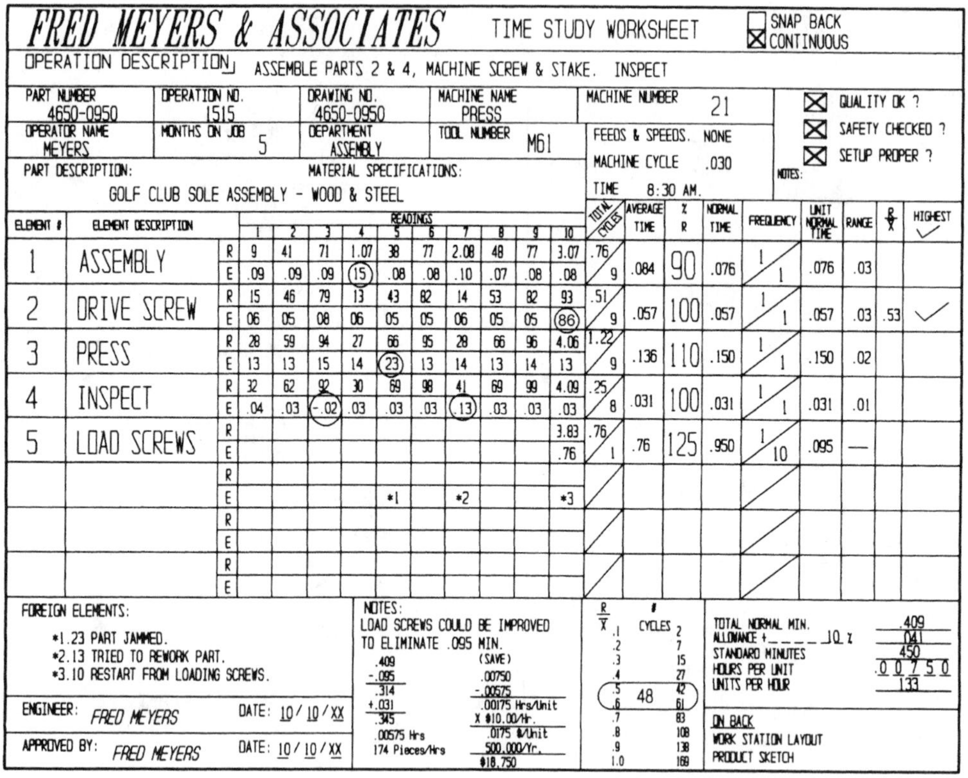

FRED MEYERS & ASSOCIATES   TIME STUDY WORKSHEET   ☐ SNAP BACK   ☒ CONTINUOUS

OPERATION DESCRIPTION: ASSEMBLE PARTS 2 & 4, MACHINE SCREW & STAKE. INSPECT

PART NUMBER 4650-0950	OPERATION NO. 1515	DRAWING NO. 4650-0950	MACHINE NAME PRESS	MACHINE NUMBER 21	☒ QUALITY OK ?
OPERATOR NAME MEYERS	MONTHS ON JOB 5	DEPARTMENT ASSEMBLY	TOOL NUMBER M61	FEEDS & SPEEDS. NONE / MACHINE CYCLE .030	☒ SAFETY CHECKED ? / ☒ SETUP PROPER ? NOTES:
PART DESCRIPTION: GOLF CLUB SOLE ASSEMBLY - WOOD & STEEL		MATERIAL SPECIFICATIONS:		TIME 8:30 AM.	

ELEMENT #	ELEMENT DESCRIPTION		READINGS 1–10	TOTAL CYCLES	AVERAGE TIME	% R	NORMAL TIME	FREQUENCY	UNIT NORMAL TIME	RANGE	R/X̄	HIGHEST ✓
1	ASSEMBLY	R	9 41 71 1.07 38 77 2.08 48 77 3.07	.76 / 9	.084	90	.076	1/1	.076	.03		
		E	.09 .09 .09 (15) .08 .08 .10 .07 .08 .08									
2	DRIVE SCREW	R	15 46 79 13 43 82 14 53 82 93	.51 / 9	.057	100	.057	1/1	.057	.03	.53	✓
		E	06 05 08 06 05 05 06 05 05 (86)									
3	PRESS	R	28 59 94 27 66 95 28 66 96 4.06	1.22 / 9	.136	110	.150	1/1	.150	.02		
		E	13 13 15 14 (23) 13 14 13 14 13									
4	INSPECT	R	32 62 92 30 69 98 41 69 99 4.09	.25 / 8	.031	100	.031	1/1	.031	.01		
		E	.04 .03 (-.02) .03 .03 .03 (.13) .03 .03 .03									
5	LOAD SCREWS	R	3.83 .76	.76 / 1	.76	125	.950	1/10	.095	—		
		E	.76									
		R	*1 *2 *3									
		E										

FOREIGN ELEMENTS:
*1.23 PART JAMMED.
*2.13 TRIED TO REWORK PART.
*3.10 RESTART FROM LOADING SCREWS.

ENGINEER: FRED MEYERS   DATE: 10/10/XX
APPROVED BY: FRED MEYERS   DATE: 10/10/XX

NOTES:
LOAD SCREWS COULD BE IMPROVED
TO ELIMINATE .095 MIN.

```
  .409          (SAVE)
 -.095          .00750
  .314         -.00575
 +.031         .00175 Hrs/Unit
  .345       X $10.00/Hr.
 .00575 Hrs    .0175 $/Unit
 174 Pieces/Hrs  500,000/Yr.
               $18,750
```

R/X̄	# CYCLES
.1	2
.2	7
.3	15
.4	27
.5	42 / 48
.6	61
.7	83
.8	108
.9	138
1.0	169

TOTAL NORMAL MIN. .409
ALLOWANCE + ____ 10 % .041
STANDARD MINUTES .450
HOURS PER UNIT .00750
UNITS PER HOUR 133

ON BACK
WORK STATION LAYOUT
PRODUCT SKETCH

FIGURE 4-5 Time study example: continuous form.

Long-cycle time study may be used for either very long jobs (31 minutes or more) or 8-hour studies, or for jobs where the elements are often performed out of sequence.

The 8-hour time study is used to find out what causes an operation's poor performance. Figure 4-7 shows an analysis of what caused a quart canning line to be down. Each line on the analysis represents 1 hour. Each line is divided into 60 parts (one minute each). An X through five minutes means the machine was down for five minutes. The number above the X indicates the reason for the downtime. At the end of the study, the analysis shows how many times the machine stopped for each reason and the total downtime for that reason. This total downtime can be converted easily to total dollars, which gives an indication of how much money we can spend to solve each problem.

Work Sampling

Work sampling is the same scientific process used in Nielsen ratings, Gallup Polls, attitude surveys, and federal unemployment statistics. We observe people working and draw conclu-

ELEMENT #	ELEMENT DESCRIPTION Started 7:00 AM Ended 3:30 PM	ENDING WATCH READING	ELEMENT TIME	% R	NORM. TIME
1	Shift Start up--No production	7:05	5.0	100	5.0
2	Run	7:06	1.0	100	1.0
3	Stopped--operator forgot something	7:07½	1½	0	----
4	Run	7:14	6½	100	6.5
5	No Lids	7:16½	2½	0	----
6	Run	7:19	2½	100	2.5
7	Check Temperature	7:20½	1½	70	1.05
8	Run	24	3½	100	3.5
9	Box Jammed in Former	25	1	110	1.1
10	Run	28½	3½	100	3.25
11	Bad Can	29	3/4	120	.9
12	Palletizer Jam-Bad Pallet	31	2	110	2.2
13	Run	33	2	100	2
14	Bad Box in Former	7:33½	½	130	.65
15	Run	7:41	7½	100	7.5
16	Bad Box in Former	7:41½	½	140	.7
17	Run	7:51	9½	100	9.5
18	Bad Box in Former	7:54	3	120	3.6
19	Run	7:56	2	100	2.0
20	No Lids in Machine	8:00½	4½	0	----

FRED MEYERS & ASSOCIATES LONG CYCLE TIME STUDY WORK SHEET

PART NO. Quart Line
OPERATION NO. Line
DATE/TIME 10/10/xx
BY I.E. Meyers

OPERATION DESCRIPTION: 4 people (loader, Machine, Cartons, Unloader) Automatic Quart Line Cann
MACHINE; TOOLS, JIGS: #1--300 CANS/Minute
MATERIAL: Motor oil any weight

FIGURE 4-6 Time study example: long-cycle time study worksheet, page 1 of 8.

sions. Everyone who has ever worked with someone else has done work sampling; you have an opinion of how hard this other person works:

1. "Every time I look at him, he's working"; or

2. "He's never working"; or

3. Somewhere in between.

Supervisors, using informal work sampling, are forming attitudes about employees all the time.

Industrial technologists can walk through a plant and state, "This plant is working at 75% performance." They should continue to say ±10% or so, depending on how many people

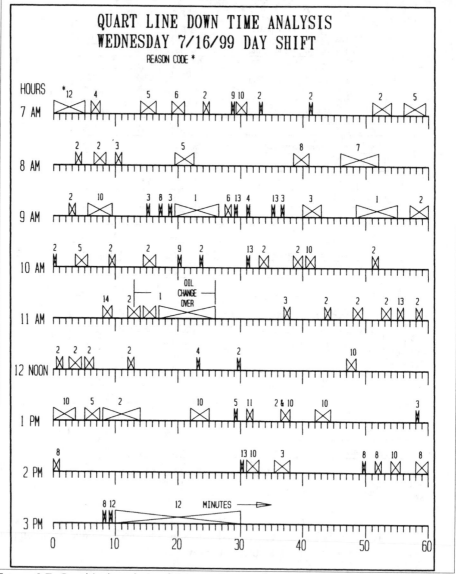

FIGURE 4-7 Graphical analysis of 8-hour time study.

they observed (number of samples). You could walk through a plant of 250 people one time and count people who are working and those who are not working and calculate the performance of that plant within ±10%. Industrial engineering consultants often start their consulting proposal with such statistics. Consultants expect to find 60% performance in plants without standards, but that is an average. A specific plant may have better management and be averaging between 70 and 75%.

Setting standards using work sampling is not very difficult. The industrial engineer samples a department and finds the following statistics:

TASK	NO. OF OBSERVATIONS	% TOTAL	HOURS WORKED	PIECES PRODUCED	PIECES/HOUR[a]
Assemble	2,500	62.5%	625	5,000[b]	8
Idle	1,500	37.5%	375	—	
Total	4,000	100.0%	1,000[c]		

[a]Pieces per hour = $\dfrac{\text{pieces} = 5,000}{\text{hours} = 625}$ = 8 pieces per hour.

[b]From supervisor (number finished products put in the warehouse).

[c]From payroll (hours paid during our study).

Eight pieces per hour is not quite the time standard. We haven't added allowances. How much time is in the 625 hours for breaks, scheduled or unscheduled? How much time is in there for delays? None. Actual hours worked is 625. All other nonwork time is part of the 375 hours that we throw away. We could add an appropriate amount of extra time to cover personal time, fatigue time, or delay time. This extra time is called *allowances*. Ten percent extra time is considered normal. A time standard of 7.3 pieces per hour would be appropriate.

Chapter 11 includes a step-by-step procedure for conducting a full work sampling study.

Standard Data

Standard data should be the objective of every motion and time study department. Standard data is the fastest and cheapest technique of setting time standards, and standard data can be more accurate and consistent than any other technique of time study. Starting with many previously set time standards, the industrial technologist tries to figure out what causes the time to vary from one job to another on a specific machine or class of machine. For example, walking time would be directly proportional to the number of feet, paces, yards, or meters walked. There might be two curves on the graph: obstructed and unobstructed.

A second example is counting playing cards. Time for counting cards would be directly proportional to the number of cards counted. Can you think of any other reasons for the time to vary?

There are several ways of communicating the time standard to future generations of factory workers, supervisors, and engineers:

1. Graph (see Figure 4-8)

2. Table

3. Worksheet (again, see Figure 4-8)

4. Formula.

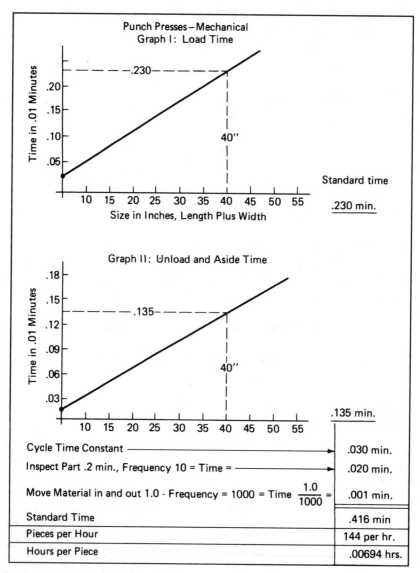

FIGURE 4-8 Standard data worksheet example.

Metal-cutting machines are examples of the need for and use of formulas. A blueprint calls for the drilling of a hole through the steel plate. Three pieces of information are required:

1. What size is the hole?
2. What is the material?
3. What tool do we use?

With this information, we look up the feeds and speeds in the *Machinery's Handbook*.[4] Feeds and speeds are communicated as follows:

Speed	500 feet/min
Feed	0.002 inches per revolution.

By substituting this information into three simple formulas, we determine the time standard.

Other machines, like welders, have simpler formulas, such as 12 inches per minute. The machine manufacturers are a good source of standard data.

Expert Opinion Time Standards and Historical Data

An expert opinion time standard is an estimation of the time required to do a specific job. This estimate is made by a person with a great experience base. Many people say, "You can't set time standards on my work." A good response would be, "You are right, but I know someone who can—you can!" The one-of-a-kind nature of many staff and service jobs makes setting time standards with the more traditional techniques unprofitable. Engineering, maintenance, and some office workers never do the same thing twice, but goals are still needed (time standards). Maintenance work is controlled by work order. Why not ask an expert how long this requested work will take? In well-managed companies, new maintenance projects will not be approved until the job is estimated. These time standards would be used to schedule and control maintenance work, just as you would schedule and control the work performed by a machine operator.

The expert in an expert opinion time standards system is usually a supervisor. In larger departments, a specialist may be used. For example, in the maintenance department, this person would be called a maintenance planner. The expert would estimate every job and maintain a backlog of work. The backlog of work would give the department time to plan the job, thereby performing that job more effectively. Expert opinion time standards and the backlog control system are discussed in detail in Chapter 13.

Historical data is an accounting approach to expert opinion time standard systems. A record is kept of how much time was used on each job. When a new job comes along, it is compared to a previous job standard. These standards are then used in a labor performance control system. The problem with historical time standards is that they do not reflect the time the job should have taken. Inefficiency is built into such a system, but a bad standard is better than no standard at all.

Table 4-3 will help you choose the correct technique for setting time standards.

---

[4]*Machinery's Handbook*, The Industrial Press, New York.

Table 4-3 Which Time Standards Technique Do We Use?

CYCLE TIME	VOLUME OF PRODUCTION		
	HIGH 1,000s	MEDIUM 100s	LOW 10s
Long	Work sampling	Work sampling	Expert opinion
		Stopwatch	Work sampling
			History
Medium	Work sampling	Stopwatch	Expert opinion
	Stopwatch	Work sampling	History
	PTSS		Stopwatch
Short	PTSS	PTSS	Stopwatch
		Stopwatch	Expert opinion

Note: Standard data is the ultimate time standard technique and can be used in all situations. Standard data should be the goal of all time study departments.

QUESTIONS

1. What is a macromotion study?
2. What are the four ways an operation can be improved?
3. What are the six charts in micromotion studies?
4. Who is famous for micromotion studies?
5. Can motion studies be applied to work cells?
6. What is normal pace?
7. Why must the operator being studied be experienced?
8. How many standard minutes in a class period?
9. What is assembly line balance?
10. Why is cost an important part of work measurement?

CHAPTER 5

Techniques of Methods Design: The Product Flow Macromotion Study

INTRODUCTION

Prior to studying individual jobs, the technologist should study the overall flow of a product as it moves through the facility. Understanding as much as possible about the present condition prepares us to make improvements on that condition. In the case of a product to be manufactured, we take that product apart and study the manufacturing sequence of each part and the sequence of assembly of the parts into subassembly, finished product, and packed out. Techniques, discussed in this chapter, are designed to record all the information required to build a manufacturing facility complete with the proper number of people, machines, and tools. These techniques will lead the technologist to ask the right questions to improve even the newly conceived plan, and they force consideration of the best method of doing a job.

The techniques covered in this chapter will be used for overall product flow study. When a technician finishes a technique, he or she turns around and does it again—only better. As a new employee, once you have applied these techniques, you should know as much about your company's manufacturing systems as anyone.

COST REDUCTION FORMULA

As you can see, the cost reduction formula is not a mathematical formula but a formula (procedure) for thinking about cost reduction. Before a cost can be reduced, we need to understand that cost. The first column in Table 5-1 is a list of six questions that we ask to understand any

Table 5-1 Cost Reduction Formula.

ASK THESE QUESTIONS	FOR EACH	TO SEEK THESE RESULTS
Why	Operation	Eliminate
What	Transportation	Combine
When	Storage	
Who		
Where	Inspection	Reroute
How	Delay	Simplify

cost. In the cost reduction formula, we ask why, what, when, who, where, and how for each operation performed on every part. Every time we move (transport) a part, we ask why, what, when, who, where, and how again, so that we understand every move made in the process of making a product. Column two of Table 5-1 shows that we ask the six questions of column one about every operation, transportation, storage, inspection, and delay. The five things listed in column two are the only things that can happen to a part while in our manufacturing plants.

Once we have asked the six questions (column one) about all five things that can happen to a production part (column two), we know what needs to happen in our manufacturing plant to make a part. If we study every part of a product, we will know exactly how we manufacture our product. Once we know how the product is manufactured, we question every step to eliminate steps, combine steps, change sequences of steps, and, finally, simplify. Column three of Table 5-1 represents the substance of the cost reduction formula. We ask the following questions:

1. Can I eliminate this step?
2. Can I combine this step with another step or steps?
3. Can I rearrange the steps to make the flow shorter and/or smoother?
4. Can I simplify the step?

These four questions must be asked in this order because the elimination step can save the most, while the simplifying step will produce the smallest percentage of savings. For example, why try to simplify a step that could be (should be) eliminated?

The eliminated step could be any operation, transportation, storage, inspection, or delay. Delays are probably the easiest to eliminate, and lean manufacturing demands that we do this. Reducing lot size would be an example of simplification. Elimination of operations is more difficult and rare, but we must always try. The question, "Why is this operation, transportation, storage, inspection, or delay necessary?" can lead to elimination.

Combination of steps is best described by the example of a machine center. A golf club iron needs a 2-inch hole drilled in the hosel (shaft). This hole needs to be spot faced, reamed, and tapped (treaded). These four operations (drilling, spot facing, reaming, tapping) have been run on four different machines moving containers of 500 clubs around the shop. A new machine with six machine heads located around a powered indexing table allows one operator to take a completed golf club head out of a fixture and place a new club head into the fixture and then activate the machine. When the last of the six machines finishes its cycle, the round table indexes and all the machines and the operator repeat the cycle. The new machine combined all four operations into one, eliminating the transportation, inspection, storage, and delays between operations. Combining of operations into work cells is a big part of the lean manufacturing concept.

Changing the sequence of operations to create a smoother, shorter product flow can be accomplished in two ways:

1. Change the sequence of operations to agree with the layout.
2. Change the layout to agree with the sequence of operations.

The first choice is the cheapest, but moving machines around may be cost-justified.

The last item in column three of Table 5-1 reflects the question, "Can I simplify the cost?" This is almost always possible: We can reduce the cost of any job. With an operation, we can always move material and equipment closer to the point of use, thereby reducing the time required to reach to them and move them back to the point of use. If we move operations closer, we simplify the move; if we reduce the number of parts being stored or delayed, we are simplifying. Another way of looking at simplifying is to downgrade the complexity of a motion (this is detailed in Chapter 8).

The first four techniques of methods analysis studied in this chapter (flow diagram, operations chart, process chart, flow process chart) will assist us in defining the steps—*all* the steps (operation, transportation, storage, inspection, and delay)—so that we can learn as much about them as possible, improve the methods, and create a cost reduction. The last three techniques of motion study covered in this chapter (work cell chart, routing sheet, assembly line balance) are special situation techniques.

FLOW DIAGRAMS

The flow diagram (see Figure 5-1) shows the path traveled by a part from receiving to stores to fabrication to subassembly to final assembly to packout to warehousing to shipping. Each path is drawn on a layout of the plant. Several parts can be shown on one flow diagram, but many pages may be required. Transparency overlays are useful.

The flow diagram will point out problems with such things as the following:

1. Cross Traffic

Cross traffic is where flow lines cross. Cross traffic is undesirable, and a better layout would have fewer intersecting paths. Anywhere traffic crosses is a problem because of congestion and safety considerations. Proper placement of equipment, service, and departments will eliminate most cross traffic.

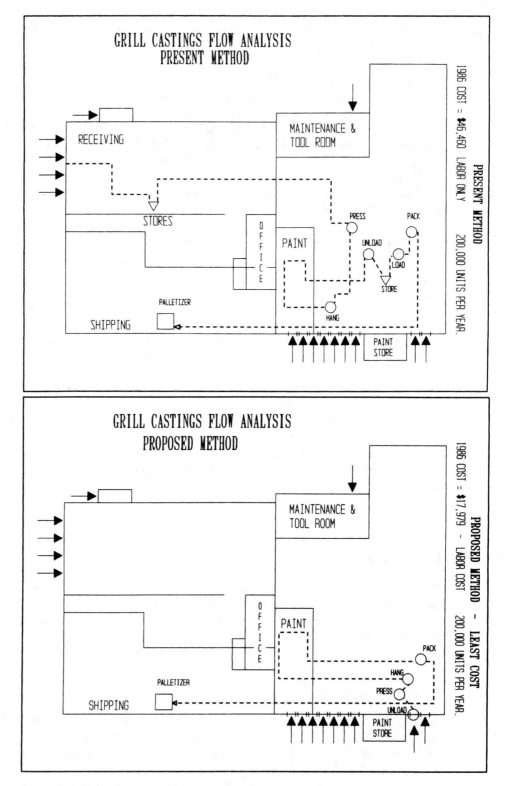

FIGURE 5-1 Flow diagram: The path taken by a part as it flows through a plant.

2. Backtracking

Backtracking is material moving backward in the plant. Material should always move toward the shipping end of the plant. If it is moving toward receiving, it's moving backward. Backtracking costs three times as much as flowing correctly. For example, five departments have flow like this:

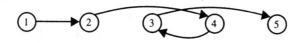

How many times did material move between departments 3 and 4? Three times: twice forward and once backward. If we rearranged this plant and changed departments 3 and 4 around, we would have straight-through flow with no backtracking:

We travel less distance. In the former example, we traveled six blocks (a block is one step between departments next to each other). In the straight-line flow, we traveled only four blocks—a 33% increase in productivity. The cost reduction formula calls this "reroute."

3. Distance Traveled

It costs money to travel distance. The less distance traveled, the better. The flow diagram is developed on a plant layout, and the layout can be easily scaled and distance of travel calculated. By rearranging machines or departments, we may be able to reduce the distances traveled.

4. Procedure

Because flow diagrams are created on plant layouts, no standard form is used. There are few conventions to restrict the designer. The objective is to show all the distances traveled by each part and to find ways of reducing the overall distance.

The flow diagram is developed from routing sheet information, assembly line balance, and blueprints. The routing sheet specifies the fabrication sequence for each part of a product. The sequence of steps required to make a part is practical and has some room for flexibility. One step may come before or after another step, depending on conditions. The sequence of steps should be changed to meet the layout, if possible, because that requires only a paperwork change. But if the sequence of operations cannot be changed and the flow diagram shows backtracking, it may be necessary to move equipment. Our objective will always be to make a quality part the cheapest, most efficient way possible.

STEP-BY-STEP PROCEDURE FOR DEVELOPING A FLOW DIAGRAM

Step 1: The flow diagram starts with an existing or proposed scaled layout.

Step 2: From the route sheet, each step in the fabrication of each part is plotted and connected with a line, color code, or other method of distinguishing between parts.

Step 3: Once all the parts are fabricated, they will meet, in a specific sequence, at the assembly line. The position of the assembly line will be determined by where the individual parts came from. At the assembly line, all flow lines join together and travel as one to packout, warehouse, and shipping.

A well-conceived flow diagram will be the best technique for developing a plant layout.

Plastic overlays of plant layouts are often used to develop flow lines for flow diagrams. The flow lines can be drawn with a grease pencil and can be grouped by classes for plants with a lot of different parts. It doesn't take a large product to make the fabrication departments of a plant layout look like a bowl of spaghetti. Using several plastic overlays will simplify the analysis.

A new industrial technologist will learn much from the creation of a flow diagram, and an experienced technologist will always find ways to improve the flow of material. Savings at this stage of methods analysis can be substantial—that is why we start with overall material flow.

NOTES ABOUT THE EXAMPLES

The examples shown in Figure 5-1 are from a barbecue gas grill manufacturing plant. The present method (top diagram) shows flow starting at the top left of the layout and flowing to stores and press shop. The proposed method (bottom diagram) shows the receiving of the gas grill casting arriving at the back door, being stored on the trailer until needed by production. The press was moved to the back door, eliminating much travel and handling.

The flow diagram works well in conjunction with an operations chart, and we continue our example in Figure 5-3a and b.

THE OPERATIONS CHART

The operations chart (see Figure 5–2) has a circle for each operation required to fabricate each part, to assemble each to the final assembly, and to pack out the finished product. Every production step required, every job, and every part is included.

Operations charts show the introduction of raw materials at the top of the chart, on a horizontal line.

The number of parts will determine the size and complexity of the operations chart.

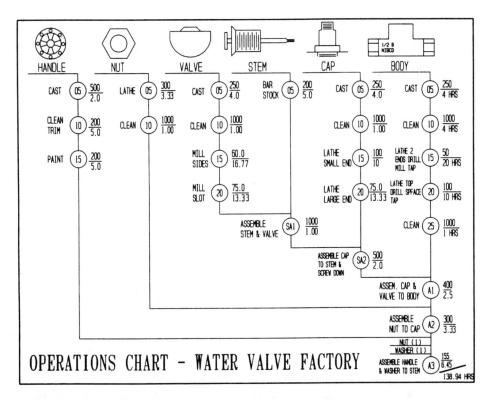

FIGURE 5-2 Operations chart for a water valve factory: complete example.

Below the raw material line, a vertical line will be drawn connecting the circles (a step in the fabrication of that raw material into finished parts).

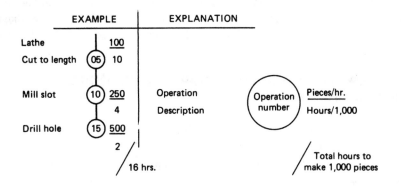

Once the fabrication steps of each part are plotted, the parts flow together in assembly. Usually, the first part to start the assembly is shown at the far right of the page. The second part is shown to the left of that, etc., working from right to left.

Example:

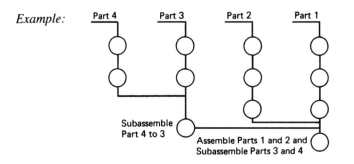

Some parts require no fabrication steps. These parts are called *buyouts.* Buyout parts are introduced above the operation at which they will be used.

Example:

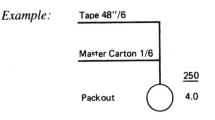

In this operation, we are going to place six products into a master carton and tape it closed.

The operations chart shows much information on one page. The raw material, the buyouts, the fabrication sequence, the assembly sequence, the equipment needs, the time standards, and even a glimpse of the plant layout, labor costs, and plant schedule can all be derived from the operations chart. Is it any wonder that industrial engineers consider this one of their favorite tools?

The operations chart is different for every product, so a standard form is not practical. The circle is universally accepted as the symbol for operations—thereby the origin of the chart's name. There is more convention in operations charting than in flow diagramming, but designers should not be too rigid in their thinking.

STEP-BY-STEP PROCEDURES FOR PREPARING AN OPERATIONS CHART

Step 1: Identify the parts that are going to be manufactured and those that are going to be purchased complete.

Step 2: Determine the operations required to fabricate each part and the sequence of these operations.

Step 3: Determine the sequence of assembly, both buyout and fabricated parts.

Step 4: Find the base part. This is the first part that starts the assembly process. Put that part on a horizontal line at the far right top of the page. On a vertical line extending down from the right side of the horizontal line, place a circle for each operation. Beginning with the first operation, list all operations down to the last operation.

Step 5: Place the second part to the left of the first part, and the third part to the left of the second part, and so forth until all manufactured parts are listed across the top of the page in reverse order of assembly. All of the fabrication steps are listed below the parts, with a circle representing each operation.

Step 6: Draw a horizontal line from the bottom of the last operation of the second part to the first part just below its final fabrication operation and just above the first assembly operation. Depending on how many parts the first assembler puts together, the third, fourth, etc. parts will flow into the first part's vertical line, but always above the assembly circle for that assembly operation.

Step 7: Introduce all buyout parts on horizontal lines above the assembly operation circle where they are placed on the assembly.

Step 8: Put time standards, operation numbers, and operation descriptions next to and in the circle.

Step 9: Sum total the hours per 1,000 units and place these total hours at the bottom right under the last assembly or packout operation.

Some parts will flow together before they reach the assembly line. This could involve welding parts together or assembling a bag of parts. This is called *subassembly* and is treated just like the main assembly, except that it is done before the parts reach the vertical line on the far right top of the page. Bag packing is a good example—all parts are usually buyouts and could be placed at the bottom left of your operations chart, like this:

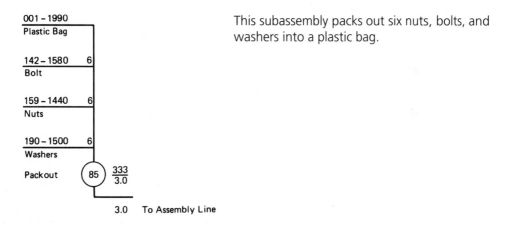

This subassembly packs out six nuts, bolts, and washers into a plastic bag.

PROCESS CHART

The process chart (see Figures 5-3a through 5-3c) is used to show all the handling, inspection, operations, storage, and delays that occur to one part as it moves from the receiving department through the plant to the shipping department. Conventional symbols have been

FRED MEYERS & ASSOCIATES PROCESS CHART

☒ PRESENT METHOD ☐ PROPOSED METHOD DATE: 5/6 PAGE 1 OF ___

PART DESCRIPTION: 2,000 UNITS/SHIFT

GRILL CASTING 75102

OPERATION DESCRIPTION: FROM RECEIVING TO SHIPPING

SUMMARY	PRESENT NO. TIME	PROPOSED NO. TIME	DIFF. NO. TIME	ANALYSIS:		FLOW
○ OPERATIONS				WHY	WHEN	DIAGRAM
⇨ TRANSPORT				WHAT	WHO	ATTACHED
☐ INSPECTIONS				WHERE	HOW	(IMPORTANT)
D DELAYS						
▽ STORAGES				STUDIED BY: F. MEYERS		
DIST. TRAVELED FT.		FT.	FT.			$7.00 PER HR.

STEP	DETAILS OF PROCESS	METHOD	OPERATION	TRANSPORT	INSPECTION	DELAY	STORAGE	DISTANCE IN FEET	QUANTITY	TIME IN .00001	COST PER UNIT	TIME/COST CALCULATIONS
1	RECEIVING UNLOAD TRUCK	PALLET FORK	●	⇨	☐	D	▽		120	31	.0025	2 MIN/PALLET
2	MOVE TO STORES	FORK TRUCK	○	➡	☐	D	▽	125'	120	23	.0016	2.5 MIN/PALLET
3	STORAGE	RACK	○	⇨	☐	D	▼	40,000				30 DAYS $3.00 EACH
4	MOVE TO MACHINE	FORK TRUCK	○	➡	☐	D	▽	625'	120	55	.0039	12.5 MIN/PALLET
5	WAIT AT MACHINE		○	⇨	☐	◗	▽					30 MINUTES
6	PUNCH		●	⇨	☐	D	▽			532	.0372	188 PER HR.
7	WAIT		○	⇨	☐	◗	▽			62	.0043	30 MINUTES
8	MOVE TO PAINT	FORK TRUCK	○	➡	☐	D	▽	200'	120	3	.0002	4.0 MINUTES
9	WAIT		○	⇨	☐	◗	▽	INV. ●				
10	PLACE ON CONVEYOR	HAND	●	⇨	☐	D	▽			595	.0417	336 PER HR.
11	TO PAINT	CONVEYOR	○	➡	☐	D	▽	10'			FREE	
12	HANG ON LINE		●	⇨	☐	D	▽			298	.0209	336 PER HR.
13	CLEAN-PAINT-BAKE	CONVEYOR	◐	➡	☐	D	▽	400'	INDIRECT		FREE	
14	UNLOAD		●	⇨	☐	D	▽			298	.0209	336 PER HR.
15	STACK		●	⇨	☐	D	▽			298	.0209	336 PER HR.
16	MOVE TO STORAGE	HAND	○	➡	☐	D	▽	20'		298	.0209	336 PER HR.
17	STORE FOR LINE		○	⇨	☐	D	▼	INV. ●				

FIGURE 5-3A Process chart: record of everything that happens to a part as it flows through a plant.

used to describe the process steps (Table 5-2). These symbols have been accepted by every professional organization working with motion and time study.

Process charts are always done in pairs, present and proposed. When doing the present chart, all observed steps are logged. This provides a record of the way the existing method is being performed. It is extremely important that all of the steps be accurately observed. The observer investigates until he or she completely understands the reason for each line by asking who, what, where, why, when, and how. Then the proposed chart is begun. Each line is subjected to elimination, combination, and sequence changing. When the proposed chart is taking shape, each line is again reviewed for simplification. Care must be taken to provide enough step-by-step details so that a future analyst can understand the operation described in order to duplicate or improve the process.

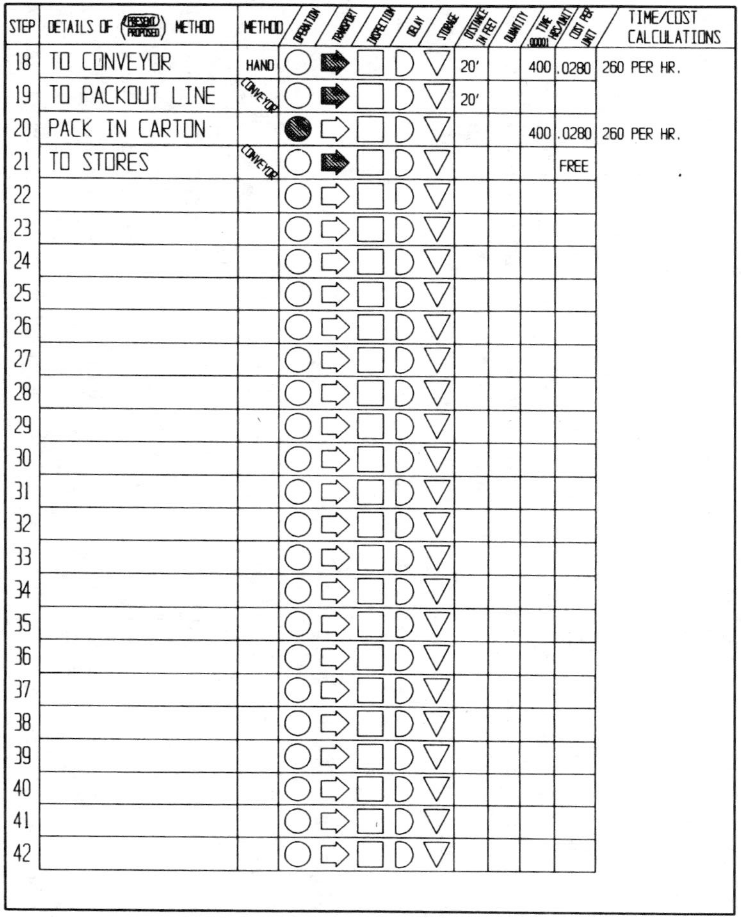

Figure 5-3c (Process Chart)

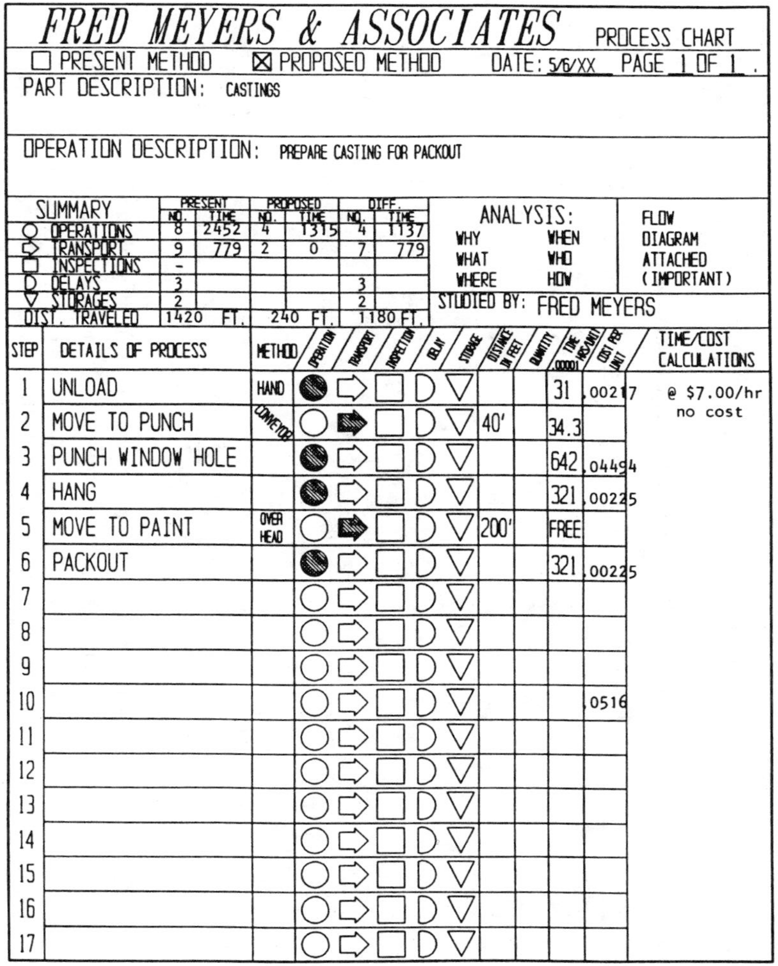

FRED MEYERS & ASSOCIATES — PROCESS CHART

PRESENT METHOD ☐ PROPOSED METHOD ☒   DATE: 5/6/XX   PAGE 1 OF 1

PART DESCRIPTION: CASTINGS

OPERATION DESCRIPTION: PREPARE CASTING FOR PACKOUT

SUMMARY	PRESENT NO.	PRESENT TIME	PROPOSED NO.	PROPOSED TIME	DIFF. NO.	DIFF. TIME
◯ OPERATIONS	8	2452	4	1315	4	1137
⇨ TRANSPORT.	9	779	2	0	7	779
☐ INSPECTIONS	–		–			
D DELAYS	3		3		3	
▽ STORAGES	2		2		2	
DIST. TRAVELED	1420 FT.		240 FT.		1180 FT.	

ANALYSIS: WHY / WHAT / WHERE   WHEN / WHO / HOW

STUDIED BY: FRED MEYERS

FLOW DIAGRAM ATTACHED (IMPORTANT)

STEP	DETAILS OF PROCESS	METHOD	DISTANCE IN FEET	QUANTITY	TIME (.0000)	COST PER UNIT	TIME/COST CALCULATIONS
1	UNLOAD	HAND			31	.00217	@ $7.00/hr
2	MOVE TO PUNCH	CONVEYOR	40'		34.3		no cost
3	PUNCH WINDOW HOLE				642	.0454	
4	HANG				321	.00225	
5	MOVE TO PAINT	OVER HEAD	200'		FREE		
6	PACKOUT				321	.00225	
7							
8							
9							
10							
11							
12							
13						.0516	
14							
15							
16							
17							

FIGURE 5-3c Process chart: an improvement over Figures 5-3a and b.

Figure 5-3b (Process Chart, continued)

STEP	DETAILS OF (PRESENT/PROPOSED) METHOD	METHOD	DISTANCE IN FEET	QUANTITY	TIME (.0000)	COST PER UNIT	TIME/COST CALCULATIONS
18	TO CONVEYOR	HAND					
19	TO PACKOUT LINE	CONVEYOR	20'		400	.0280	260 PER HR.
20	PACK IN CARTON		20'				
21	TO STORES	CONVEYOR			400	.0280	260 PER HR.
22					FREE		
23							
24							
25							
26							
27							
28							
29							
30							
31							
32							
33							
34							
35							
36							
37							
38							
39							
40							
41							
42							

FIGURE 5-3b Process chart: present method (continued).

Table 5-2 Process Chart Symbols.

SYMBOL	DESCRIPTION	INDICATES	MEANING
○	Circle	Operation	Performing work on a part of a product
□	Square	Inspection	Used for quality control work
⇨	Arrow	Transportation	Used when moving material
▽	Triangle	Storage	Used for long-range storage
D	Big D	Delay	Used when storage of less than a container

One way of classifying process charts is by the type of subject. The observer must be careful that the same subject type is being charted. Some of the common subject types include the movement of materials, the movement of an operator, the distribution of a form, the flow of information, the movement of a vehicle or material handling device. The process chart will be evaluated based on the reduction of the number of delays, storages, inspections, operations, transportations, and the distance moved in transportation as recorded in the process chart summary. Care must be taken to ensure that each step in the present chart has not been converted into more or fewer steps in the proposed chart.

Process charting lends itself to a standard form. A properly designed form will lead the designer to ask questions of each step. (The questions were asked earlier in this chapter, in the section titled "Cost Reduction Formula," on p. 67. Please review them.)

A blank form for process charting has been included at the back of this book for your future use. Examples of a completed before-and-after study are included in Figures 5-3a through 5-3c. Read and study them. Try to visualize what is happening to this part. Have all your questions been answered? The process chart communicates the same information as the flow diagram.

NOTES ABOUT THE EXAMPLES

Figures 5-3a and 5-3b show the present method of producing a charcoal grill casting. There are 21 steps in the process. Figure 5-3c shows the proposed method of producing the same part. There are only six steps in the process. We eliminated 15 steps, 4 operations, 2 delays, 2 storages, and 7 transportations. The transports that are left are by conveyor and free of labor. Figure 5-1 shows the flow diagrams for producing charcoal grills (present and proposed methods), and Figures 5-3a through 5-3c are process charts for the same parts. The flow diagram is the picture, and the process chart is the text. The two techniques are often used together to sell a proposed method to management. In the examples, the proposed idea saved over $26,000 for this one part, and there were four parts that saved about the same amount. Every part of a product needs to be studied like this because even the manufacturing of the smallest parts can be improved. There were nearly 100 different parts in this study. The charcoal grill manufacturing company saved $250,000 per year, with an expenditure of $325,000—a 77% return on investment. The company paid its cost back in less than 16 months. Study the two before-and-after examples to see if you can follow what changed.

STEP-BY-STEP DESCRIPTION FOR USING THE PROCESS CHART

Let's look at a step-by-step example of process charting (see Figure 5-4).

① ☐ Present Method   ☑ Proposed Method

A check mark in one of the two boxes is required. A good industrial engineering practice is always to record the present method so the improved (proposed) method can be compared to it. Costing the present and proposed methods will be required to justify your proposal, especially if any costs are involved. Recording and advertising cost reduction dollars saved is a smart idea.

② Date _____ Page _____ of _____

Always date your work. Our work tends to stay around for years, and you will someday want to know when you did this great work. Page numbers are important on big jobs to keep the proper order.

③ Part description

This is probably the most important information on the form. Everything else would be useless if we didn't record the part number. Each process chart is for one part, so be specific. The part description also includes the name and specifications of the part. Attaching a blueprint to the process chart would be useful.

④ Operation description

In this block, you record the limits of the study (for example, from receiving to assembly). Also, any miscellaneous information can be placed here.

⑤ Summary

The summary is used only for the proposed solution. A count of the operations, transportation, inspection, delays, and storages for the present and proposed methods is recorded and the difference (savings) is calculated.

The distance traveled is calculated for both methods and the difference calculated. The time standards in minutes or hours are summarized and the difference calculated. This information is why we did all the work of present and proposed process charting; it is the cost reduction information. We will come back to Step 5 after Step 15. ˙

FRED MEYERS & ASSOCIATES   PROCESS CHART

☐ PRESENT METHOD ①☐ PROPOSED METHOD   DATE: ②   PAGE __ OF __ .

PART DESCRIPTION: ③

OPERATION DESCRIPTION: ④

SUMMARY	PRESENT NO. TIME	PROPOSED NO. TIME	DIFF. NO. TIME	ANALYSIS:	FLOW ⑦
○ OPERATIONS				WHY ⑥ WHEN	DIAGRAM
⇨ TRANSPORT		⑤		WHAT WHO	ATTACHED
☐ INSPECTIONS				WHERE HOW	(IMPORTANT)
D DELAYS					
▽ STORAGES				STUDIED BY:	
DIST. TRAVELED FT.	FT.	FT.			

STEP	DETAILS OF PROCESS	METHOD	OPERATION / TRANSPORT / INSPECTION / DELAY / STORAGE	DISTANCE IN FEET	QUANTITY	TIME PRC.0001	COST PER UNIT	TIME/COST CALCULATIONS
1			○⇨☐D▽					
2			○⇨☐D▽					
3	⑧	⑨	○⇨⑩D▽	⑪	⑫	⑬	⑭	⑮
4			○⇨☐D▽					
5			○⇨☐D▽					
6			○⇨☐D▽					
7			○⇨☐D▽					
8			○⇨☐D▽					
9			○⇨☐D▽					
10			○⇨☐D▽					
11			○⇨☐D▽					
12			○⇨☐D▽					
13			○⇨☐D▽					
14			○⇨☐D▽					
15			○⇨☐D▽					
16			○⇨☐D▽					
17			○⇨☐D▽					

FIGURE 5-4 Process chart: the step-by-step form.

⑥ Analysis

The questions why, what, where, when, how, and who are asked of each step (line) in the process chart, and *why* is first. If we don't have a good *why,* we can eliminate that process chart step and save 100% of the cost. Questioning each step is how we come up with the proposed method. With these questions, we try to

1. Eliminate every step possible, because this produces the greatest savings. However, if we can't eliminate the step, we try to

2. Combine steps to spread the cost and possibly eliminate steps between. For example, if two operations are combined, delays and transportations can be eliminated. If trans-

portations are combined, many parts will be handled as one. However, if we can't eliminate or combine, maybe we can

3. Change the sequence of operations to improve the product flow and save many feet of travel.

As you can see, the analysis phase of process charting gives the process meaning and purpose. We will come back to Step 6 after Step 15.

⑦ Flow diagram attached (important)

Process charting is used in conjunction with flow diagramming. The same symbols can be used in both techniques. The process chart is the words and numbers, whereas the flow diagram is the picture. The present and proposed methods of both techniques must be telling the same story; they must agree.

Your name goes in the section labeled "Studied By." Be proud and put your signature here.

⑧ Details of process

Each line in the flow process chart is numbered, front and back. One chart can be used for 42 steps. Each step is totally independent and stands alone. A description of what happens in each step aids the analyst's questions. Using as few words as possible, describe what is happening. This column is never left blank.

⑨ Method

Method usually refers to how the material was transported—fork truck, hand cart, conveyer, hand, etc.—but methods of storage could also be placed here.

⑩ Symbols

The process chart symbols are all here. The analyst should classify each step and shade the proper symbol to indicate to everyone what this step is.

⑪ Distance in feet

This step is used only with the transportation symbol. The sum of this column is the distance traveled in this method. This column is one of the best indications of productivity.

⑫ Quantity

Quantity refers to many things:

a. Operations: The pieces per hour would be recorded here.

b. Transportation: How many were moved at a time.

c. Inspection: How many pieces per hour if under time standard and/or frequency of inspection.

d. Delays: How many pieces in a container. This will tell us how long the delay is.

e. Storage: How many pieces per storage unit.

All costs will be reduced to a unit cost or cost per 1,000 units, so knowing how many pieces are moved at one time is important.

⑬ Time in hours per unit (0.00001)

This step is for labor costing. Storage and delays will be costed in another way: inventory carrying cost. This column will be used only for operations, transportations, and inspection. Time per unit is calculated in two ways:

a. Starting with pieces per hour time standards, say 250 pieces per hour, divide 250 pieces per hour into 1 hour, and you get 0.00400 hours per unit. On our process chart, we place 400 in the time column, knowing that the decimal is always in the fifth place.

b. The material handling time to change a tub of parts at a work station with a hand truck is 1.000 minutes, and we have 200 parts in that tub. How many hours per unit is our time standard?

$$\frac{1.000 \text{ min/container}}{200 \text{ parts/container}} = 0.005 \text{ min/part}$$

$$\frac{0.005 \text{ min/part}}{60 \text{ min/hr}} = 0.00008 \text{ hr/part.}$$

⑭ Cost per unit

Hours per unit multiplied by a labor rate per hour equals a cost per unit. For example, consider the aforementioned two problems using a labor rate of $15.00 per hour:

a. $0.00400 \times 15.00 = \$0.06$ per unit
b. $0.00008 \times 15.00 = \$0.0012$ per unit.

The cost per unit is the backbone of process charting. We are looking for a better way, so the method with the overall cheapest way is the best method.

⑮ Time/cost calculations

Technologists are required to calculate costs on many different things, and how costs were calculated tends to get lost. This space is provided to record the formulas developed to determine the costs so they do not have to be redeveloped over and over again.

⑤ Returning to *Summary*

Once all steps in the present-method process chart have been completed, the summary is completed by

a. Counting all the operations, transportations, etc.
b. Adding up the unit time for all steps
c. Adding up the distance traveled.

⑥ Returning to *Analysis*

Once the present method is recorded and costed, the search for a better method starts. It is not appropriate to start the analysis phase until after the completed present method has been recorded. During the analysis phase, a proposed method is being developed and the industrial engineer goes through the process all over again. Figure 5-3c is an improved process chart for the gas grill housing in Figures 5-3a and 5-3b. How much was saved?

Do you remember what was said in an earlier chapter about successful people doing what others do not like to do? Working hard was one of those things. There is no easy way to finding the best method and selling that method to management—only hard work using the tried-and-true techniques of motion and time study. The process chart is one of the best techniques and one of the most widely used.

FLOW PROCESS CHART

The flow process chart (see Figure 5-5) combines the operations chart with the process chart. The operations chart used only one symbol—the circle, or operations symbol. The flow process chart uses all five process chart symbols. Another difference is that buyout parts are treated like manufactured parts. No standard form exists for flow process charting.

The flow process chart is the most complete of all the techniques, and when completed, the technologist will know more about the plant's operation than anyone in that plant.

Step-By-Step Procedure for Preparing a Flow Process Chart

Step 1: Start with an operations chart.

Step 2: Complete a process chart for each part.

Step 3: Combine the operations chart and the process chart, working in all the buyouts.

WORK CELL LOAD CHART

The work cell load chart is different from the previous techniques in that it is not for a complete part or product, but for only a few operations. We could end up with a chart for a complete part, but that is not the goal of a work cell chart. For example, we may cast or forge a part elsewhere, machine the part in a cell, and chrome plate it in a third area. Once we determine what the cell is going to do, we need to conduct a micromotion study of all the operations involved. This will be discussed in Chapter 6.

A work cell is a collection of equipment required to make a single complicated part. This equipment is placed in a circle around an operator or operators. Figure 5-6 shows a typical work cell layout. The operator (most often a single operator) takes a part from the in basket and moves that part around the circle of equipment. The equipment is usually automatic machines that only need to be loaded, activated, and then unloaded. Once the machine is

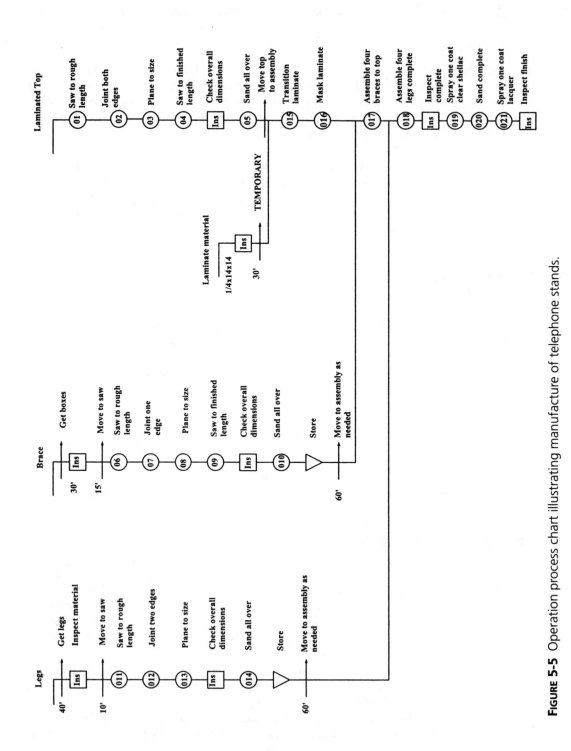

FIGURE 5-5 Operation process chart illustrating manufacture of telephone stands.

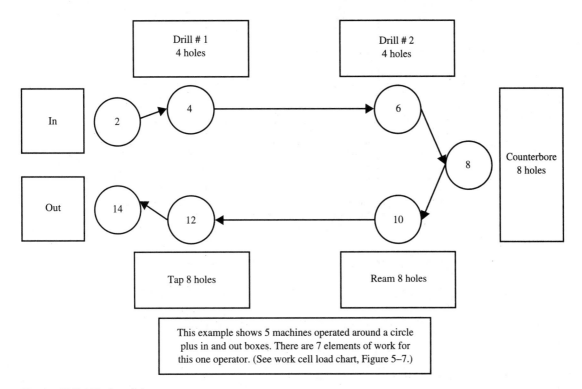

FIGURE 5-6 Work cell layout.

loaded and activated, the operator moves the just-completed part from the first machine to the second, where the operator removes the previous part and loads the next part. This process continues around the cell: taking parts out of one machine, putting new parts back into this machine, then activating that machine until arriving at the last machine, where the part is removed, inspected, and placed in the finished parts basket. Work cells are being developed at a very fast rate because they eliminate all storage between operations, eliminate most of the moving time between operations, and eliminate delays spent waiting for the next machine; this leads to cost reductions, work-in-process reductions (reduced inventory), and reductions in manufacturing-in-process time. The work cell concept considers operator time and utilization to be more important than machine utilization.

Work cell load charts are a special operations chart used for multimachine situations (see Figure 5-7 for an example). Work cell load charts will visually show the operator time, machine time, and walking time required to run a work cell to produce one part per cycle using many machines. The result shows us the total cycle time, operator utilization, and machine utilization. Because they are visual, work cell load charts help people see problems and make improvements on the operation by properly loading the operator(s) and/or machines.

Part Number 1675   Part Name crank   Date 5/13/xx   Engineer Justin M.

| Operation Number | Operation Description | Time in 0.001 Minutes | | | Time in 0.025 Minutes |
		Manual	Machine	Walk	
2	Get new part	54	0	45	
4	Drill #1	49	455	25	
6	Drill #2	55	470	35	
8	Counterbore	75	289	35	
10	Ream	111	115	35	
12	Tap	175	300	25	
14	Inspect & aside	55	0	25	
	TOTAL	0.574	1.629	.225	
①	②	③	④	⑤	⑥

Total cycle time = 0.8 minutes + 10% allowance = 0.88 minutes

Pieces per hour = 68

Hours per piece = 0.01467

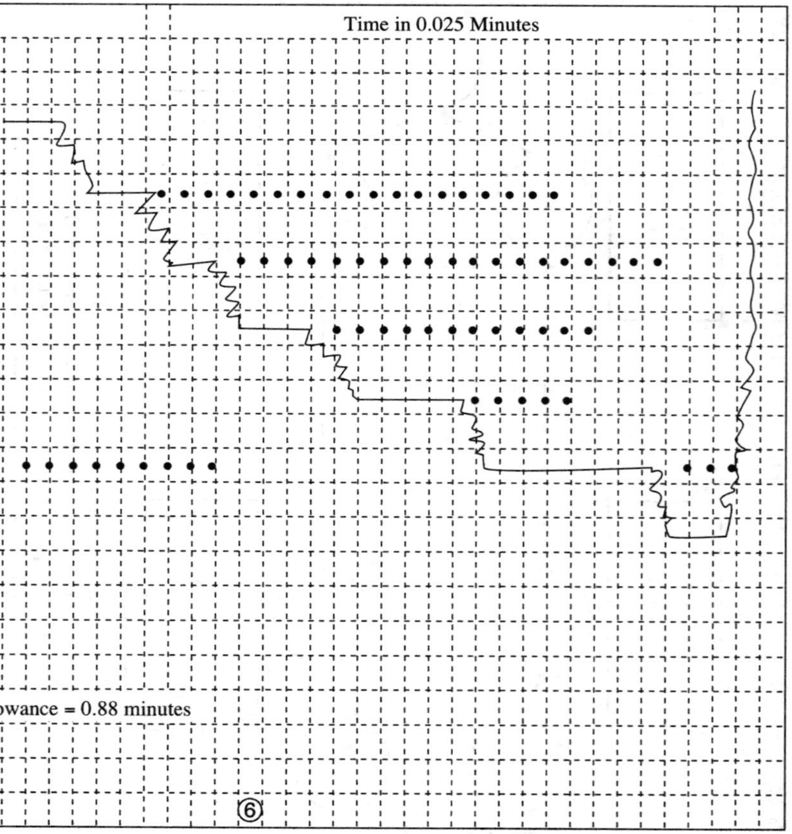

FIGURE 5-7 Work cell load chart.

Step-By-Step Procedure For Preparing a Work Cell Load Chart

The circled numbers in this discussion correspond to those in Figure 5-7.

① Operation sequence: This is just a numerical sequence of steps. Good procedure uses numbers that leave room for expansion, like 2, 4, 6, 8 or 5, 10, 15, 20. This will allow the insertion of new operations between existing operations without having to renumber everything.

② Operation description: This will include machine names and operation descriptions being performed. It should be descriptive enough to communicate to others what is being accomplished, so they can follow the sequence of operations.

For 3 to 5: Time standards in decimal minutes. These times were developed by time study techniques to be discussed later.

③ Manual time: The time it takes the operator to load, unload, inspect, and do anything else the operator is required to do. This time is totally under the control of the operator.

④ Machine time: Once the operator activates the machine, the machine does its job automatically, and the operator goes on to the next operation. This machine time is usually calculated using feeds and speed formulas (discussed in Chapter 10). Machine times are usually out of the control of the operator.

⑤ Walk time: Walk time is the time it takes an operator to move from one machine or operation to the next. The time standard for walking has already been determined to be 0.005 minutes per foot, and it can be easily calculated from a work station layout. For example, it would take 0.050 minutes for the average person to move 10 feet. This statistic is based on the basic time standard of walking 3 miles per hour, plus a 25% allowance for obstruction and turning.

⑥ Operation accumulation time graph: This is the visual meat of the form. The time data is plotted on the chart using three standard symbols:

1. _____ = a solid line = manual time = operator time
2. •••••••• = dotted line = machine time = automatic time
3. ᴡᴡᴡᴡ = zigzag line = walk time to next operation.

The resulting graph visually shows the work load on a time scale and can be used to better balance the operator and machines. Also, the total cycle time results in a time standard. With analysis and a little imagination, an improvement can be attained.

ROUTE SHEET

The route sheet, or routing sheet (Table 5-3), is used by many manufacturers to communicate the routing instructions (sequence of operations and machines) to the various departments that will perform the work. If our product has 25 fabricated parts (parts we make), we will

Table 5-3 Routing Sheet.

① PART NAME	BOX BODY	② PART NUMBER 1600		DATE 10/22/xx	③ QUANTITY 1000	
	A	B	C	⑥ Machine Number	E	F
	④ Operation Number.	⑤ Operation and Tooling Description	Machine Name		Pieces per Hour	Hours per Piece
1				D		
2						
3	5	Cut to length	Shear	12	1200	0.00083
4	10	Cut to width.	Shear	12	400	0.0025
5	15	Notch corners	Press	65	300	0.00333
6	20	Punch four holes	Press	65	300	0.00333
7	25	Form two short legs	Press break	55	250	0.004
8	30	Form two long legs	Press break	55	250	0.004
9	35					
10	40					
11						
12						
13						
14						
15						
16						
17						
18						
19						
20						

have 25 route sheets. The route sheet is much like the process chart, except that it concentrates on operations. The route sheet lists the operations required to make a part in the sequence of operations to make that part. The route sheet often accompanies an order of parts around a plant from operation to operation. The route sheet tells the production operators about the following:

1. Part name
2. Part number—for control
3. Quantity of parts needed
4. Operation numbers—in sequence
5. Operation description—describes what needs to be done
6. Machine number—if available

QUESTIONS

1. Why do we start our methods study with the broad view?
2. Describe flow diagrams, operations charts, process charts, and flow process charts. When would you use each technique?
3. What is the most important reason for using any of these techniques?
4. Describe the use of the cost reduction formula.
5. What is cross traffic and backtracking? What should be our attitude toward these?
6. Why is distance traveled important?
7. What are the five process chart symbols? Give examples.
8. Develop a flow process chart for the water valve plant shown in Figure 5-2.
9. What is a work cell?

Techniques of Micromotion Study: Operations Analysis

The techniques of motion study discussed in the last chapter identified the operations required to produce a product, and they gave us guidelines for our micromotion study. (See the operations chart in Figure 5-2, for example.) Chapter 5 was called the broad view, or macromotion study, because it included all the required operations. In this chapter, we are interested in individual operations, and the techniques help us understand the smallest detail of each operation so we can make small improvements. Jobs come in every conceivable form:

1. Individuals working alone with simple hand tools and fixtures.
2. Individuals working at a machine (called operator/machine).
3. Individuals working many machines (called multimachine cells), sometimes the same kind of machine, sometimes different types of machines. One person operating several different kinds of machines is an example of one work cell.
4. Groups of people working together (called gangs, cells, or assembly and pack out).

But they have much in common. One-half of all human work is reaching for and moving items. The other half of work is a small group of body motions like grasping, walking, and positioning. The pioneers in motion study have developed predetermined time standards for all body motions. We need only to develop a work station design and a motion pattern to analyze the motions and thus determine how long they will take. Every one of the four types of work listed here can be analyzed the same way. The most complicated work is just a larger series of simple elements. First, break down the job into elements of work, then analyze each element.

The first two types of work listed are the more traditional, common types, and we will spend much more time on them (Chapters 6 and 7). The last two are also very important, so much so that they are the subject of Chapter 10. The techniques to be studied in Chapter 6 all need time standards, but we will wait until Chapter 8 to discover how Predetermined Time Systems (PTS) develop their time requirements.

The techniques of operation analysis, motion economy, work station design, motion patterns, and predetermined time standards system (PTSS) will allow us to develop a work station design and work method for achieving the least production cost possible. There is no limit to cost reduction possibilities. With the right tools, you can be a cost reduction champion.

The techniques of micromotion study charts discussed in this chapter are

1. Operations analysis
2. Operator/machine
3. Multimachine
4. Gang
5. Left-hand/right-hand
6. SIMO

These techniques have several common factors:

1. Each activity is broken into elements. An activity is one unit of production. For example, if one operator were to operate three pieces of machinery, there would be four activities: one operator and three machines. Under operator activity, we may have multiple elements of work for each machine, while the machine activity would have two elements—working or idle.

2. Time is measured linearly. A scale is drawn down the side of an activity, and the unit of measurement (usually .01, or one-hundredth, of a minute). The elements are listed sequentially and divided by horizontal lines. Time values were historically gathered by viewing and counting motion picture frames. Individuals may use standard data, predetermined time values, time study, or operations analysis. Cycle times or estimates would not be sufficiently accurate.

3. All operations analysis techniques can use the same form. Only the number of activities varies. In the foregoing four-activity example, we would need two of the standard forms side by side. The operations analysis chart would use only one half of one form.

4. All these charting techniques are visual and are good sales tools. The length of the operation is shown by a scale on the chart, and the present method can be held next to the proposed method, with the choice being obvious.

OPERATIONS ANALYSIS CHART

The operations analysis chart (see Figure 6-1) is used to describe a single activity, usually one operator using only tools and equipment that are totally operator-controlled. This is the simplest of all the charts discussed in this chapter, because it has only the one activity; however, the process is the same process used for more difficult charts. The single activity

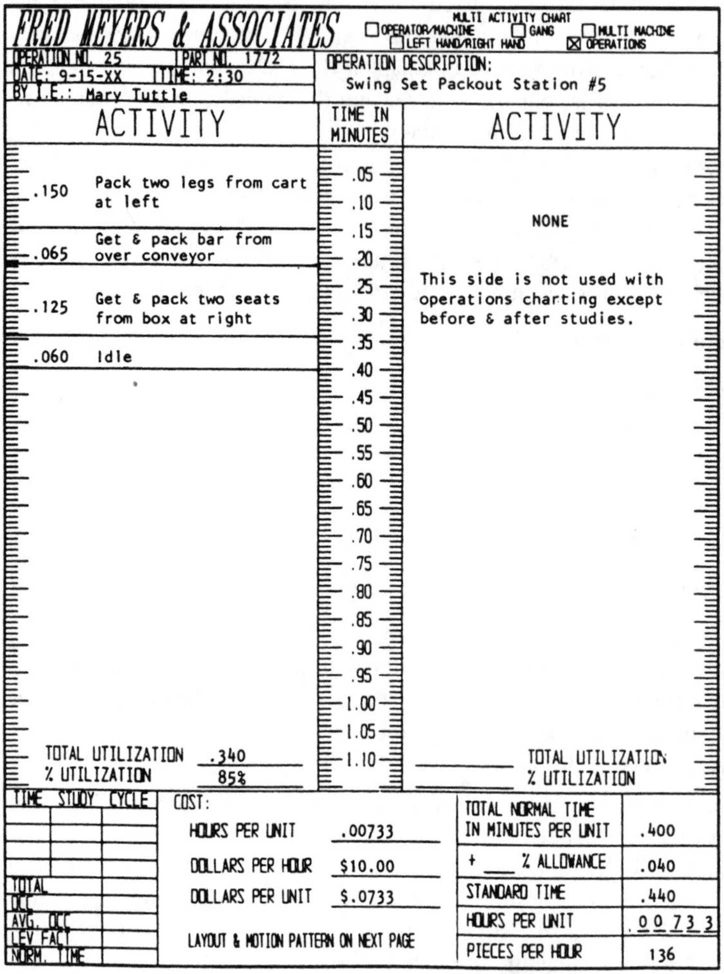
FRED MEYERS & ASSOCIATES

MULTI ACTIVITY CHART
☐ OPERATOR/MACHINE ☐ GANG ☐ MULTI MACHINE
☐ LEFT HAND/RIGHT HAND ☒ OPERATIONS

OPERATION NO. 25 PART NO. 1772
DATE: 9-15-XX TIME: 2:30
BY I.E.: Mary Tuttle

OPERATION DESCRIPTION:
Swing Set Packout Station #5

ACTIVITY		TIME IN MINUTES	ACTIVITY
.150	Pack two legs from cart at left	.05 .10	NONE
.065	Get & pack bar from over conveyor	.15 .20	
.125	Get & pack two seats from box at right	.25 .30	This side is not used with operations charting except before & after studies.
.060	Idle	.35 .40	
		.45 .50 .55 .60 .65 .70 .75 .80 .85 .90 .95 1.00 1.05 1.10	
TOTAL UTILIZATION	.340		TOTAL UTILIZATION
% UTILIZATION	85%		% UTILIZATION

TIME STUDY CYCLE / TOTAL / OCC. / AVG. OCC. / LEV FACT / NORM. TIME

COST:

HOURS PER UNIT	.00733
DOLLARS PER HOUR	$10.00
DOLLARS PER UNIT	$.0733

LAYOUT & MOTION PATTERN ON NEXT PAGE

TOTAL NORMAL TIME IN MINUTES PER UNIT	.400
+ ___ % ALLOWANCE	.040
STANDARD TIME	.440
HOURS PER UNIT	.00733
PIECES PER HOUR	136

FIGURE 6-1 Operations analysis chart.

is broken up into elements (an element of work is one unit of work that cannot be divided realistically), and these elements are timed. The method of timing is not important at this stage, but decimal minutes are used.

EXAMPLE: A packout operator on an assembly line is required to pack the following parts into a carton passing on the assembly line:

PART NUMBER	QUANTITY/SET	DESCRIPTION	TIME
1	2	Leg	.150
2	1	Bar	.065
3	2	Seat	.125

The assembly line runs at .400 minutes/set. The 100% station on the assembly line (normally only one station) is fully loaded and will have .400 minutes of work.

The operations analysis chart will show the aforementioned information on a vertical time scale, with the size of each element in direct proportion to the amount of time that element takes.

OPERATOR/MACHINE CHART

The operator/machine chart (see Figures 6-2 and 6-3) is twice as complicated as the operations analysis chart. The operator/machine chart has two activities—the operator and the machine. The operator/machine chart is much more useful because it shows the

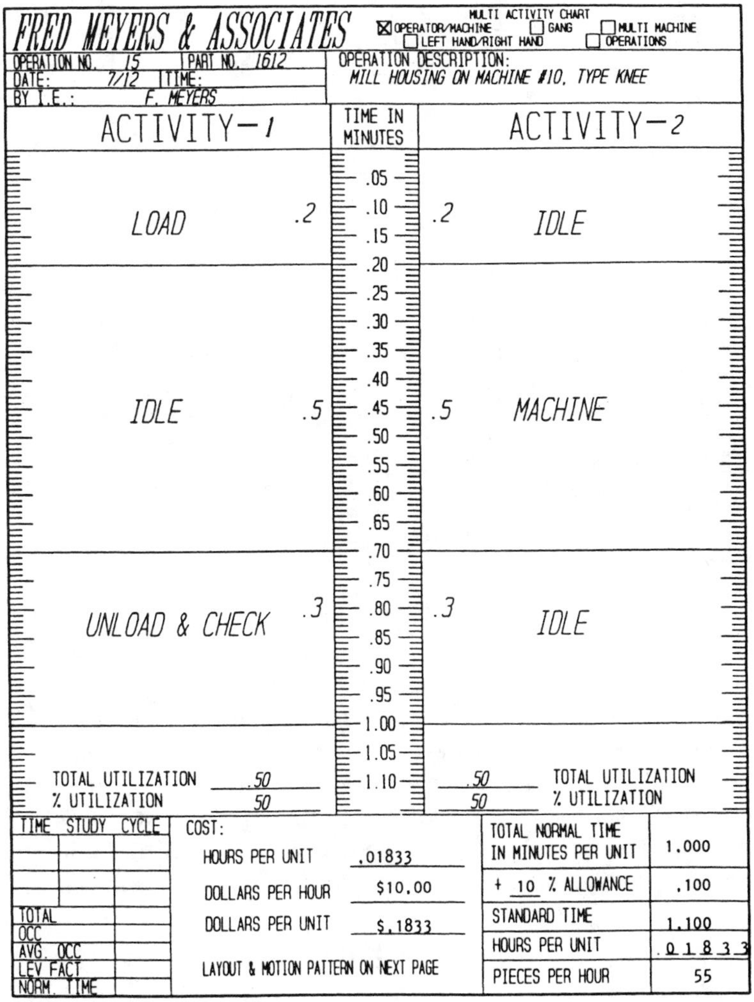

FIGURE 6-2 Operator/machine chart: two activities working together.

FRED MEYERS & ASSOCIATES

MULTI ACTIVITY CHART
☒ OPERATOR/MACHINE ☐ GANG ☐ MULTI MACHINE
☐ LEFT HAND/RIGHT HAND ☐ OPERATIONS

OPERATION NO.	5	PART NO.	1612	OPERATION DESCRIPTION:
DATE:	7/12	TIME:		MILL VALVE BODY ON MACHINE #16
BY I.E.:	F. MEYERS			

ACTIVITY—1	TIME IN MINUTES	ACTIVITY—2
LOAD	.05	IDLE
	.10	
	.15	
	.20	
PRE POSITION & ASIDE	.25	MACHINE TIME
	.30	(MILL SURFACE)
	.35	
	.40	
	.45	
	.50	
IDLE	.55	
	.60	
UNLOAD	.65	IDLE
	.70	
	.75	
	.80	
	.85	
	.90	
	.95	
	1.00	
	1.05	
TOTAL UTILIZATION .60	1.10	.50 TOTAL UTILIZATION
% UTILIZATION 86		71 % UTILIZATION

TIME STUDY CYCLE		COST:			
.75	3.80	6.80	HOURS PER UNIT	.01295	TOTAL NORMAL TIME IN MINUTES PER UNIT	.700
1.55	4.50	7.50			
2.20	5.25		DOLLARS PER HOUR	$10.00	+ 11 % ALLOWANCE	.077
3.05	6.05				
TOTAL		7.50	DOLLARS PER UNIT	$.1295	STANDARD TIME	.777
OCC		10			
AVG. OCC		750			HOURS PER UNIT	.01295
LEV FACT		95%	LAYOUT & MOTION PATTERN ON NEXT PAGE		
NORM. TIME		.712			PIECES PER HOUR	77

FIGURE 6-3 Operator/machine chart: an improvement over Figure 6-2.

interrelationship of the operator and the machine. Both the machine and the operator work intermittently, and this chart shows what each is doing at every moment in time. Each activity is reduced to a series of elements (an element is a unit of work that cannot be subdivided). These elements of work are placed in order down one side of the chart, and the other activity's elements are placed down the opposite side of the chart. Each element must be aligned according to time, so the same moments are across from each other.

Notes on Figure 6-2 (Present Method)

Activity 1 is the operator (left side), and Activity 2 is the machine (right side). The first element of the operator is the load element, and the first element of the machine is an idle element. These two elements end when the operator pushes the start button to activate the machine. At this moment, the operator starts the idle element while the machine starts its work element. When the machine stops cutting, the machine work element ends; at the same moment, the operator's idle element ends. This also starts the last elements: The operator unloads the machine and places the part aside, and the machine is again idle. These elements end when the operator puts down the finished part and starts reaching for the next part, which starts the whole process again. Both activities took 1.00 minutes, and both activities actually worked only 50% of the time. Is this effective use of our company's resources—people and machinery? No, it is not.

Notes on Figure 6-3 (Proposed Method)

This is the same job as in Figure 6-2, but during the machine element of the machine activity, the operator pre-positioned and placed the finished part aside. Hold the present method and the proposed method side by side. See the improvement? Any doubt about which is the best procedure? If 1 million of these parts are produced each year, what is the savings?

Present method	$.1833 per unit	.1833
Proposed method	$.1295 per unit	.1295
Savings	$.0538 per unit	.0538
Times 1,000,000 units per year equals $53,800/year		

Step-By-Step Procedure for Producing Operations Analysis Charts in This Chapter

See Figure 6-4 for an example of a multiactivity chart.

Step ①: Identify the problem: operations analysis, operator/machine, gang, multimachine, left-hand/right-hand chart. The only differences are the number of activities and the kind of activities. The examples given in this chapter will help identify which technique to use for each task. Check the appropriate box once the problem has been identified.

Step ②: Operation number: The operations required to manufacture every part are numbered to identify them. The operations numbers can be anything the systems designer wants, but usually the first operation number is 05 followed by 10, 15, 20, etc. A few parts would require more than 20 operations, but for those parts, operation numbers between the one divisible by 5 may be used. The system of skipping numbers gives room for expanding the operations if a new operation is required between 15 and 20, so this new operation could be 17 or 18—your choice.

Step ③: Part number: Every product is made up of parts. To keep control of these parts, each part is assigned a part number. There are many techniques for setting up part numbers, and most companies try to make part numbers meaningful and useful.

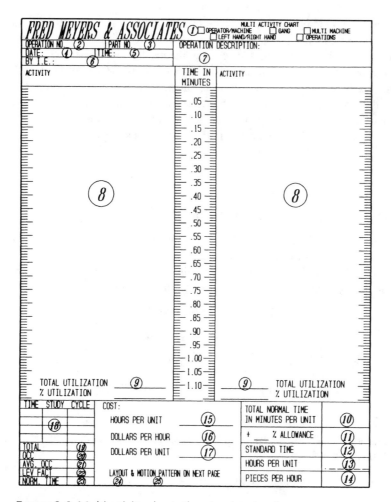

FIGURE 6-4 Multiactivity chart: the step-by-step form.

One useful system is an eight-digit system—XXXX = XXXX. The first four digits identify the finished product, while the last four digits identify the part. In addition, the last four digits can identify parts groups, such as

9XXX	Packaging
8XXX	Plastics
7XXX	Paints
6XXX	Sheet metal parts
5XXX	Castings
	etc.

Any industrial engineering work done on any part should carry a part number and operation number, or this work will be useless. After a short period of time, the busy engineer will not remember what the work was for.

Step ④: Date: Always include the date on your work. This includes the year. Your work will stay around for years, and subsequent studies will be compared to older ones. How will you know the sequence of studies unless you have the date?

Step ⑤: Time: Time may affect the study. For example, a machine or process may work better or worse in the morning before it warms up. You may not know this at the time of your study, so just in case, record the time of the study.

Step ⑥: "By I.E.": I.E. refers to industrial engineer, your professional service. Your name as the analyst, technologist, or engineer goes in this space. A study without a name is worthless.

Step ⑦: Operation description: The operation description should be as complete as possible. You must be able to communicate what is happening in such detail that the next generation of engineers can understand what you did. Every senior industrial technologist has experienced a time when he or she couldn't figure out what was done six months ago for lack of an operation description. Don't save time by being sloppy and incomplete; do it right the first time—do it now.

Step ⑧: Activity name: The activity is the guts of the chart. The activity name is the operator's job description name or machine name. Under each activity name, the elements of the job are listed in the exact order they are performed. The size of the elements depends on the time required. The more time required, the larger the element size. The time is measured on the linear scale at the side of each activity. In the center is the cumulative time. Time can be determined by any of the techniques studied later in this book, but do not worry about this now.

Step ⑨: Total utilization, % utilization: This comprises two pieces of information. First, total utilization is how much time this activity was working. This time does not include idle time. The second piece of information is percent utilization, which indicates how well the activity (operator or machine) is being used. The percent utilization is calculated by dividing the total utilization of the activity by the total time of that activity. In Figure 6-2, for example,

$$100 \times \frac{.5 \text{ minutes total utilization}}{1.0 \text{ minutes total time}} = 50\% \text{ utilization.}$$

Step ⑩: Total normal time in minutes per unit: Total normal time is how long it takes each activity to complete a cycle and includes both working and idle time. The total normal time will be the same for all activities of an operation. In Figure 6-2 of the operator/machine chart, the total normal time was 1.00 minutes.

Step ⑪: +___% allowance: Allowances are the time we add to normal time to make our time standards realistic. Allowances include time for personal time, fatigue (breaks), and unavoidable delays. We will study allowances in the time study chapter (Chapter 9), but until then, we will use a 10% allowance because it is the most common. So, we have +10% allowance; 10% of the total normal time in minutes per unit is placed here. In our example, 10% of 1.00 minutes is .10 minutes.

Step ⑫: Standard time: Standard time is normal time plus allowances. The difference between normal time and standard time is allowances.

Step ⑬: Hours per unit: Hours per unit is calculated by dividing standard time by 60 minutes/hour. In our example,

$$\frac{1.10 \text{ minutes/unit}}{60 \text{ minutes/hour}} = .01833 \text{ hours/unit}.$$

Multiply .01833 hours per unit by 1,000 units to determine the hours per 1,000 units—18.33 hours/1,000. Hours per 1,000 is a more meaningful number than hours per unit.

.01833 hours per unit

or

18.33 hours per 1,000

Which is the most meaningful to you?

Step ⑭: Pieces per hour: Pieces per hour is calculated by dividing the hours per unit into 1 or the hours/1,000 into 1,000. In our example,

$$\frac{1}{.01833} = 55 \text{ pieces/hour} \qquad \frac{1,000}{18.33} = 55 \text{ pieces/hour}.$$

Step ⑮: Hours per unit: This is the same number calculated in Step 13 and is the first ingredient of cost.

Step ⑯: Dollars per hour: The dollars per hour refers to the operator's wage rate in $/hr. This can be more complicated than shown in Figure 6-3, but a useful method is the departmental average labor rate plus fringe benefit rate. All the employees' salaries of a department are added together and divided by the number of employees to get the average hourly rate. The cost of employee benefits, such as vacation, holiday, and insurance, is calculated yearly and converted to a percentage of the hourly rate; 33.3% is a common fringe number. If our hourly average rate is $15.00/hour, fringe benefits (15.00 × 33.3%) equals $5.00, for a total cost of $20.00 per hour.

Step ⑰: Dollars per unit: The dollars per unit is the labor cost for one unit. To calculate the dollars per unit, multiply (Step 15 × Step 16) hours per unit times dollars per hour. This calculation can be complicated by having more than one operator; in that case, cost would double. Dollars per unit is our measure of desirability. We want this measure to be as low as possible.

Step ⑱: Time study cycles: The time study cycles block is a small time study form and allows the analyst a place to check his or her work. There are 12 blocks in which to place stopwatch time study observation readings. When the operator finishes a unit, the industrial technologist will start the stopwatch and record the ending time of the next 12 parts. The watch could be left running (continuous time study) or reset each time a part is run (snap-back time study). These two techniques are discussed in the time study chapter, Chapter 9.

Step ⑲: Total: On a continuous time study, the total is the last reading. The last reading would be the total time for running 12 parts. On the snap-back time study, each reading is the time for 1 part; therefore, the technician needs to add the 12 readings together to get the total time.

Step ⑳: Occurrences: This is the number of occurrences or cycles. If the technician studied all 12 cycles, then this figure is 12. But if the time study technician studied less than 12 cycles, the number of cycles studied goes in this block.

Step ㉑: Average occurrence: The average occurrence is the average time per part. It is calculated by dividing the total time (Step 19) by the occurrence (Step 20). Average time per occurrence is simply the arithmetic average of the cycles checked.

Step ㉒: Leveling factor: The leveling factor is a complicated subject that is discussed fully in Chapter 9. At this time, the best definition is that the leveling factor is the time study technician's opinion of the speed or tempo of the operator, or, in simpler terms, how fast the operator is working. Normal is 100% performance; therefore, if someone is working less than normal, we would put a leveling factor of less than 100. An operator working faster than normal would be leveled at over 100%. (An example is shown in Step 23.)

Step ㉓: Normal time: Normal time is calculated by multiplying the average occurrence time (Step 21) by the leveling factor (Step 22).

EXAMPLES:

AVERAGE OCCURRENCE	LEVELING FACTOR (%)	NORMAL TIME IN MINUTES
1.00	120	1.20
1.00	80	0.80
1.00	100	1.00

Note that the leveling factor makes a big difference in normal time; therefore, a time study analyst needs much practice in leveling. The normal time in Step 23 needs to be compared to the total normal time in Step 10. If they are close, the analyst will feel confident that this time standard is a good one.

Step ㉔: Layout: A top-view drawing of the work station layout is very important to a complete description of the operation (Step 7). Chapter 7 discusses work station layout, so further discussion is postponed until then.

Step ㉕: Motion pattern: The motion pattern is the path made by the hands in the process of making one part. The motion pattern is the modern technique that replaces the cyclograph shown in Chapter 2. The motion pattern is also a major discussion of Chapter 7, so further discussion is postponed until then.

The operator/machine chart is the most popular of the multiactivity charts, but all charts are similar. Only deviations from the aforementioned procedures are discussed in the future charts and will be titled "Procedure Deviations."

GANG CHART

A gang chart (see Figure 6-5) is used when two or more people work together and their activities intertwine. A two-person team gang chart would look just like an operator/machine chart. When three or more operators are involved, more columns are needed on the standard form. This is accomplished by taping two or more pages side by side, building as many columns as needed, as shown in Figure 6-5. All multiactivity charts have idle time. Our goal is to minimize this idle time and to ensure proper crew size.

Procedure Deviations

1. Hours per unit will be multiples of one operator's time. Take the number of operators working together times the hours per unit for one person.
2. Pieces per hour will still be calculated from the one-person hours per unit, just the same as in the operator/machine chart.
3. Cost must reflect the total crew size, so hours per unit must reflect all the operators.

Notes About Figure 6-5

Swing sets are packed out in boxes 10' long, 18" wide, and 8" high, and weigh 125 pounds. Finished swing sets must be removed from the packout line continually, stacked on hand carts (using two people because of weight and length), rolled to a storage location, and restacked on the floor to a 12' height. The activities are

1. Two people remove the sets and stack them on carts,
2. Carts are rolled to the stacks and an empty cart is brought back,
3. Two people remove the sets from a roller cart and hand them to a stacker,
4. The person on top of the stack straightens the sets to align and cross tie each tier.

The big question is, How many cart roller operators do we need?

Time Study Information

1. Sets are coming out of production at a rate of 5 per minute. The operators are busy 70% of the time. Sets are stacked 10 per cart. Ten sets weigh 1,250 pounds plus cart weight.
2. A cart can be rolled to the warehouse and back in 1.6 minutes.
3. Sets can be stacked in .18 minutes per set or faster, depending on stack height.
4. Once boxes are placed on a stack, a third person straightens and aligns them in .10 minutes per box.

FRED MEYERS & ASSOCIATES — MULTI ACTIVITY CHART

OPERATION DESCRIPTION: UNLOAD FINISHED SWING SETS FROM PACK OUT LINE & STACK IN WAREHOUSE

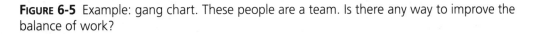

FIGURE 6-5 Example: gang chart. These people are a team. Is there any way to improve the balance of work?

MULTIMACHINE CHART

When an operator is asked to run more than one machine, the multimachine chart is used (see Figure 6-6). How many machines can one person operate? The multimachine chart can show us. The operator/machine chart is expanded so that each machine has a column of its own. The longer the cycle, the more machines a person can operate. The multimachine chart

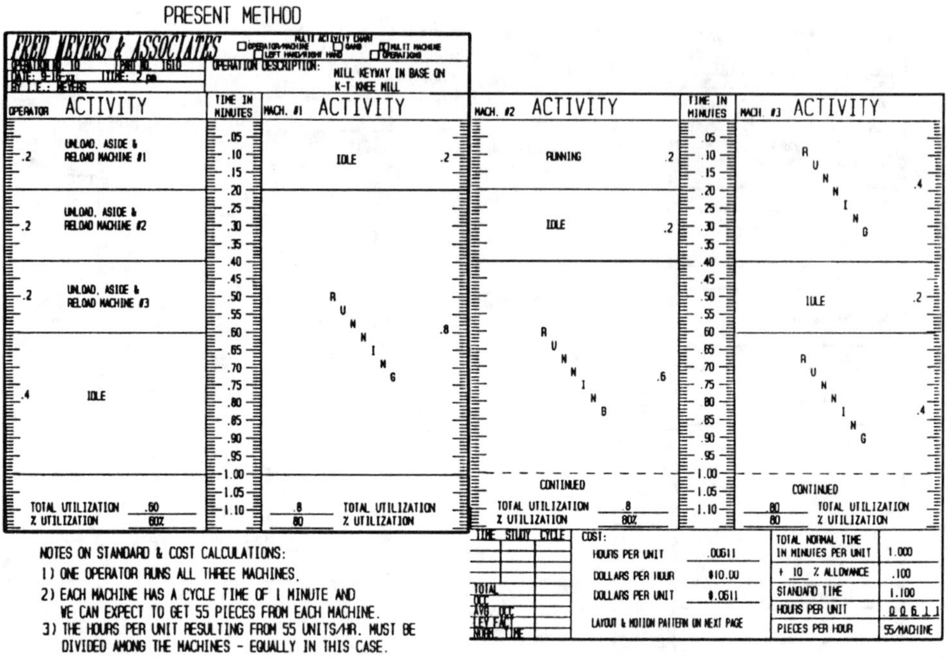

PRESENT METHOD

FIGURE 6-6 Example: multimachine chart. All three machines are the same and do the same job. Different parts/machines are possible. How can this method be improved?

is very similar to the gang chart in appearance. In our example (Figure 6-6), how many machines could our operator run?

Procedure Deviations (Steps 2 and 3)

1. More than one part and one operation are possible. Be sure to include all part numbers and operation numbers.

2. Total normal time (Step 10) will be the same for each activity. Total normal time is the operator's time, and when an operator runs three machines, that time must be split among the machines in proportion to the cycle times of the machines. This split is done by dividing the hours per unit (Step 13) by the number of machines. The hours per unit can be multiplied by the percentage of the operator's productive time required for each machine.

LEFT-HAND/RIGHT-HAND CHART

The left-hand/right-hand chart (see Figure 6-7) is very different from the previous charts because it is for one operator only. It is also different from the operations chart because it treats each hand as an activity. Each hand's activity is broken into elements and is plotted in

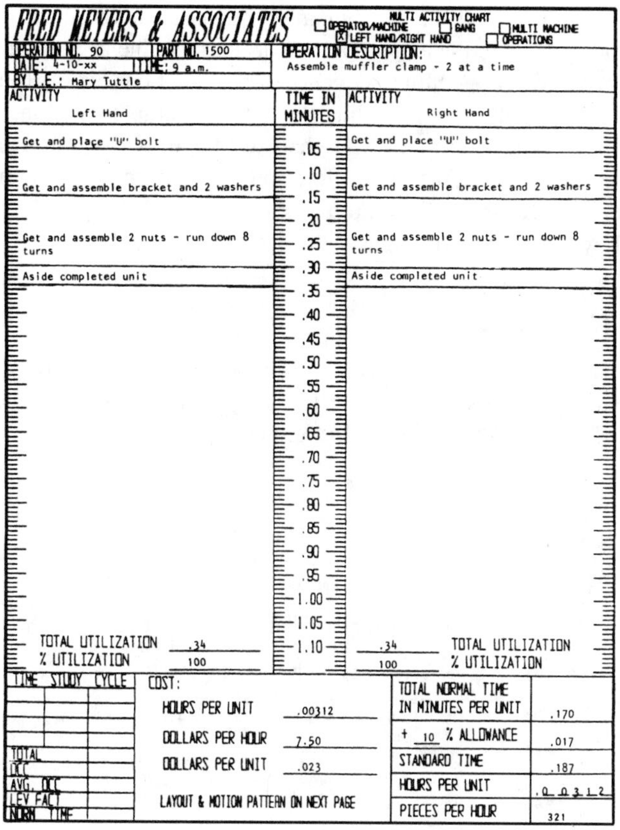
FRED MEYERS & ASSOCIATES

MULTI ACTIVITY CHART: ☐ OPERATOR/MACHINE ☐ GANG ☐ MULTI MACHINE ☒ LEFT HAND/RIGHT HAND ☐ OPERATIONS

OPERATION NO. 90 | PART NO. 1500 | DATE: 4-10-xx | TIME: 9 a.m. | BY I.E.: Mary Tuttle

OPERATION DESCRIPTION: Assemble muffler clamp - 2 at a time

ACTIVITY Left Hand	TIME IN MINUTES	ACTIVITY Right Hand
Get and place "U" bolt	.05	Get and place "U" bolt
	.10	
Get and assemble bracket and 2 washers	.15	Get and assemble bracket and 2 washers
	.20	
Get and assemble 2 nuts - run down 8 turns	.25	Get and assemble 2 nuts - run down 8 turns
Aside completed unit	.30	Aside completed unit
	.35	
	.40	
	.45	
	.50	
	.55	
	.60	
	.65	
	.70	
	.75	
	.80	
	.85	
	.90	
	.95	
	1.00	
	1.05	
TOTAL UTILIZATION .34	1.10	.34 TOTAL UTILIZATION
% UTILIZATION 100		100 % UTILIZATION

TIME STUDY CYCLE: TOTAL OCC / AVG. OCC / LEV FACT / NORM TIME

COST:

HOURS PER UNIT	.00312
DOLLARS PER HOUR	7.50
DOLLARS PER UNIT	.023

LAYOUT & MOTION PATTERN ON NEXT PAGE

TOTAL NORMAL TIME IN MINUTES PER UNIT	.170
+ 10 % ALLOWANCE	.017
STANDARD TIME	.187
HOURS PER UNIT	.00312
PIECES PER HOUR	321

FIGURE 6-7 Example left-hand/right-hand chart for the assembly of two muffler clamps.

the column adjacent to the other hand, each moment aligned exactly across from the other. This chart is useful in showing idle time by either hand. When one hand is idle or being used as a fixture (just holding a part), this time is shown as blank space and is referred to as a "one-arm bandit." One-hand operations are inefficient and must be eliminated. The left-hand/right-hand chart will give utilization rates on both hands. Full (100%) utilization of both hands is impossible, but by being creative a technician can approach full utilization. There are no procedural deviations from the operator/machine chart standard procedure. The PTSS form in Chapter 8 has been developed to improve on the left-hand/right-hand form.

SIMO CHART

The SIMO chart (simultaneous motion chart) was invented by the Gilbreths when they found the left-hand/right-hand chart to be lacking in one important detail. If a movement of four inches occurred, reading the left-hand/right-hand chart would only tell which hand was used. When evaluating human effort and resultant fatigue, it was important to determine which *body member* was used. One of the Gilbreths many principles of motion economy is that the body member requiring the least effort be used. Thus, the left-hand/right-hand chart was given a new column, which recorded the body member being used. The column is next to the center time line and begins with the easiest motion at the time line and the most difficult farthest out. The column is colored in, with a sequence of easy motions having little area colored in and a sequence of difficult motions having the entire column colored in.

The body members represented on the SIMO chart are finger, wrist, forearm, full arm, and shoulder. Finger motions require the least effort and shoulder motions the most effort. Subsequent to the work of the Gilbreths, the problems of sustained complex finger motions became apparent. Such repetitive motions appear only in a SIMO analysis.

Completing the SIMO Chart

The chart is filled out according to the steps for every multiactivity chart on page 00. As Step 8 is completed, each activity is analyzed for the body member used and the appropriate column is filled in. The completed chart will look like Figure 6-8.

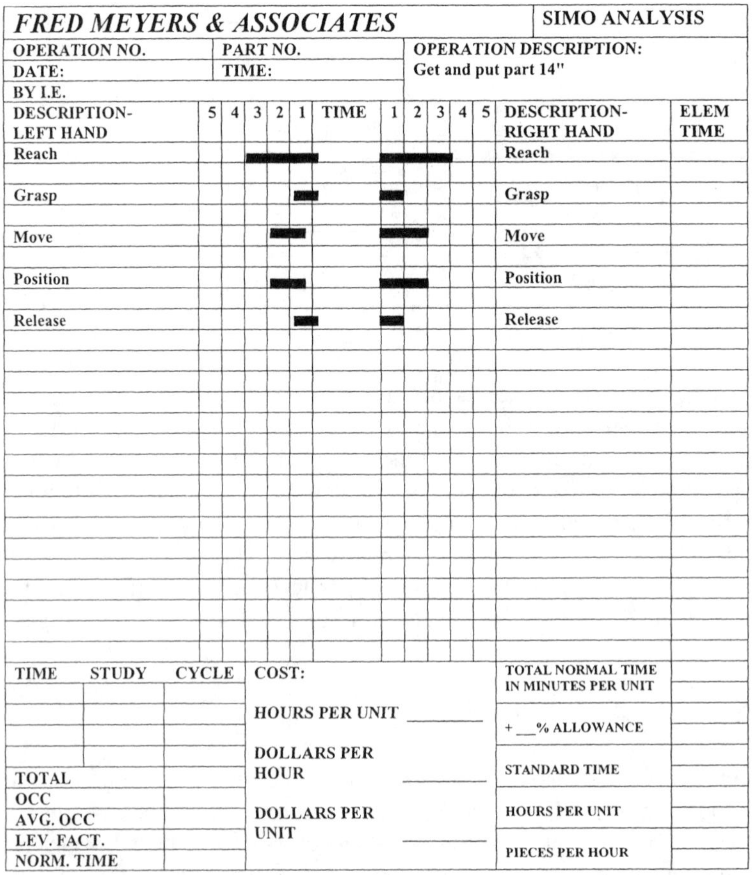

FRED MEYERS & ASSOCIATES												SIMO ANALYSIS	
OPERATION NO.			PART NO.				OPERATION DESCRIPTION: Get and put part 14"						
DATE:			TIME:										
BY I.E.													
DESCRIPTION- LEFT HAND	5	4	3	2	1	TIME	1	2	3	4	5	DESCRIPTION- RIGHT HAND	ELEM TIME
Reach			■					■				Reach	
Grasp				■				■				Grasp	
Move			■					■				Move	
Position				■				■				Position	
Release			■					■				Release	

TIME	STUDY	CYCLE	COST:	TOTAL NORMAL TIME IN MINUTES PER UNIT	
			HOURS PER UNIT _____	+ ___ % ALLOWANCE	
TOTAL OCC			DOLLARS PER HOUR _____	STANDARD TIME	
AVG. OCC			DOLLARS PER UNIT _____	HOURS PER UNIT	
LEV. FACT.					
NORM. TIME				PIECES PER HOUR	

FIGURE 6-8 SIMO chart, factory work.

Evaluating the SIMO Chart

The SIMO chart must be evaluated for two conditions: potential stress on trunk and potential repetitive stress on fingers and wrists. If the job has a considerable amount of type 4 and type 5 motions (shoulder and full arm), then a thorough study of the weight involved, the heights of the beginning and ending motions, the distance away from the body, and frequency of the activity in the task should be evaluated to ensure the operator has sufficient recovery time. Lifting 60-pound bags from a skid to a conveyor is shown in Figure 6-9. If the activity has a considerable number of type 1 or type 2 motions, the task should be studied for potential repetitive motion stresses. Figure 6-10 shows the movement of a package while scanning with a barcode reader.

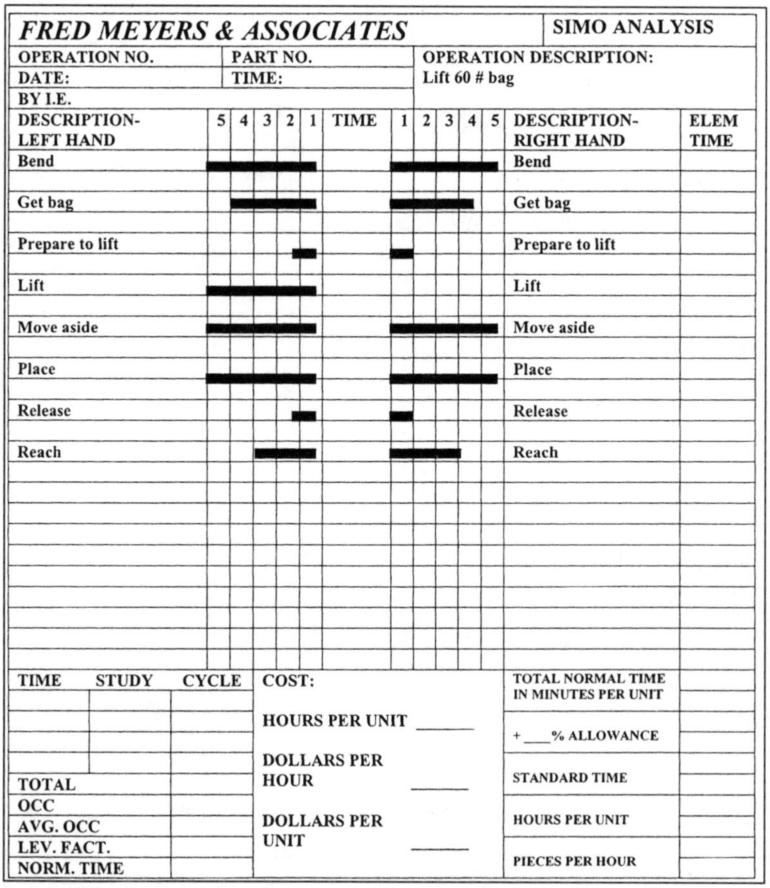

FRED MEYERS & ASSOCIATES							SIMO ANALYSIS						
OPERATION NO.		PART NO.				OPERATION DESCRIPTION:							
DATE:		TIME:				Lift 60 # bag							
BY I.E.													
DESCRIPTION-LEFT HAND	5	4	3	2	1	TIME	1	2	3	4	5	DESCRIPTION-RIGHT HAND	ELEM TIME
Bend												Bend	
Get bag												Get bag	
Prepare to lift												Prepare to lift	
Lift												Lift	
Move aside												Move aside	
Place												Place	
Release												Release	
Reach												Reach	

TIME	STUDY	CYCLE	COST:	TOTAL NORMAL TIME IN MINUTES PER UNIT	
			HOURS PER UNIT _____	+ ___ % ALLOWANCE	
TOTAL			DOLLARS PER HOUR _____	STANDARD TIME	
OCC					
AVG. OCC			DOLLARS PER UNIT _____	HOURS PER UNIT	
LEV. FACT.					
NORM. TIME				PIECES PER HOUR	

FIGURE 6-9 SIMO chart, heavy lifting.

FRED MEYERS & ASSOCIATES								**SIMO ANALYSIS**				
OPERATION NO.				**PART NO.**			**OPERATION DESCRIPTION:**						
DATE:				**TIME:**			Knit by hand						
BY I.E.													
DESCRIPTION-LEFT HAND	5	4	3	2	1	**TIME**	1	2	3	4	5	**DESCRIPTION-RIGHT HAND**	**ELEM TIME**
Hook					▄		▄					Unhook	
Pull					▄		▄					Reach	
Set					▄		▄					Hook	
Unhook					▄		▄					Pull	
Reach					▄		▄					Set	
Hook					▄		▄					Reach	
Pull					▄		▄					Hook	
Set					▄		▄					Pull	
Unhook					▄		▄					Unhook	

TIME STUDY CYCLE		**COST:**		**TOTAL NORMAL TIME IN MINUTES PER UNIT**	
		HOURS PER UNIT _____			
		DOLLARS PER HOUR _____		+ ____ **% ALLOWANCE**	
TOTAL				**STANDARD TIME**	
OCC		**DOLLARS PER UNIT** _____		**HOURS PER UNIT**	
AVG. OCC					
LEV. FACT.				**PIECES PER HOUR**	
NORM. TIME					

FIGURE 6-10 SIMO chart, knit.

EXAMPLE: Assemble U Bolt

Figure 6-11 is a picture of the work station and a drawing of the product. (Figure 6-11 is the back of Figures 6-7 and 6-8.) This picture and drawing help the technologist describe the operation.

QUESTIONS

1. Which is the most common use of the multiactivity chart?
2. Why don't we use the left-hand/right-hand chart more?
3. Review the cost reduction calculation for Figures 6-2 and 6-3 for 1 million units per year.

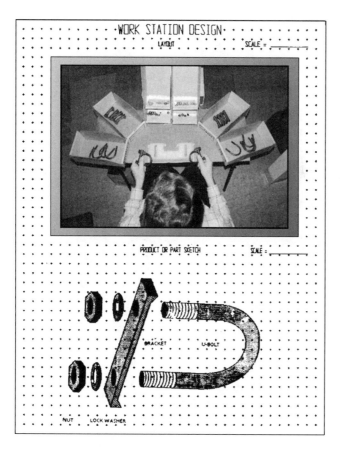

FIGURE 6-11 Work station design is the back of the time study form.

4. What is an activity? How many activities are there when two operators operate five machines?

5. When one operator runs five machines, each producing 200 pieces per hour, what are the standards?
   _____ pieces/hour
   _____ hours/piece

6. When 10 people work together on a gang job, produce 500 units per hour, and earn $10.00 per hour each, what is the cost of one unit?

7. Do a left-hand/right-hand chart of your project. Compare this to your PTSS after Chapter 8.

Motion Study: The Work Station Design

The work station design in this chapter is discussed from the motion and time study perspective, not the manufacturing engineering perspective, which would include strength and power considerations. We are interested in the effort and time considerations of the design. This chapter is not meant to be the ending point of work station design, but the beginning of motion and time study. Neither motion study nor time study can be performed if no work station design exists. In some cases, we may be required to select equipment and physically determine the requirements of a work station. In those cases, we are performing the manufacturing engineering job, and other training or experience is required. Some companies have one person responsible for everything needed to get a new product into production. That person would be required to design work stations as well as set the time standards to produce a specific number of products per day. This project would include a production start date, and management would expect the project engineer to deliver on time and to do whatever was required to start the new product.

Many of the previous techniques required work station designs. The work station design is a drawing, normally top view, of the work station, including equipment, materials, and operator space. The design of work stations has been an activity for industrial engineers and technologists for nearly a century. During this time, the profession has developed a list of principles of motion economy that all new technologists should learn and apply. When the principles of motion economy are applied properly to the design of a work station, the most efficient motion pattern results.

This chapter is divided into three sections:

Work station design
Principles of motion economy
Motion patterns.

WORK STATION DESIGN

The first question a new work station designer asks is, Where to start? The answer is very simple—anywhere. No matter where you start in work station design, another idea will come along and make the starting point obsolete. Where to start depends a great deal on what is to be accomplished at the work station. The cheapest way to get into production is the best rule for the starting point. The cheapest way means just that—the simplest machines, equipment, and work stations. Any improvement on this cheapest method must be justified by savings. Therefore, you are free to start anywhere, and then improve on the first method.

Our first work station example is a simple assembly station for the muffler clamp assembly shown in Figure 7-1. A work table of 36″ × 36″ × 42″ high with the parts placed on top would be the simplest equipment.

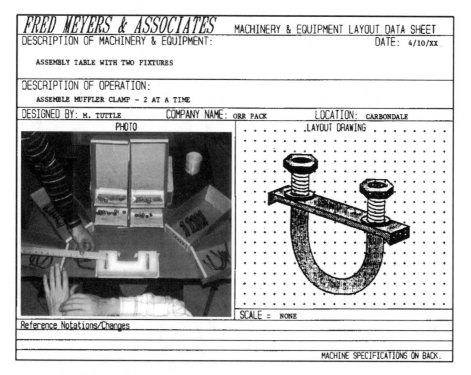

FIGURE 7-1 Muffler clamp assembly.

The following information must be included in the design (see Figure 7-2):

1. Work table.
2. Incoming materials (bolt, clamp, washers, and nuts; packaging and quantity must be considered).
3. Outgoing material (finished product).
4. Operator space and access to equipment.
5. Location of waste and rejects.
6. Fixture and tools.
7. Scale of drawing.

A three-dimensional drawing would show a great deal more information, and a talented technician would attempt this. (Costing and improving this work station design is accomplished in Chapter 8, where we cost and improve this job.)

Our second example of work station design is a machine operation (see Figure 7-3). The needs of the station design are the same as in the foregoing list, but the equipment (machines, jigs, and fixture) will be added.

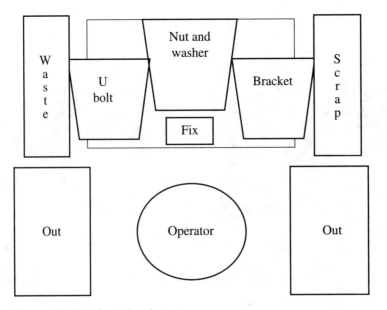

FIGURE 7-2 Work station layout.

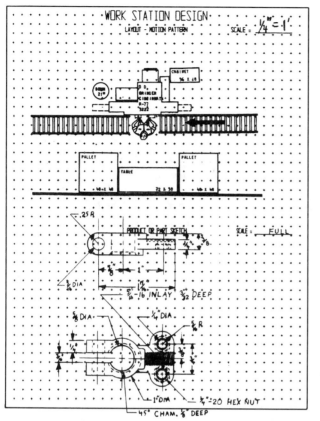

FIGURE 7-3 Machine operation.

PRINCIPLES OF MOTION ECONOMY

Industrial engineers and technologists have been continually developing guidelines for efficient and effective work station design. These guidelines are collectively called the principles of motion economy. These principles have become part of every motion and time study book. The principles are often used together in creative ways. The only limit to improved work station design is the technologist's creativity.

Effectiveness is doing the right things (the job), and efficiency is doing the things right (method). So effectiveness and efficiency mean doing right things right. Effectiveness is important to consider first because doing a job that is not necessary is bad, but making a useless job efficient is the worst sin. Efficiency, or doing things right, is a goal of motion and time study.

The principles of motion economy should be considered for every job. Sometimes principles will be violated for good reasons. These violations and reasons should be written down for future use. You will have to defend yourself to every new technologist—be prepared.

Hand Motions

The hands should operate as mirror images (see Figure 7-4). They should start and stop motions at the same time, they should move in opposite directions, and both should be working at all times.

If the hands are reaching for two parts at the same time, the bins should be placed at an equal distance from the work area and the same distance from the center line of the work station. To design a normal work station, the technologist should place all parts and tools between the normal and maximum reach, but make the reaches as short as possible (see Figure 7-5).

One hand reaching for only one part leaves the question, What will the other hand do? To keep both hands working at all times is a challenge and can be accomplished most easily by doing two parts at a time (one with the left hand and one with the right). Holding parts in one hand while assembling other parts to it is a poor use of the holding hand. It is said that the most expensive fixture in the world is the human hand (see Figure 7-6). In work station design, we don't consider people right-handed or left-handed, unless hand tools are used. Then we consider everyone right-handed.

Basic Motion Types

Ballistic Motions Ballistic motions are created by putting one set of muscles in motion and not trying to end those motions by using other muscles. Throwing a part in a

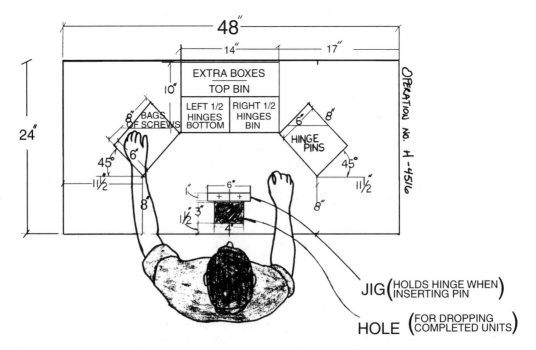

FIGURE 7-4 Design work stations to promote mirror image motion patterns.

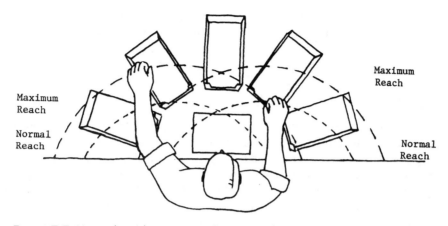

FIGURE 7-5 Normal reach versus maximum reach.

tub and hitting a panic button on a machine are good examples. Ballistic motions should be encouraged.

Controlled or Restricted Motions Controlled or restricted motions are the opposite of ballistic motions and require more control, especially at the end of the motion. Placing parts carefully is an example of a controlled motion. Safety and quality considerations are the best justification for controlled motions. But if ways of substituting ballistic motions for controlled motions can be found, cost reduction can result.

Continuous Motions Continuous motions are curved motions and are more natural. When the body has to change direction, speed is reduced, and two separate motions result. If direction is changed less than 120°, two motions are required. Reaching into a

FIGURE 7-6 The most expensive fixture: the human hand.

box of parts lying flat on the table is an example requiring two motions—one motion to the lip of the box and another down into the box. If the box were placed on an angle, one motion could be used. (We look at this principle again when we discuss gravity later in this chapter.)

Location of Parts and Tools

Have a fixed place for everything and have everything as close to the point of use as possible. Having a fixed place for all parts and tools aids in habit forming and speeds up the learning process. Have you ever needed a pair of scissors, and when you looked where they were supposed to be, they were gone? How efficient were you in the next few minutes? A tool maker's tool box is arranged so the tool maker knows where every tool is and can retrieve it without looking. That should be our goal in every work station we design.

The need for placing parts as close to the point of use as possible will be evident in the next chapter. However, for now we can just say that the farther you reach for something, the more costly that reach will be. Creativity is required to minimize reaches. We can tier parts: Instead of having one row of parts across the top of the work station, maybe three rows of parts, one over the other, would be better (see Figure 2-2). We can hang tools from counterbalances over the work station (see Figure 7-7). We can use conveyers to move parts into and out of the work station.

Release the Hands of as Much Work as Possible

As stated earlier, the hand is the most expensive fixture a designer can use. So we must provide other means of holding parts. Fixtures and jigs are designed to hold parts so operators

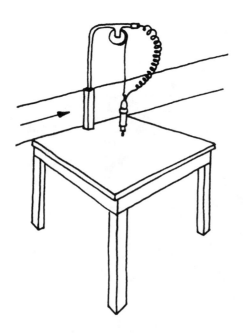

FIGURE 7-7 Counterbalance: pre-positions tool close to the point of use.

can use both hands (see Figure 7-8). Foot-operated control devices can be designed to operate equipment to relieve hands of work. Conveyers can move parts past operators, so operators don't have to get or put aside the base unit (see Figure 7-9). Powered round tables are also used to move parts past an operator.

Fixtures can be electric, air, hydraulic, and manually operated. They can be clamped with little pressure or tons of pressure, and they can have any shape, as dictated by the part. A hex nut can be placed in a hex-shaped hole that has no clamping need but will be held firm because of the part and fixture shape. Fixture design is easy, and only your knowledge of the part and necessary processes are required to design fixtures. Many tooling vendors would be happy to supply you with fixture-building materials.

Use Gravity

Gravity is free power. Use it! Gravity can move parts closer. By putting an incline in the bottom of parts hoppers, parts are moved closer to the front of the hopper. Production management allows us to spare every expense, and the use of gravity can do that. For example, consider a box 24″ × 12″ × 6″ lying flat on a table. The average part in that box (the only part the designer is interested in) is 12″ back, 6″ over, and 3″ down the exact middle of the box. If we get a 2″ × 4″ scrap board out of the trash, place it under the rear end of the box, and raise the box up 4 to 5 inches, the parts will slide down to the front of the box as the parts

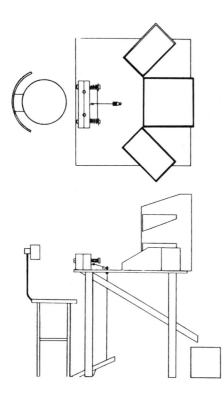

FIGURE 7-8 Fixtures and jigs relieve the hands for more useful work.

FIGURE 7-9 Conveyors move material into and out of work stations.

are used (see Figure 7-10). The reach has been reduced from 12 inches to 3 inches from the front lip of the box, a significant cost reduction.

Large boxes of parts can be moved into and out of work stations using gravity rollers and skate wheels. Parts can be moved between work stations on gravity slides made of sheet metal, plastic, and even wood.

Gravity can also be used to remove finished parts from the work station. Dropping parts into chutes that carry the parts back, down, and away from the work station can save time and work station space (see Figure 7-8). Slide chutes can carry punch press parts away from the die without operator assistance by using jet blasts or mechanical wipers, or the next part can push the finished part from the die.

Figure 7-10 Gravity will move parts closer to point of use.

Potential uses of gravity are everywhere. Try to incorporate as much gravity use into your designs as possible.

Operator Considerations—Ergonomics

Efficient operators must be allowed to work at the right height, with comfortable chairs, and enough light and adequate space to perform their tasks. Much effort has gone into trying to empower workers and including them in the important activities of regulating the processes and operations of the company. Today's employee is often expected to accept responsibility for the quality of the product, the meeting of schedules, and the reduction of costs. Yet many companies do not invest in providing appropriate environments for their employees.

People come in various sizes and shapes. There is a science called anthropometry which studies such body shapes among different population groups. Unfortunately, the findings of such studies are often ignored, with the result that totally inappropriate work stations are perpetuated. As educators, we see too many examples of college classrooms with chairs and desks designed for middle school 12-year-olds. And with no accommodation for the left-handed. Human dimension tables with 5, 50, and 95% data by age, sex, and nationality for 36 dimensions are in common use. However, particular populations may be classified in a way that false conclusions are made. One tends to think of Orientals as petite until one meets a group of Sumo wrestlers. There are examples in all cultural groups. Designing for the average size means that half the people will be too small for the plan and the other half too large. Devices must be adjustable to the comfort level of each individual.

Work station design, indeed any design, should be based upon user-centered design principles. Two tools, task analysis and user trial, are appropriate for the design process. In task analysis, the work station duties are reviewed to ensure that all of the parts, tools, work, and storage areas are appropriately placed, that the work station is sufficiently apart from the work areas of all other work stations, that the operator is comfortable, and that the transfer of materials is convenient. User trial occurs after the design is ready and ensures that the station is comfortable, functional, and efficient. During a user trial, every circumstance of use should be considered. The work station must fit the body sizes of all users. It must be hard to use incorrectly (particularly if such use may harm the worker or weaken the product). It must be comfortable to use for long periods, and it also must be easy and convenient. It must be easy to learn and have clear instructions. It must be easy to maintain, including cleaning. And it cannot interfere with sight and hearing. And finally, does the user feel relaxed after a period of use? This is a difficult list to fulfill, especially when only the concept is available. The details of good work station design are worked out by an iterative process. A good set of preliminary rules to start this process is

1. Design each opening so that the largest person may fit through it,
2. Design each surface height so that the smallest person may reach it,
3. Design so that the weakest can perform,
4. Design so the strongest will not overcome,
5. Provide easy adjustment for the range of sizes,
6. Provide independent adjustment for each human dimension,
7. Provide for alternate postures during the work period.

Work Surface   The correct standing work height is around elbow height (see Figures 7-11 to 7-13). With the forearm held parallel to the ground and the upper arm straight down, measure the elbow to the floor. This can range from 48″ for the tallest American male to 27″ for the shortest Japanese female. It is apparent that the Japanese woman cannot reach up to the 48″ level (it is above her shoulder height) and that the American male would be reaching below his fingertips. Even if the work station were designed for the variations in American males (13″) or Japanese females (8″), it would be uncomfortable for many. Designing the workstation at the highest level and using a platform for each individual may compensate for variation. Be sure the platform is sturdy, safe, and large enough to comfortably work on.

The work station is designed so that the task is actually done at that height. If the work is on top of a 20″ object, the work surface is 48 − 20, or 28″. The amount of force will also affect work height. Heavy manipulative tasks (especially requiring downward pressure) should have the hands about 8″ below the work surface, moderate force and precision about 3″ below, and delicate tasks about 3″ above the elbow (wrist support often being necessary). Lifting tasks are best done between the knuckle and elbow vertical space, and hand controls from elbow to shoulder. Work above the shoulders or below the knuckles is generally stressful, requiring additional strength and time to perform.

Any position is fatiguing if it is held for a long period of time. A seat should be provided that will permit the same elbow height and forearm attitude as standing. An alternative to this is to provide a trade-off task that can be done seated. Preferably, such a task would require different finger and wrist movements and frequencies as well.

FIGURE 7-11 Wrong work height will create problems.

FIGURE 7-12 Proper work height will produce less fatigue.

FIGURE 7-13 Design a work station for sitting and standing, but keep work height constant.

Seating Chairs have been neglected in the workplace. Some sort of strange idea that people work less efficiently while seated prejudiced many managers. The logic, extended, would improve the efficiency of the office by removing all chairs there also. In the modern factory, the comfort of the employees is important to enable them to be contributing members of the team. Consider the metal stool found in many shops. The wood seat is cracked and it is swathed with rolls and rolls of duct tape. Why use tape? Because the tape can function as a crude seat cushion.

So, the designers of ergonomic chairs enter the picture. First, all chairs are somewhat ergonomic. They have a base, a seat, and a back. Some manufacturers assume that making some of these parts adjustable can command a high price, but many of the expensive chairs violate the ergonomic principles of chair design. The principles are no secret and have been used in chairs for centuries. The seat of the chair will be most comfortable when the forces of the cushion are about equal on the human buttock. Wooden chairs often have a cutout for the body shape; cane, wicker, and canvas have been used to provide the shape. Modern cushions provide the same function. Tests using pressure pads show that conditions of even pressure applied to the largest surface provide the highest comfort scores. One of the most promising designs includes a wide saddle at the back with an upward-sloping horn at the front center. This is similar to the design of the tractor seat from the last century. Cushions should be padded enough to be comfortable—too much, and the person will sink into the seat, a condition not conducive to work. The height of the seat should be adjusted so that the thighs are approximately horizontal and the lower legs vertical, with the feet resting firmly on the floor. The back of the chair needs to provide support for the upper back and also lumbar support for the lower back. The height and slope of the upper back support depends on the individual. Once set, it is rarely changed. Lumbar support, however, will depend on the forces required for the particular task and on the condition of the lower spine from prior activities. Adjustment there may be more frequent. Having tested chairs both with extra cushioning at that point and an adjustable air bladder, the air bladder was far superior. The base of the chair requires five legs for safety (four-legged swivel chairs do fall over). And good rollers are also required if the job requires such movement, even occasionally.

Summing up the requirements for an ergonomic chair,

1. A comfortable seat with an appropriate contour and a firm cushion,
2. The ability of the seat to tip forward and to lift so that the buttock is higher than the knees,
3. A back that is adjustable for the height and angle of the spine,
4. An adjustable head and neck support,
5. A lumbar cushion that is adjustable for the curve of the spine,
6. Appropriate arm rests and wrist wrests designed for the job,
7. Appropriate base and wheel system for the work location and job.

Such a chair may not be available for a reasonable cost or even at all. An extensive test by the employees who will be expected to use the chair should be made, especially if the cost is high. The label "ergonomic chair" does not guarantee comfort, productivity, or chronic stress reduction. Indeed, the true test of any principle is the reduction of work stress by the users of the chair (see Figure 7-14). Many of the designs or even supposed optimal adjustments are favorites of some and shunned by others.

Space Allocation Operator space of 3' by 3' is generally adequate unless work-station fixtures are larger. A minimum of 3' from an aisle is required for safety. Areas for parts on either side of the operator may be necessary additions. If two people are back to back, then five feet between stations is adequate. With complex machinery, more space may be necessary for setup and repair.

Work Station Environment

No matter how motivated the worker or how effective the design of the individual work station, the climate of the plant must be considered for the health and well-being of the employees. An unhealthful climate is the responsibility of the management and can, in severe cases, result in personal civil and criminal liability in addition to the liability of the company. As this topic is far beyond the scope of this book, a brief description is all that we will provide here.

Environmental topics include illumination, vibration, noise, climate, toxicology, and safety. Only extremely severe, and unusual, conditions have immediate observable effects. In most cases, secondary—possibly chronic—effects may occur.

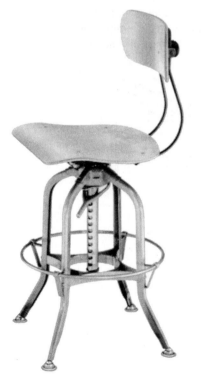

FIGURE 7-14 Chairs must be adjustable and comfortable. (Courtesy of Toledo Furniture.)

Illumination   The effects of illumination depend on three things: the individual's perception of the quantity and quality of light, the light source output, and the visual environment that is illuminated. The measure of light output is in lumens from a light source. The amount of light that can be used is the lumens/m^2, or lux, returned from the object illuminated.

   People see differently. As one ages, the amount of light required to see and complete tasks increases dramatically. An aging work force will generally mean an increase in the amount of light required to operate the company. Also, the need for glasses changes over time. In many studies, it was consistently found that about two-thirds of the people without glasses in a plant required them. Further, about one-third of the employees with glasses required lens changes. Individuals see in low vision environments with rods, a structure in the eye. Since the rods do not detect color, the whiteness of a moonlit night, for example, is enhanced. As some color is perceived, the cones, another structure in the eye, begin to work. For all repetitive industrial work, light levels should be above about 500 lux overall, well above the point where cones operate and color is detected. Insufficient light will cause eye muscle fatigue, bloodshot eyes, and headaches (see Figure 7-15). The effects are temporary.

   The perfect light source would be free, provide true color, come from the proper direction, and give the right amount of light upon demand. There is no such thing. Using sunlight requires decisions about noise, ventilation, heat loss, and many costly side effects. The amount and direction of sunlight often require supplementing and occasionally require blocking. Blinds, shades, and curtains can be costly. Artificial light requires electricity, which is about 90% of lighting costs. For large areas, including work bays, storage bays, and outdoors, there are several kinds of sodium, mercury vapor, and other high-efficiency lighting systems. For spot illumination, low-voltage halogen lamps are becoming popular. The usual choice for lighting, however, is incandescent or fluorescent lamps. Incandescent lamps generate most of

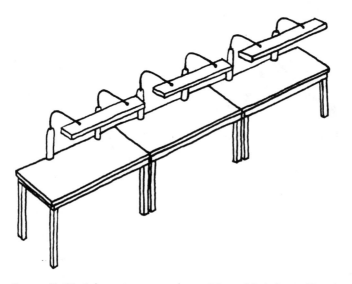

FIGURE 7-15 Adequate, properly positioned lighting will reduce fatigue.

their energy as heat and infrared energy. Thus the warm color is at the sacrifice of longevity and output. Either of these can be optimized. Low-maintenance long-life bulbs also have a lower lumen output. Fluorescent lamps provide about four times the lumens of an incandescent bulb. Replacement fluorescent bulbs that fit incandescent lamps are available.

The visual environment affects the way the work is seen. Given the same employee and the same light source, the size and shape of the room, the color of the walls, ceiling, and floor, the color and shape of equipment, even the amount of dust in the environment will affect how things are seen. Contrast can have as great an effect as intensity. Backgrounds for color inspection should be the same color as the part being inspected to make changes in surface characteristics apparent. Shapes, including print, require high contrast, and colored objects require a contrasting background color for effective inspection. Light sources affect the apparent color. Inspection tasks should specify the light source. The contrast between the light from the windows and battleship gray, industrial green, and safety yellow interiors require tiring adjustments for the eyes. Light-colored walls, floors, and ceilings increase lighting efficiency and reduce glare. Light-colored work surfaces reduce contrast with paper. Machinery moving parts should be painted one matte color while parts that are being worked on should be a contrasting color.

Glare can indirectly be reflected from a surface illuminated by too much light, and metallic or gloss finishes can enhance it. Polarization or shielding can reduce this, as can taking such measures as not polishing floors to a high gloss. Direct glare from the light source may require special shielding or orientation.

Vibration Vibration conditions in industry include segmental vibration caused by hand tools, whole-body vibration associated with large machines, and whole-body vibration of trucks and other transportation vehicles. Vibration can be reduced by redesigning at the source or by isolating the person from the source. Equipment can be mounted on springs or compression devices, materials can be substituted that have little vibration properties, amplitude can be reduced with more frequent balancing and maintenance, and machine motions, including speeds and feeds, can be modified. Persons can be isolated from vibrations by insulating materials on handles, by soft floor mats, or by springs or cushions on chairs.

The severity of vibration effects is affected by exposure time, acceleration, and frequency. The effects of whole-body vibration, most severe at frequencies of 4 to 8 Hz and above .1 G include abdominal pain, loss of equilibrium, nausea, muscle contractions, and chest pain. Effects of segmental vibration at frequencies over 8 Hz and above 1.5 G include loss of strength as well as numbness, stiffness, pain, and blanching of the fingers.

Noise Noise provides a common source of annoyance in an industrial environment. It requires employees to increase concentration, and this tends to increase fatigue. The level of noise is also increasing outside the workplace, which rather than causing a higher tolerance in the workplace, causes a greater employee desire to have shop sounds reduced. What is acceptable and expected at popular entertainment or in the commute to work is unacceptable in the workplace. Noise may affect people in the following ways:

Degrade task performance

Distract or annoy nearby persons

Interfere with communication

Contribute to temporary or permanent hearing loss.

Noise is measured by a logarithmic scale and is expressed as dBA (decibels in the A band). Because the scale is not linear, one device can be measured at 80 dBA, while two of the same devices will measure only 86 dBA. Controlling noise intensity is therefore not a matter of merely shutting down devices. A living environment, such as a home, hospital, church, or courtroom, measures about 40 dBA; a conference room about 50 dBA; a loud office, about 60 dBA. At about 80 dBA, noise is very loud but has no permanent effect. At about 90 dBA, exposure is to be limited to eight hours; at 100 dBA, there is a two-hour limit; above 115 dBA, exposure is not permitted.

From about 50 to 80 dBA, hearing problems are temporary and are related to efficiency and discomfort. In jobs where one must concentrate or listen to complex explanations, noise distractions can be a hindrance to the job. In other cases, such as repetitive assembly or heavy lifting, noise may aid in completing the task or work day. Although not recommended for areas with a background noise of over 70 dBA, music may be found to be work-enhancing. Even then, some employees will not like the type or volume of the particular music. Employees should be involved in the selection of the type of music and the schedule. The broadcasting system should produce a quality sound, at levels only slightly (3 to 5 dBA) above the background sounds. In environments with noise above 90 dBA, hearing loss is complicated. It can be spectrum- or bandwidth-specific, permanent or temporary, partial or total, and involve such complications as tinnitus. The impact of a particular noise is difficult to ascertain. However, the measurement is quite simple and the governmental penalties for violations are severe.

Noise reduction is accomplished by planning and management. There are three techniques to reduce noise: placing a barrier to reduce transmission through air or other materials, absorbing the reflected or causal noise, or reducing the volume or spectrum of the noise generator.

Evacuation alarms should be part of overall noise management planning. Evacuation alarms should last over 10 seconds, sweep tones from 500 to 2000 Hz frequencies, and be 10–12 dBA higher than ambient noise. It should be followed by a verbal message 14 dBA higher than ambient noise. Visual alarms should be part of the alarm system.

Cold or Hot Surfaces Cold surfaces, when touched by bare skin, can tear and damage it. The schoolboy dare to put a tongue on the flagpole is a fine example. Metal below 45°F and wood or plastic below freezing should be handled with gloves. Hot surfaces can burn bare skin. Metals become uncomfortable at 110°F and cause burns at 140°F. Wood and plastics can be over 200°F before burn occurs. In an environment with a hot or cold radiating material, radiated energy will increase the effects of climate.

Changes in body heat affect comfort. Specifically, body heat is controlled by airflow, temperature, and humidity. Cold conditions increase manipulative task effort; very wet climates may require special precautions and clothing to maintain health; hot, humid climates add to the demands of physical work; dry conditions put increased strain on the body. Even in a comfortable environment, surface heat or cold can cause injury.

Individual perception of comfort changes with behavior, diet, season, time of day, job stress, and expectations. Choice of clothing affects comfort. Since people are different, it must be expected that at any given temperature some people will feel too hot and some too cold. If we can assume that clothing can be adjusted for discomfort, and light work is to be done, individuals will be comfortable in temperatures between about 66 and 76°F. If the humidity is below 30%, then temperatures up to 79°F can be tolerated. Humidity readings between 70% and 20% are preferable. Increased airflow can be a benefit at higher temperatures but also a major cause of complaint as temperatures drop. Airflows of 40 feet/minute are common in the comfort zone. At velocities of 100 feet/minute, temperatures can increase to about 82°F.

As temperatures increase beyond the comfort zone, sweating increases and skin temperature rises. As workload is increased, more body heat is produced and the immediate effect is that less work can be performed. In two-hour exposures with a typical summertime humidity of 60%, the worker can do work involving above 300 kcal/hour at 80°F, up to 300 kcal at 90°F, and below 120 kcal at 100°F. Two-hour work periods at temperatures above 100°F are not recommended.

Heat Discomfort Reduction

Reduce the temperature (air conditioning)	Quickest, but often most expensive approach
Reduce the humidity (dehumidifying)	Increases the ability to sweat, permitting increased temperature tolerance
Increase air velocity (fan)	Increases evaporative heat loss
Reduce the workload (methods analysis)	Additional fixtures and holding devices are needed
Adjust the clothing (protective clothing)	Special cooling systems or fabrics are available
Work practices	Rest periods in cool (72°F) area
	Drinking water and eating salted food
	Training in recognizing and eliminating heat illness symptoms
	Acclimatization allowances

As temperatures decrease below the comfort zone, the range between comfort and freezing will affect the ability to perform dexterity tasks. Local cold discomfort is the major cause of complaint in cold environments. Although generally protected, the hands, feet, and face are often subjected to increased stresses. Exposure and reduced blood flow combine to cause such discomfort. Skills start to degrade at 60°F and can have a reduction as severe as 20% at 45°F. The solutions to reducing the overall risk increase the decrement to performance, however. Clothing is the general approach to correcting general cold exposure. This includes appropriate fabrics such as wool, designs that cover the whole body closely but permit a skin-side layer of air, windproofing layers, and hand pockets. Wet clothing increases risk, and a clothing drying cabinet for use during breaks is useful. Air velocity reduction such as wind breaks or portable tents help against air velocity. Workload should be leveled because short spans of high activity cause sweat and periods of inactivity cause discomfort.

When the core temperature and the skin temperature converge, the body becomes unable to eliminate heat and dehydration and heat stroke are possible. Work practices for protection against heat illness include

Radiant heat protective clothing

On-the-job buddy system

Work and rest task flexibility scheduling

Provision of tepid (not cold) drinking water

Scheduling tasks in cooler hours of day

Training workers in heat illness first aid.

When core temperature cannot be maintained, serious hypothermia and subsequent death from exposure can occur. At a body temperature below 95°F, severe shivering occurs. Usually local cold symptoms occur first, because in order to work hands and eyes cannot be protected. Frostbite and freezing are possible, and aggravation of vibration injury and of arthritic conditions can occur. Short-term exposure of −10°F is dangerous if gloves are not worn. Gloves with holes in the fingers are used, but most effective are frequent breaks when the hands can be warmed.

Because the physiology and psychology of the individual is important to the performance of the worker, it is a part of the engineering design process to ensure that nothing in the manufacturing system impedes the worker's performance. It is not enough that the work station be comfortable, simple, and safe. It must be perceived as such.

MOTION PATTERNS

A motion pattern is the path taken by both hands in the process of making one part or pair of parts (if making two at a time). The path for each hand must be unbroken and a complete loop. It is much like a computer program. A motion pattern is also a blueprint of the work method and a bill of material for a time standard. This second definition will prove useful in Chapter 8, on predetermined time standards systems (PTS). Work station designs and motion patterns are required before PTS can be applied, because the motion pattern (which is drawn on the work station layout drawing) is the blueprint of the work method and the bill of material for the time standard.

The work station design must be completed before the motion pattern is drawn. The first motion pattern is drawn on the work station and then redrawn on its own page to allow analysis. The motion pattern and cyclograph for the same job would look identical. Cyclographs are nice but too expensive to produce (see Figure 2-2).

Each job is made of elements. An element is one indivisible piece of work, which usually includes a reach, a grasp, a move, and an alignment/position. More simply put, an element is getting one part or tool and doing something with it. Examples of motion patterns are shown in Figures 7-16 and 7-17. Notice in Figure 7-16 that there is only one loop per hand. This is a one-element job. In Figure 7-17, there are three loops per hand; therefore, this is a three-element job. Elements are important in setting time standards.

Figure 7-16 Work station and motion pattern for pin board: from average pin location to average hole = 6 inches.

Work station designs using the principles of motion economy and the resulting motion pattern are required before a predetermined time standard can be set. The back of the PTS form has a place for these two pieces of information. Chapter 8 discusses setting the time standard on the work stations designed in this chapter. Once time standards are set, costs are calculated and all future improvements to the job are compared to these costs.

QUESTIONS

1. What is included in a work station design?
2. What is the basic rule regarding the cost of new work stations?
3. Where do you start a work station design?
4. What are the principles of motion economy that relate to
   a. Hand motions?
   b. Basic motion types?
   c. Location of parts and tools?
   d. Relieving the hands of work?
   e. Use of gravity?
   f. Operator considerations?

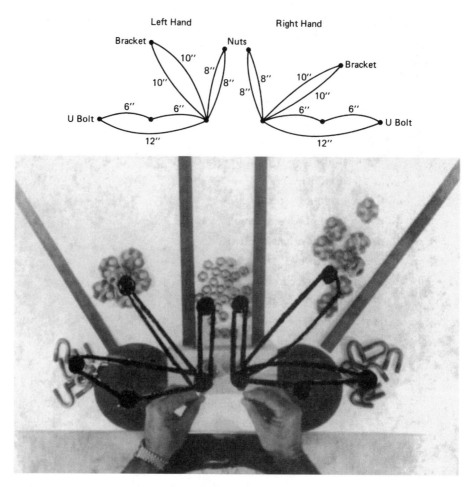

FIGURE 7-17 Motion pattern for cable clamp.

5. What is a motion pattern?
6. How is a motion pattern used?
7. Design a work station for your semester project.
8. Draw the motion pattern for your project.
9. Write a summary of how you used the principles of motion economy in your project.

Predetermined Time Standards (PTS)

INTRODUCTION

Predetermined time standards (PTS) is a modern technique of motion study and time standards development. All work has been reduced to basic motions, and each motion has been reduced to a specific time value. With PTS, we can describe the method in terms of the basic motions required to perform the operation, and then the time value for each motion is looked up on a table and recorded on the PTS form. Each element of work is totaled, and the elemental times are totaled for the time required for the total job. PTS is a general term, to be distinguished from PTSS, a system developed by Fred Meyers.

Station design and motion pattern development are the first two steps of the predetermined time standards (PTS). Once these two steps are complete, a time standard is needed for the many reasons discussed in Chapter 4, but costing the method is most important at this early stage of design. In Chapter 7 we asked where to start, and the answer was anywhere, because the first design is never the best design; it is only a starting point for optimization. All future designs will be compared to our first design, and cost is the measure of desirability.

PTS is best used before production starts. Think about it: A new product has been invented, and you have been asked to design a production line to produce 2,000 units per shift. There are no machines, no assembly stations, no operators to study. How will you set time standards so you will know how many machines to buy or to build, how many people will be needed, or how much your product will cost? The PTS technique is the answer. All new work stations start with an idea in a technician's head. PTS will allow the technician to evaluate and improve methods ideas.

PTS is the technologist's most efficient tool. It can

1. Develop the best method.
2. Establish the time standard.
3. Help develop standard data (see Chapter 10).
4. Establish labor cost.
5. Help justify better tooling.
6. Help select the best machine for the job.
7. Assist in the training of the operator in the best method.
8. Develop motion consciousness and cost consciousness in supervisors, engineers, and employees.
9. Settle grievances in connection with time standards.

Predetermined time systems are almost as old as the systematic measurement of work. With the invention of time study in the 1880s, various tables and formulas were devised to group such standards. The wide discrepancies in individual performance, the variations in observational conditions, and the occasional human error have, over a century, led to the conclusion that time study data probably comes no closer than about 10% to measuring the actual performance and often is much worse. Further, if some of the standards were at +10% and others at −10%, the effect would be as though a crew of five had an extra helper. Certainly, any time study person will tell you that the law of averages applies and that this wide variation is usually compensated for. In individual time studies, an increase in accuracy (the relationship of the individual standard to the individual performance) is often cancelled out by reduced consistency (the relationship of one standard to others with similar work).

Consistency is an inherent feature of a predetermined time system. It is built by correlating large numbers of similar motions over a range of tasks, conditions, and operators. Accuracy becomes meaningless because it is not a single task being measured, but consistency becomes important because so many similar motions were included in the average. Individual differences, however, are certainly an important time variable for both predetermined time systems and time standards. With a range in performance time, primarily due to slight variations in method, of about 2.6 to 1, rating (normalizing) to determine a "should take" rather than a "did take" time is used to find a benchmark. Because rating is done to the original predetermined time data, any analyst error will be consistent, while the stopwatch will yield varying individual point estimates.

Decision tasks are difficult to measure. As such, they are often left out of predetermined systems and may even go unrecognized in time study. Only the simplest reactions and specific defining conditions are used. Because of the high potential for variability, decision tasks are usually measured when they represent only a tiny portion of the work content.

For half a century, predetermined systems and standards were not successful, but several successful systems appeared after World War II. These systems were used by engineers with considerable specialized training in applying different methods of time codes to jobs.

METHODS TIME MEASUREMENT (MTM) FAMILY

MTM, developed by Maynard, Stegemarten, and Schwab in 1948, is probably the best-known predetermined system in use today. MTM-1 has 10 elements of micromotion. Each element is assigned a number of time-measured units (TMU), which are .00001 hours (one hundred thousandths of an hour). One minute thus equals 1,667 TMUs. The elements of MTM-1 are Reach, Move, Turn, Apply Pressure, Grasp, Position, Release, Disengage, Eye Movements, and Body Movements. There are distances, cases, and classes that modify the table values so that there are about 1,700 different values to apply. There is also a way of determining which motions can be performed simultaneously. It requires over 80 hours to become proficient in using MTM-1. The MTM-2 and MTM-3 systems are faster but less accurate than MTM-1. Each of these systems has considerably fewer variables, and a standard that would take a week to set by MTM-1 would take a day in MTM-3. There is a complete package for computerizing the MTM analysis, called 4M, as well as several systems for specialized applications such as maintenance and clerical. Upon completing a training course by a certified instructor, a student must pass an exam to be certified as an MTM practitioner (the famous blue card). Practitioners usually work for companies that use the system.

The H. B. Maynard company developed, in 1980, a predetermined time system called Maynard Operational Sequence Technique (MOST), which originated as a computerized application of MTM but in the final version had an entirely different system for data collection and coding. The time value in TMUs can be determined in MOST by adding all of the factor numbers, which is not a feature of the MTM family. Also in MOST, upon completing a line of code, a complete cycle has been performed. This is unique to MOST and is a convenience in editing time values.

MODAPTS

Modular Arrangement of Predetermined Time Standards (MODAPTS) is a predetermined time system that is most consistent for manual work tasks. It can also be used for determining the manual internal and external work content of machine-paced cycles. The person vs. machine chart can be used in conjunction with the analysis.

MODAPTS is simple enough to be learned by every team member and have its principles recognized by every employee. As such, it is useful for lean manufacturing, productivity efforts, and proactive quality activities. It provides a rapid answer to the question of individual task times and line balance issues. The system is in current use by UAW-Ford to resolve workload disputes and by Ford to estimate model costs.

The system is easy to learn because the rules for movement and terminal classes are simple, and the application of auxiliary class activities is unambiguous. Generally, the appropriate code can be applied based on visual identification. It is very useful for teams or peer group analysis or for analysis by supervisors or process engineers. Each analysis consists of movement, terminal, and auxiliary classes of activities. The movement class tells which body member moved, and the terminal class tells what type and difficulty of manipulation occurred at the end of the move. These classes of activity account for most of the time at a modern work station. Auxiliary classes include walk, sit, stand, read, decide, vocalize, and other miscellaneous activities that occur infrequently.

MODAPTS is useful to those concerned with workplace ergonomics. Because it is based on the upper body member moved rather than the distance traveled, the method description represents the pattern of each individual body member movement. Extensions of the system (ErgoMOD) have been developed in Australia and used by the Victoria Health and Safety Authority to evaluate ergonomic problems in the workplace in the enforcement of legislated ergonomic standards. Like the other predetermined time systems, MODAPTS cannot measure work effort. However, there is a system, engMOD, utilizing MODAPTS coding that calculates energy use and predicts the need for rest periods.

History of MODAPTS

In 1966, G. C. (Chris) Heyde, an experienced work measurement engineer and consultant took a fresh look at the predetermined time systems that had been widely available for several decades and decided to take a radically different approach. Rather than trace the distance an object moved and correlate distance and movement, he would attempt to evaluate movements based upon the observed body member. In particular, the movements of the fingers, hand, lower arm, upper arm, and shoulder would each be analyzed separately. This is the system that came to be known as MODAPTS (MODular Applications of Predetermined Time System).

The results of Heyde's study found that the body member observed moving determined with high probability the distance that was moved and the distance moved determined with high probability the body member that was moved. Each body member usually moved its set distance. Members closer to the finger moved shorter distances, and members closer to the shoulder moved most of the really long distances. Even more interesting to the work measurement analyst is the fact that Heyde also found that these body members moved in multiples of each other. Thus, a system of viewing the human body member at work and calculating a task time came into existence.

In refining this system, the Australian team that developed MODAPTS tried to devise a method for determining task times that was even simpler to apply and calculate. It is because of MODAPT's ease of calculation and observational characteristics that it is advocated for use in the evaluation of digital simulation. A task time requiring a minute can be calculated and documented in less than an hour by a person with knowledge of the job and less than a week's training.

MODAPTS: Identifying and Coding Human Motion

Motion classes represent nearly all of the time spent in performing work, play, or, indeed, any form of active endeavor. Each human is constantly moving things or reaching for things. Depending on the task, our fingers, hands, or arms are in motion. It only takes one unit of time (MOD) to close the fingers. If the fingers must move further than about an inch, another body member also moves. If the move is about 2 inches, the person has probably moved only the palm. Typing with the wrist on a pad is this type of motion; it takes 2 MODS. Most of the time, things are a short distance from the assembly area and the reach is about 6 inches, which is completed in 3 MODS. Longer reaches with the full arm extended can take up to 7 MODS.

Terminal classes occur at the beginning and end of virtually all motion classes. These involve puts or gets and represent differences in control at the beginning or end of each motion. Both puts and gets can be done with no difference in control and have no time allowance. There are also events that require a little attention and some requiring a great deal of concentration.

Auxiliary classes include all other factors, motions, calculations, and decisions. Once an allowance for weight is added to the motion and terminal time, over 70% of the work is covered. By adding MODS for body motions and mental activities, virtually the entire task can be measured in MODS.

EXAMPLE: Remembering that, depending on method, skill, and motivation, a considerable difference can occur between humans, the leveled time for a MOD is about 129 milliseconds per MOD. In Table 8-1, the worker moves a small box to a scale, weighs and sets it aside, then returns the hands to the table edge.

The Use of MODAPTs at Ford-UAW

MODAPTS is not unfamiliar to the transportation industry. It is used throughout Ford and UAW-Ford as a method of establishing line task times. It has also been used for ergonomic purposes, and it has been used to establish corporate labor book performance.

Evaluating Production Line Task Load In the Ford labor contract, there is a clause that names MODAPTS as the system to be used for resolving productivity issues in the assembly plants. In the implementation, a number of union representatives have been trained and currently evaluate work independent of Ford industrial engineers. According to representatives from both sides, this system has reduced the number and severity of grievance issues about productivity.

Estimating Labor Book Performance At Ford for the past decade, the corporate labor book standards have been set using MODAPTS. Because the standards have been found credible, easy to visualize, and of sufficient detail, the estimates of standard time are accepted as benchmarks by the various plants.

Table 8-1

MODAPTS CODE	DESCRIPTION	DISTANCE (INCHES)
M3G1 M2P2	Get package and put on scale	12
M2P0	Move hand aside	6
R3	Read 2 digits	6
M2G1 M2P0	Get and set package aside	6
M3P0	Move hand to start	12

Ergonomic Analysis Several years ago, Ford and some consultants extended MODAPTS in order to look at particular ergonomic problems in the automotive industry. That several alternative elements could be developed for the system was possible because within the MODAPTS system the times are based on observed movements, unlike other predetermined time systems where times are based on distance and difficulty.

ErgoMOD Measures Increased Task Difficulties

ErgoMOD is a system that provides an analysis of the difficulty of work. The length of time needed to perform a task is only one measure of performance difficulty. Within the range of human capability, there are tasks that can be performed by nearly all persons, and some that can be performed by only a few. Dr. Judith Farrell of Melbourne, Australia has developed a system (FWAPS) in which the categories used to distinguish between the length of time (MODAPTS) are further divided into categories that involve insignificant amounts of time but are major categories for differentiation of task difficulty. The combined use of MODAPTS and ErgoMOD extends the contribution of each beyond that which either can make alone. MODAPTS is used to describe the elements or steps of the task and to display the time needed for each step; ErgoMOD records the status, at each step, of the posture or action characteristic of interest. Thus, the period of time (number of steps) over which a given ErgoMOD posture or action is sustained and how often it is repeated can be seen.

In MODAPTS the number of classes of action within categories such as arm movement, grasp, or placement of components is small, actions that require a similar length of time being grouped together in the same class. A MODAPTS analysis provides a step-by-step sequence of actions of which a task is comprised, and many of the actions are identified by the body part used, but the categories are not designed to distinguish between actions using different muscle groups. When dealing with pain associated with repetitive work, it is important to know whether the work is being spread, in turn, among a number of muscles or whether the fibers of one or more particular muscles are contracting in the same way. Further, MODAPTS does not have a code for maintaining control of an item once it has been picked up, yet static muscle work is known to be a factor in chronic pain associated with repetitive work. Neither does MODAPTS record the worker's posture, yet maintaining posture is, in itself, a form of static muscular work.

ErgoMOD was designed to provide descriptors to record the presence of actions with characteristics difficult to perform by people with known musculoskeletal problems, including partial paralysis, uncoordination, and musculoskeletal pain. The frequency with which a difficult action must be performed determines whether the person for whom it is difficult should apply for or continue in a job that requires it. In the case of soft-tissue musculoskeletal pain associated with repetitive work, the frequency with which any action is repeated and the duration over which the repetitive motion pattern is sustained are as important as the characteristics of the actions themselves. Also important are the length of the break from the sustained unvaried actions and the extent to which those actions performed during these breaks differ from those within the unvaried series. MODAPTS is ideally suited to keep a record of those factors of repetition.

EngMOD as a Measure of Fatigue And Recovery

EngMOD provides an analysis of energy usage during manual work. It is based on the concept that each task requires a certain amount of energy and that this expended energy can be summed in a manner similar to the summing of task times. Although there are a number of different variables, the process of summing is similar. Paul Carey and Chris Heyde devised this system to be applied during a MODAPTS study of warehouse operations. The system aids in determining appropriate rest allowances, long an enigma in the field of work measurement. The system is based on the average energy use during a period of heavy physical work, which was determined from physiological research during the late 1960s. It includes a method of calculating the amount of time required for extra allowances and an energy bank concept, with varying amounts of reserves depending on the amount of energy being used.

Although any job can be analyzed for energy usage, recovery time only becomes important when the time required to recover the energy exceeds the task time. For practical purposes, this usually is significant only when there are vertical body motions, the object is over 16 kg, or the object is over 8 kg and moved more than .75 meters. All calculations require a measurement of the person's energy use when not moving heavy objects as well as of the object itself. Calculations are not made on the basis of individual MODAPTS elements, but on the average of a completed physical task. Recovery and fatigue are macrocycle conditions. Thus, walking up 2 steps does not require even .01 of the rest required from walking up 200 steps. Energy use is calculated on the average of physically fit persons skilled at heavy lifting, not the average working population. When the average of 87.5 joules per second is exceeded, the individual starts to use energy reserves from the bank.

Energy use can be calculated directly (in joules/sec) from the sum of three factors: the MODAPTS movement of a body member, the get or put of an object, and the mass, distance, and direction of a carried object. Although calculations could be carried out in joules/sec, a number comparable to the MOD can be developed. Because 1 MOD takes about 1/7 sec, the amount of energy that can be expended in 1 MOD is 12.5 joules (87.5/7). Actual expenditures for both the individual and load can be converted into this energy MOD.

Through the use of tables, energy usage for each element of work calculated by MODAPTS can be determined and compared to the MODAPTS time values. Values that have larger energy than time values are causing fatigue; those that have lower values are contributing to recovery. Some examples of the differences between energy and time expenditure are shown in Table 8-2.

Table 8-2

ACTIVITY	MOTION-TIME	ENERGY-TIME
Bend and arise; shallow movement	14	14
Bend and arise; deep movement	14	20
High reach up, and recover position	12	8
Walk, per pace	5	3
Go up stairs, per step	5	10
Go down stairs, per step	5	3

PREDETERMINED TIME STANDARDS SYSTEM (PTSS)

The Predetermined Time Standards System (PTSS) was developed by Fred Meyers to provide a time system that was quick to learn and easy to apply. The application of this system is described in detail in this chapter. It may be studied for two reasons: First, it provides the general student with an overview of application practices unavailable when studying a commercial predetermined time standard. Second, upon learning its application, the student will find it easier to learn the application of more complex systems.

The motion pattern is the blueprint for PTSS. Each line on a motion pattern is either a reach or a move. Larger stations could use body motion, but this is not good station design. (We discuss body motions later in this chapter.) At the end of reaches, there are grasps. Grasps are shown as big dots on the motion pattern. At the end of a move are alignments/positions, releases, or sometimes other grasps. These are also shown as big dots on the motion pattern. A well-defined motion pattern makes PTSS easy. Time spent on motion pattern development will save twice as much in PTSS analysis and development.

The motion pattern is the bill of materials for a time standard. The finished motion pattern consists of dimensional lines representing reaches and moves, plus big dots representing alignments/positions and grasps. These lines and dots form the sequence of motions required to do a job, and when we assign time to each motion, we end up with a time standard.

We discuss PTSS as follows:

1. Review of the table and definition of terms
2. The form
3. A 13-step procedure
4. Example problems

PTS Table

The PTS table (see Figure 8-1) is the source of time standards for all motions. The left-hand side of the table is the time standards in thousandths of a minute (.001). The decimal point has been omitted; however, if we are consistent and do all work in thousandths of a minute, we will have no problem, at the appropriate time, putting the decimal point back where it belongs.

On the right side is the motion pattern construction table (which tells you what motions you can't do at the same time). On the bottom right side are definitions of the terms. Our discussion starts with the definition of each term and a discussion of what causes the time to vary. The table needs to be kept at hand when studying these terms.

Reach: Symbol R Reach is the basic motion used to move the hand to a location or destination, for example, reaching for a part or a tool. We may even be carrying a tool (a pencil, for example), but if our basic motivation is to get to another part or tool, the motion is called a reach. Most of the time, the hand will be empty.

FRED MEYERS & ASSOCIATES

PREDETERMINED TIME STANDARDS
NORMAL TIME IN .001 MINUTES

REACHES & MOVES   R or M

2"/.001 + .003 MAXIMUN 48"
+25% FOR EACH 10 # OVER 5

GRASPS - G

G	CONTACT GRASP	1
G1	LARGE PARTS 1" OR OVER	3
G2	MEDIUM PARTS 1/4-1"	6
G3	SMALL PARTS UNDER 1/4"	9
G4	REGRASP	4
RL	RELEASE	-

POSITIONS & ALIGNMENTS   AP

AP1	LESS THAN 1/4"	5
AP2	LESS THAN 1/32"	10
AP3	LESS THAN 1/64"	20

BODY MOTIONS

SF	STATIC FORCE	5
EF	EYE FIXATION	5
ET	EYE TRAVEL	10
FM	FOOT MOTION	5
LM	LEG MOTION	SAME AS REACH
B	BEND AT KNEES	15
AB	ARISE FROM BEND	15
S	STOOP TO FLOOR	30
AS	ARISE FROM STOOP	30
T1	TURN ONE FOOT DISPLACED	10
T2	TURN TWO FEET DISPLACED	20
ST	SIT OR STAND	20
W-F	WALK FEET	4
W-P	WALK PACES	10

GOALS: ELIMINATE
COMBINE
CHANGE SEQUENCE
DOWNGRADE

MOTION PATTERN CONSTRUCTION TABLE

RIGHT HAND

LEFT HAND

		AP1	AP2	AP3	G1	G2	G3
AP1	-10	-	1	1	-	1	1
	+10	1	2	2	1	2	2
AP2	-10	1	1	2	1	2	2
	+10	2	2	3	2	3	3
AP3	-10	1	2	3	1	2	2
	+10	2	3	4	2	3	4
G1	-10	-	1	1	1	1	1
	+10	1	2	2	2	2	2
G2	-10	1	2	2	1	2	2
	+10	2	3	3	2	3	3
G3	-10	1	2	2	1	2	2
	+10	2	3	4	2	3	3

CODE 1 — 1st MOTION FULL VALUE / 2nd MOTION 1/2 VALUE

CODE 2 — 1st MOTION FULL VALUE / 2nd MOTION FULL VALUE

CODE 3 — 1st MOTION FULL VALUE PLUS AN R2 or M2 / 2nd MOTION FULL VALUE

CODE 4 — 1st MOTION FULL VALUE PLUS AN ET / 2nd MOTION FULL VALUE

THE BASIC MOTION EMPLOYED TO:

REACH: MOVE THE HAND TO A DESTINATION
MOVE: MOVE AN OBJECT TO A DESTINATION
GRASP: SECURE SUFFICIENT CONTROL TO PERFORM THE NEXT MOTION
ALIGNMENT/POSITION: ALIGN, ORIENT & ENGAGE ONE OBJECT WITH ANOTHER
STATIC FORCE: EXERT FORCE WITH NO MOVEMENT
RELEASE: RELINQUISH CONTROL
EYE FIXATION: FOCUS THE EYES & DETERMINE CERTAIN READILY DISTINGUISHABLE CHARACTERISTICS WITHIN A 16" DIAMETER
EYE TRAVEL: SHIFTING THE EYES

FIGURE 8-1 PTSS table: time for each motion.

The cause for time to vary is logical: The farther you reach, the more time it takes. In the motion picture analysis of work, it was discovered that the hand moves 2 inches in 1/1,000 of a minute. Motion pictures of work were taken at 1,000 frames per minute. When the film is reviewed at one frame per second, the hand can be seen moving in 2-inch increments across the screen. At the beginning and end of a reach, time is required to accelerate and decelerate, about .002 minutes each. If you could eliminate either of these accelerations or decelerations, you could save .002 minutes.

The formula for reaching is 2 inches per .001 plus .003 minutes, up to 48 inches. Beyond 48 inches, the body bends at the same time the hand moves, creating a compound motion that saves time but fatigues the operator. Reaches over 36 inches are very tiring and should be eliminated.

EXAMPLES:

	CODE	REACH	TIME IN .001	EXPLANATION
1.	R1	1″	4	$1 \div 2 + .003$[a]
2.	R15	15″	11	$15 \div 2 + .003$
3.	R36	36″	21	$36 \div 2 + .003$
4.	R50	50″	27	$48 \div 2 + .003$

[a]Never split .001.

The one complexity to add is that when the reach begins or ends in motion, .002 minutes is subtracted for each.

EXAMPLES:

	CODE	REACH	TIME IN .001	EXPLANATION
5.	R10m	10″	6	$10 \div 2 + .003 - .002$
6.	mR20	20″	11	$20 \div 2 + .003 - .002$

The lowercase "m" at the end of the reach indicates ending in motion. Example 5 reads reach 10 inches ending in motion, while example 6 reads beginning in motion reach 20 inches. Unless otherwise noted, reaches begin and end at a stop. Most of our reaches (95%) will be like examples 1 through 4. Any time the hand changes directions under 120°, the hand comes to a stop.

The hand may reach in a curved motion, but all measurements are made as if the motions were flat. This makes measuring the distances at the work station or on the work station drawing much easier. Just measure point to point.

Move: Symbol M Move is the basic motion used to move an object to a location or destination. Moving a part from the bin to the fixture, or moving a tool from the table to where it is needed, are good examples of moves.

There are three reasons for move time to vary:

1. Distance measured in inches
2. Beginning or ending in motion
3. Weight or force required.

The first two causes for move time variations are exactly the same as those for reaches. In fact, the time is exactly the same for reaches and moves if the item being moved weighs less than 5 pounds. For those items (tools or parts) that weigh over 5 pounds, we add 25% more time for every 10 pounds over 5 pounds. If both hands are used, the weight is divided by 2.

POUNDS	% ADDITIONAL TIME
5–15	25
15–25	50
25–35	75
35–45	100
45–55	125

PTSS EXAMPLES:

	SYMBOL	MOVE IN INCHES AND POUNDS	TIME IN .001	EXPLANATION
1.	M18	18″	12	18/2 + .003
2.	M6m	6″	4	6/2 + .003 − .002
3.	M20-20#	20″-20#	20	(20/2 + .003) 150%
4.	M17-50#/2	17″-25#	21	(17/2 + .003) 175%
5.	M11-7#	11″-7#	12	(11/2 + .003) 125%

Note: When weight is a factor, no adjustments are made for beginning or ending in motion.

Reaches and moves account for 50% of all operator-controlled work. At this point, you can set time standards for 50% of all work. Try these examples:

LH	TIME	RH
R16	_____	R12
M17	_____	M15
R15m	_____	R15m
mM21	_____	mM21
M5-50/2	_____	M5-50/2
M50	_____	M50
R36	_____	M24

Grasps: Symbol G Grasp is the basic motion used to secure sufficient control of an object to perform the next motion. After a reach to a tool or part, we must secure sufficient control to move that part or tool back to the point of use. There are five types of grasps:

1. Contact Grasp: Symbol G A contact grasp requires no closing of the fingers; it is merely touching something at the end of a reach. Touching a piece of paper before moving it off a desk is an example of a contact grasp. A contact grasp is the fastest motion in all PTSS—.001 minutes. When something must be moved without picking it up, a contact grasp is used.

2. Large Parts Grasp: Symbol G1 This grasp is used when picking up something that measures at least 1 inch at the point of grasp. The time includes time for the fingers to separate one part from other parts, to close the fingers around that part, and to apply sufficient pressure to gain control.

3. Medium Parts Grasp: Symbol G2 This grasp is used when picking up parts between $\frac{1}{4}$ inch and 1 inch at the point of grasp. The time for a G2 includes the same motions as a G1, but more time is added because smaller parts are more difficult to handle than larger parts.

4. Small Parts Grasp: Symbol G3 This grasp is used when picking up parts under $\frac{1}{4}$ inch at the point of grasp. The time for a G3 is the largest time for all grasps because it concerns the smallest dimensions. G3s have time to search, separate, close fingers, and apply pressure to pick up a part or tool. (Think about picking up a straight pin.)

G1s, G2s, and G3s are the normal grasps used in work. The grasps all include time for separation of parts. If this separation could be eliminated, two-thirds of the time could be eliminated, and then G1 = 1, G2 = 2, and G3 = 3. When one tool or part is by itself, it can be picked up using less time. This is a fine point, and failure or inability to design a station and methods so the operator doesn't have to separate tools or parts will not adversely affect the operator's performance.

5. Regrasp: Symbol G4 A regrasp is used in many different situations and is called the *contingency element*. A contingency is defined as an accidental happening; in PTS, a G4 is some extra time allowed for a contingency.

The first use is as the name implies: regrasping a part previously under control. For example, a screwdriver, when turned 180°, cannot be turned any more by hand until it is regrasped. A wrench cannot be rotated past 180° until it is regrasped. A nut cannot be run down past 180° until it is regrasped. If a nut is screwed down on a bolt three turns, six regrasps would be needed.

A second use of G4 is palming and unpalming. When more than one part is picked up out of a box of parts, the first part must be palmed to free the fingers before the second part is picked up. If four parts are needed, there would be four G1s, G2s, or G3s and three G4s. The last part would remain in the fingers, be moved back to the fixture, and be used. When the second part is needed, it is unpalmed and put back in the fingertips where it can be used. There will normally be a palming and an unpalming. Tools are also palmed and unpalmed. Think of a pencil that must be used every cycle. It could be kept in the hand, even if the fingers are needed for something else; it is palmed when not needed and unpalmed when needed.

The third use of G4 is the contingency element. If parts are very small (under $\frac{1}{64}''$), very large (over 10 pounds), fragile, slippery, or dangerous, extra time may be needed. Any grasp or alignment/position with one of these conditions requires one G4 for every condition.

Release: Symbol RL Release is used when control is relinquished. When a tool is set down, when a part has been used, and when a finished product is put aside, the ending element is release. Release has no time value. It is used for description only. If any situation arises where the analyst feels time is needed on the release because of safety, quality, or other considerations, add a G4. The key to good time standards is to be realistic and use common sense.

Alignment and Position: Symbol AP Alignment is the basic motion required to bring an object to a point or a line. Position is the basic motion required to align and engage one object with another. Alignment and position are the two motions required to do something with the parts in one's hands. The time values are very similar; therefore, they have been combined in PTS. The position includes only 1 inch of engagement; therefore, if a 6-inch engagement is needed, an AP would be followed by an M5. There are three APs.

AP1 is an alignment or position with less than $\frac{1}{4}$-inch tolerance. If an alignment can be within $\frac{1}{4}$ inch, an AP1 is used. If the two parts in a position have $\frac{1}{4}$ inch or less of difference between the dimensions, an AP1 is used. If more than $\frac{1}{4}$ inch of clearance is available, no AP is needed, and at the end of a move, only a release is needed.

AP2 is the same as AP1, only with closer tolerance—$\frac{1}{32}$-inch tolerance for alignments and positions. If a pin and hole had a $\frac{1}{32}$-inch difference or less, it would be an AP2.

AP3 is the same as AP1 and AP2, but with still closer tolerance—$\frac{1}{64}$ inch or less. There is nothing less than AP3.

The tolerance for positions is at the point of engagement, as shown in Figure 8-2. Figure 8-2 is an example of an AP3 and will take .020. The point of engagement is where the top of the block meets the leading edge of the pin when engaged. For cost reduction, the pin will be redesigned to add a chamfer or a countersink (as in Figure 8-3). A $\frac{1}{32}$-inch chamfer and/or a $\frac{1}{32}$-inch countersink will reduce this action to an AP2 and will take out .010 minutes. A 50% savings results. Now, if we chamfered and countersunk $\frac{1}{8}$ inch on each, the clearance at the point of engagement will be $\frac{1}{4}$ inch, and an AP1 results. Now the time is reduced to .005 minutes.

AP1s and AP2s are not too costly, but AP3s are very time consuming and should be eliminated. Product redesign can eliminate AP3s by downgrading, but tool design and fixturing can totally eliminate the AP1s, AP2s, or AP3s. Let a press put it in. This is especially true of AP3s. Alignments have been almost totally eliminated by fixtures. If a hole is needed in the middle of a steel plate, the tolerances are only .001 inch. The tool designer will provide a fixture with two back pins and one side pin. The operator will move the part into the fixture up against the back two pins and then move it a few inches sideways to the third pin. Now it is located and no APs were used, just moves. The operator now reaches for the two palm buttons to activate the machine.

Static Force: Symbol SF A static force is the basic motion used where no movement is involved but a pressure must be built up. Static force is a tightening or breaking-loose

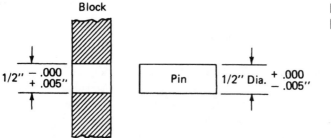

Block

FIGURE 8-2 Position pin in hole.

1/2″ $\begin{array}{c} - .000 \\ + .005″ \end{array}$

Pin

1/2″ Dia. $\begin{array}{c} + .000 \\ - .005″ \end{array}$

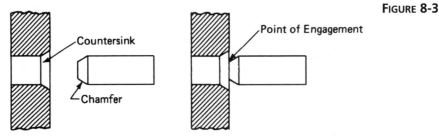

FIGURE 8-3

element. If the method requires a tight nut, bolt, or screw, the G4s required to run the part down are followed by an SF. If a bolt has to be loosened, a static force preceded the G4s. Static forces are used to cut wire, press parts together, and loosen faucets (the list will grow with experience). Static force is characterized by a hesitation or lack of motion that takes time. This time must be accounted for with an SF.

Eye Fixation: Symbol EF An eye fixation is the basic motion used when focusing the eyes on an object to determine certain readily distinguishable characteristics within a 16-inch diameter. The time for eye fixation is normally cancelled by other motions, and the time is included only when the hands stop so that the eyes can see something that couldn't be seen if the hands and object were moving. Eye fixations are uncommon in production work and common in inspection work. The new technologist should be careful not to use eye fixations when not required. The eyes are always working, and they see most everything without stopping the hands. The best test for when to use an eye fixation is to ask the question, "Does the hand have to stop for the operator to see this?" The eye fixation is a red flag to a chief industrial engineer, so the technologist must be prepared to state why the eye fixation was used.

Eye Travel: Symbol ET Eye travel is the basic motion required to shift the eyes from one position to another. Eye travel is required when more than one eye fixation is required and the distance is over 16 inches apart. The same cautions discussed for eye fixations are true with eye travel—they should be used only when everything else stops. Eye travel is part of the motion pattern construction table and is discussed further later in this chapter.

Eye fixations and eye travel are the inspection elements. If the quality control procedure calls for checking this and that, the technician counts up the number of items to be checked and allows for that many EFs plus the travel between them if the items are over 16 inches apart.

There are 12 body motions. Our attitude toward body motions is negative, and such motions should be eliminated. Many existing jobs include body motions; therefore, the present method will require the setting of standards using these body motions. The proposed method would eliminate these motions and save a lot of money. The body motions are used in pairs and are discussed that way here.

Foot Motion (FM) and Leg Motion (LM) The foot motion is the basic motion used to pivot the foot at the heel; the leg motion is the basic motion used to pivot the leg at the knee or hip. Both motions are measured in inches and are calculated just like reaches and moves. Foot motion is a useful and efficient body motion as long as the distances are minimized. Leg motions should be eliminated or at least reduced to foot motions. Microswitches can be activated with a 1-inch foot motion and are easily cost-justified. Pressing the gas pedal on the car is a foot motion; moving to the brake pedal is a leg motion. Some people ride with their foot over the brake at all times. (Do you think they are industrial engineers?)

Bends (B) and Arise from Bend (AB) A bend involves reaching down (to about knee level) to pick up a part or tool. Once at knee level, the operator will pick up or put down a part or tool and then arise from bend. Bending and arising from bends move the hands about 24 inches, which takes .015 minutes; therefore, bends and arising from bends comprise .015 minutes. Bending the body down to around the knees several times per minute will quickly fatigue the worker. For this reason, bends and arising from bends should be eliminated. Parts that come to a work station in 4′ × 4′ shop tubs will require a bending motion to pick up the average part (the one in the exact middle of the tub). A methods improvement would bring the parts to the work station on a conveyer at waist level. This would eliminate the bending and the arise from bending, a savings of .030 minutes.

The position of the average part is always used because only one cost or standard per operation is allowed. The top parts on a tub will be faster, and the bottom parts will be slower, but the one in the middle is the average time. PTSS is better than time study because of this fact.

Stoop (S) and Arise from Stoop (AS) Stooping is the basic motion used to reach down around the ankles, and arise from stoop is coming back upright. When parts or tools must be picked up from the floor, stoop and arise from stoop are required. Stooping is extremely fatiguing and must be eliminated. Bringing parts to the work station on elevated material-handling devices is the answer. The cost of the material-handling device will be paid for by the reduction in time resulting from the improvement. Stoop is the maximum reach (48 inches) plus a 10% fatigue allowance.

Turning: Symbol T1 or T2 Turning is the basic motion of displacing one foot up to 110°. If an operator must turn sideways, one foot is displaced, and this is called a T1 (one foot displaced). If the operator must turn around, two feet must be displaced, and this is called a T2. Work stations that require operators to turn are not properly designed, and opportunities to reduce costs are available. A T2 moves the part 34 inches, and the time is .020 minutes. A T2 also takes two paces, and that time is .020 minutes.

Sit and Stand: Symbol ST Sitting and standing motions are considered from a shop chair (about 36-inch seat height). No work station designer should design a job where the person has to stand up each time and sit down again during the same cycle, but someone may have in the past. That is the only reason for knowing how much time sitting and standing should take. All work stations should allow the employee to work sitting and standing alternately to reduce fatigue, but this does not require the operator to get down and up in the same cycle. Normally, no time is allowed for the operator to change positions for comfort. This time is in the allowance addition (discussed in Chapter 9).

Walking: Symbol W__F, W__P Walking can be measured in feet (symbol F) or paces (symbol P); therefore, W10F is read, "walk 10 feet," and W5P is read, "walk 5 paces."

W10F equals $10 \times .004$ minutes, or .040 minutes, and W5P equals 5×10, or .050 minutes. Walking time standards are based on the universally accepted standard of 264 feet/minute, or 3 miles/hour and rounded off, since we don't split .001. ($\frac{1}{264} = .00378$ minutes.) A pace is 30 inches.

A special instance of walking, with the symbol W__FO or W__PO, is walking around a work station that may be obstructed, or a sideways move (called a sidle), which takes about 25% more time. Each foot will be .005 minutes, and each pace will be .013 minutes. Walking is not part of efficient work station design, but work cells requiring the operators to move a part around a circle of machines include sidling as a substantial part of the job. Walking three paces sideways to the next machine is written as W3PO and takes .039 minutes to perform. If it were two paces, it would take only .026 minutes, resulting in a savings of .013 minutes per cycle, a saving of $3.60 per day forever. This is what micromotion study is all about.

EXAMPLE: Set a time standard for the job illustrated in Figure 7-16. You are ready to set your first standard.

ANSWER:

L.H.	TIME	R.H.	FREQ	
R6	6	R6		
G2	6			
	6	G2		
M6	6	M6		
AP1	5	AP1		
	29		15	.435
			+10% allow	.044
			Standard time	.479
			Pieces/hour	125
			Hours/piece	.00800

Notes: Each of the motions above places two pins in the board. To place 30 pins into the pin board, we must do the motions 15 times.

Two pins in .029 minutes $\times$ 15 cycles = .435 minutes

(You might ask why we used 2 lines for G2. This question will be answered later.)

Motion Pattern Construction Table

The top right-hand side of the PTSS table is called the motion pattern construction table. It tells us what we can do at the same time and what we can't do. Reaches and moves can be done two at a time for just the time cost of one reach or move. Reaches and moves are not on the motion pattern construction table; those symbols not included on the table can be done two at a time. Only alignments, positions, and grasps are included. Anytime the work method asks the operator to perform two alignments, positions, or grasps at the same time, we must look to the motion pattern construction table for guidance.

The motion pattern construction table is arranged like a mileage map. The intersection of two motions is a box with two numbers in it. These numbers are code numbers. The top number is used when the hands are working within 10 inches of each other (for example, picking up a part with each hand in the same container). The bottom number is used when the hands are working over 10 inches apart. When the hands are over 10 inches apart, more time is required, and the motion pattern construction table tells us how much more time.

Until this point, all the numbers on the PTSS table have been in .001 minutes. The motion pattern construction table is in code. Four codes are used.

Code 1: First Motion Full Value, Second Motion Half Value
If the motions are not on the chart, or there is no number in the box, both motions can be done at the same time; on the form, both motions would be on the same line. A Code 1 would be on two lines. For example, grasp two large parts, hands under 10 inches apart.

	LH	TIME	RH	
Grasp part	G1	3		
		2	G1	Grasp part

The time does not go on the same line, because it does not happen at the same time. Also, note that half of 3 is 2, because we don't split .001 minutes.

Code 2: First Motion Full Value, Plus Second Motion Full Value
This code requires each hand to work separately. For example, place two parts in holes over 10 inches apart.

LH	TIME	RH
AP2	10	
	10	AP2

Note that either hand can go first; the motion pattern would tell you what each hand is doing, and common sense is the only guide to which works first. Normally, it doesn't make any difference.

Code 3: First Motion Full Value, Plus a 2-inch Reach, or a Move 2 Inches, Plus Second Motion Full Value
After we have grasped a small part in one hand, our other hand must finish its reach to the other part; therefore, it would be a 2-inch reach followed by the second grasp. For example, grasp two small parts over 10 inches apart.

LH	TIME	RH
G3	9	
	4	R2
	9	G3

As another example, if two parts are being assembled, then the second motion would be preceded by an M2.

LH	TIME	RH
AP3	20	
	4	M2
	20	AP3

The R2 or M2 is just finishing the reach or move. When the next motion is a grasp, an R2 is used. When it is aligning or positioning, the M2 is used. Both motions have the same time value (.004).

Code 4: First Motion Full Value, Plus an Eye Travel, Plus Second Motion Full Value When two complicated motions are performed at the same time and the first motion is completed, the eyes must move before that second motion can be performed. Code 4s are always over 10 inches apart; therefore, they are the bottom number in the box. For example, grasp a small part and position a tight fit over 10 inches apart.

LH	TIME	RH
G3	9	
ET	10	ET
	20	AP3

Notice that the hands do not have to be doing the same thing, but they cannot do a G3 and an AP3 at the same time.

Now that the definitions have been given, as well as the instructions for their use, practice is needed. The PTSS form has been designed to help the technologist develop the method and standard. The next step in understanding PTS is to understand the form.

The Form

Figure 8-4 shows a PTSS form with circled numbers in each block. In this section, we describe what goes in each block. In an earlier section, we discussed in detail much of the information asked for in the top block: part number, operation number, etc. That discussion should be reviewed, because it will not be repeated here.

① A *Operation Number*   ① B *Part Number*

Both the operation number and part number are needed to identify the specific job being worked on. If either number is omitted, the study will be valueless.

② A *Date*   ② B *Time*

The date, including the year, is important to keep track of the sequence of methods. PTSS studies tend to stay around for years, and it is important for further study to know how old a study is.

③ *By I.E.*

The industrial technologist will commonly be given the title of industrial engineer by his or her company. The industrial technologist's name goes in this block.

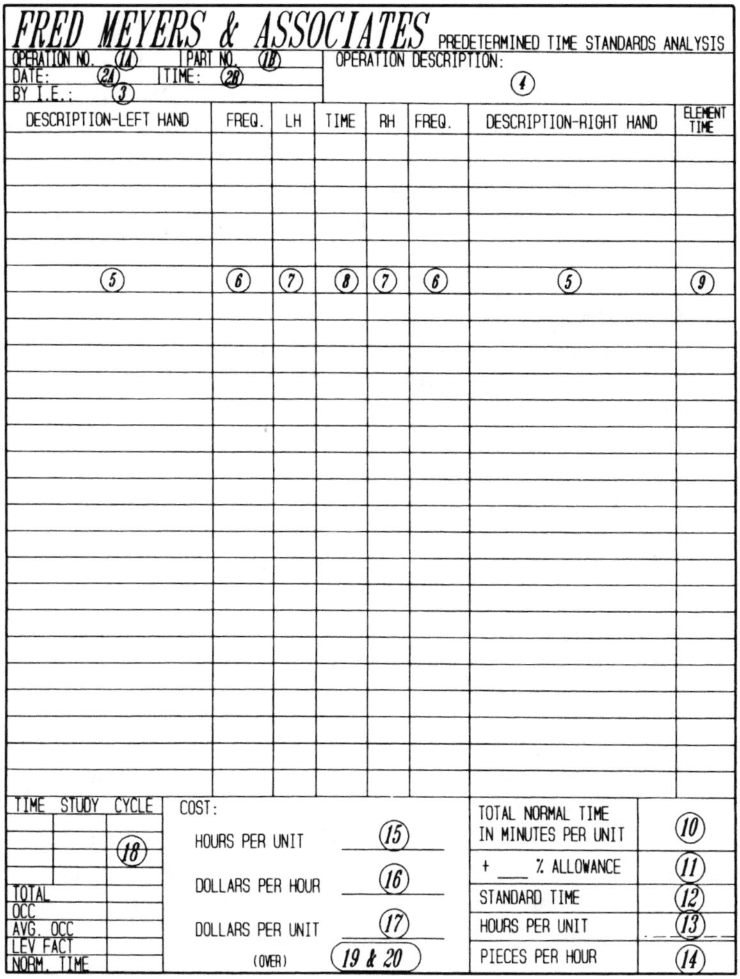

FIGURE 8-4 PTSS form: the step-by-step form.

④ *Operation Description*

This is a brief but accurate description of the work being performed at this station. Key words like *assemble, weld, mill, packout,* etc., are important.

⑤ *Description—Left Hand or Right Hand*

This is where the description of the motion goes. The elements have been reduced to the simplest motions, and a description of that small piece of work should be easy. This description is important for understanding.

⑥ *Frequency*

Frequency is on both the left-hand and right-hand sides of the form. This column is extremely useful and will save you time. Frequency asks how many times you want to perform this motion. For example, if a nut is required to be turned down eight revolutions, 16 G4s would be needed. Instead of writing 16-G4s one after the other, 16 is placed in the frequency column and multiplied by the time value for 1 G4. The total (64) is placed in the time column. At other times, an element may be repeated. Instead of rewriting the total element, the technologist only has to place a 2 in the RH frequency column next to the total element time, then multiply the total element time by the 2, and enter the total in the element time ⑨ column. An inspection element may not be repeated every cycle. In this case, we calculate the total time required to inspect one part and divide that time by the frequency of inspection needed by the quality control department: 1/10 means to inspect 1 out of every 10 parts; therefore, 1/10 of the time would be included in the time standard. This philosophy is used for material handling, clean-up, loading parts, etc.

⑦ *LH, RH*

The left-hand, right-hand columns are for symbols. The symbols are meaningful and reduce the number of words needed in the description column. An M16-50/2 means move 16 inches with a 50-pound object using two hands; therefore, this doesn't need to be part of the description. What is needed in the description would be the part name being moved and the destination. A G3 means pick up a small part, but the part name is missing.

⑧ *Time*

There is only one column for time because many motions can be done two at a time (one in each hand). If the motions are too complicated to do two at a time, two lines must be used. The motion pattern construction table will be our guide to what is too complicated to do at the same time. The time is in .001 of a minute, but the decimal point is omitted in this column for simplicity. Every line is a moment in time, and if two motions cannot be performed at the same time, they cannot be on the same line. The motion times are totaled for every element under the last motion. A release is a good ending point.

PTSS EXAMPLE:

DESCRIPTION—LEFT HAND	FREQ.	LH	TIME	RH	FREQ.	DESCRIPTION— RIGHT HAND	ELEMENT TIME
Get braces and							
pack parts #1 and 2							
To part #1		R48	27	R48		To part #2	
Grasp 2	2	G2	12				
			8	R2	2		
			12	G2	2	Grasp 2	
Regrasp 1st		G4	4	G4		Regrasp 1st	
To box		M24	15	M24		To box	
In box		RL	—	RL		In box	
			78				.078

⑨ *Element Time*

The element time is the total of all the motions of one element of the job. The number of elements is equal to the number of legs on a motion pattern. An element usually includes a reach, grasp, move, alignment/grasp, and a release for each hand. The motion times for each of these were totaled in the time column. Our previous example showed doing this element twice, so two times the element is placed in the element time column. Up to this point, no decimal point was needed. Now the decimal point is placed in the third place from the right—1.307, for example.

⑩ *Total Normal Time*

The total normal time is simply the total of the element times. If there are three elements, there will be three element times. These elements are added together for the total normal time. Normal time is defined as the amount of time required for a person to perform a specific task at a normal pace. There are no allowances in normal time.

⑪ *+__% Allowance*

Plus 10% allowance will be used until our discussion in Chapter 9. Allowances are that extra time allowed in each cycle for getting tired, personal needs, and unavoidable delays. Ten percent of the previous number (total normal time) is placed here.

⑫ *Standard Time*

Standard time is normal time ⑩ plus allowances ⑪.

⑬ *Hours/Unit*

Hours per unit equals standard time ⑫ divided by 60 minutes per hour.

⑭ *Pieces/Hour*

Pieces per hour equals the whole number 1 divided by hours/unit ⑬. ⑬ and ⑭ are $1/x$ of each other. This is called the *reciprocal*.

EXAMPLES:

⑩ NORMAL TIME	+	⑪ 10% ALLOWANCE	=	⑫ STANDARD TIME	⑬ HOUR/UNIT	⑭ PIECES/HOUR
0.200		.020		0.22	.00367	273
0.520		.052		0.572	.00953	105
0.725		.073		0.798	.01330	75
1.000		.100		1.100	.01833	55

⑮ *Hours/Unit*

This is the same as ⑬ (now being used for costing).

ⓖ *Dollars per Hour*

This is the the wage rate of the operator. Most often it is an average wage rate of a department. The loaded labor rate means that the labor rate has fringe benefits included.

ⓗ *Dollars per Unit*

The dollars per unit is the labor cost of one unit of production using the method described. Dollars per unit is used to compare methods.

ⓘ *Time Study Cycle*

The time study cycle is a small time study form. It has room for 12 cycles. (Time study is discussed in Chapter 9.) What we want here is normal time to compare to ⓾, total normal time. An industrial technologist can feel comfortable if the time study normal time agrees with the PTS normal time. Often, however, the work station is not available for time study until months after the PTSS standard has been set. However, when it is available, a time study should be made.

ⓙ *Layout*

The time standard cannot be set without a layout. Layout of the work station comes first, and a large area on the back of the PTSS form has been assigned for this purpose. (In Chapter 7, we discussed work station design.)

⑳ *Motion Pattern*

The motion pattern is the second step of PTSS. After the work station layout is completed, a motion pattern is drawn. The motion pattern is the blueprint for the PTSS method and the bill of materials for the time standard.

This 20-item list is a reference for what goes in each block. Refer to it when you are not sure what is needed. The step-by-step procedure discussed next will assist you in understanding the PTSS form.

Thirteen-Step Procedure for Developing a PTSS Study

Step 1: Select an operation to study. A new product, a request from management, or an industrial person who has a cost reduction idea are all possible sources of new projects to study. There are always opportunities to study work.

Step 2: Collect data. The data needed to design a work station includes blueprints, bill of materials, and production volume. The technologist needs to understand what work must be accomplished. During the process design stage of a new product, route sheets must be developed for each part and the sequence of assembly and packout developed. Each of these operations (as shown on an operations chart) requires a work station design. The more information the designer has, the better job he or she can do.

Step 3: Lay out work station. We discussed work station layout in Chapter 7. Nothing can be done until this step is complete. If the work station layout changes, a new time standard will result.

Step 4:   Develop a motion pattern. We discussed motion pattern design in Chapter 7. The motion pattern is shown on the station layout, and the motion pattern becomes the blueprint for the motion study analysis.

Step 5:   Break the job down into its smallest motions. The motion pattern developed in Step 4 is the blueprint (see Figures 8-5 and 8-6). The motion pattern shown in Figure 8-5 is for assembling two U bolts. The motion pattern has broken the job down into three elements—three loops. Note that the cyclograph (shown in Figure 8-6) looks just like a motion pattern. This cyclograph is for a six-element job.

Step 6:   Break the job down into elements. The motion pattern shown in Figure 8-5 has three elements. Note the three loops for each hand. Each loop is a part. Since both hands work at the same time, this is a three-element job. An element of work cannot be subdivided between operators. If this were part of a larger assembly, one part (or element) could be taken from one person and given to another as the volume of output needs changes. Each element will be subtotaled on the PTSS form and carried over to the elemental time column with the properly placed decimal—three places.

Step 7:   Calculate allowances. Allowances are extra time added to the time standard to allow for coffee breaks, personal time, and unavoidable delays. A typical allowance is 10% in addition to the work content time. At this point, a constant 10% is added to the normal time for allowances. In Chapter 9, we discuss allowances in more detail.

Step 8:   Calculate the time standard. A time standard consists of three numbers:

1. *Standard time:* Normal time plus allowances equals standard time in minutes.

2. *Hours per unit:* Hours per unit is the standard time (in minutes) divided by 60 minutes/hour. This standard is always communicated in five decimal places (.00001 hours). Because this number is not meaningful to most people, hours per 1,000 units is often used. In that case, two decimal places are used—0.01 hours per 1,000.

3. *Pieces per hour:* Hours per unit are divided into the whole number 1 to calculate pieces per hour. This is the number most people think of when talking about time standards.

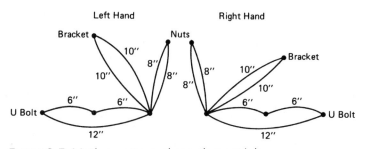

FIGURE 8-5 Motion pattern: three-element job.

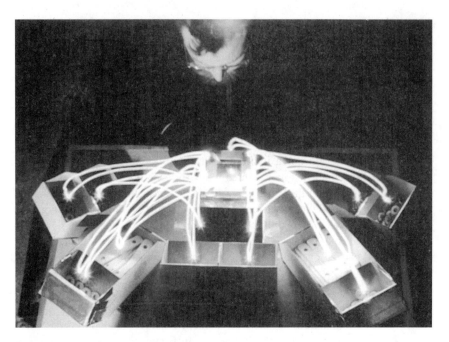

FIGURE 8-6 Cyclograph: six-element job.

Step 9: Calculate cost. The hours per unit multiplied by the hourly rate of the operator equals the unit labor cost. This labor cost is used to compare all future methods improvements.

Step 10: Perform a time study comparison. When a work station exists, a stopwatch time study can be done and used to compare the PTSS time standard. Time study is the subject of Chapter 9, but the PTSS time form includes a small time study form at the lower left. When the PTSS normal time agrees with the stopwatch normal time, the technologist can be confident that the time standard is correct. If they don't agree, the technologist will look for problems that were overlooked in either the time study or the PTSS.

Step 11: Improve for cost reduction. Once the first 10 steps are complete, we have a method and a cost. There is always a better way, so the technologist will search for a better method. The technologist will try to

1. Eliminate motions,
2. Combine motions,
3. Change the sequence of motions,
4. Downgrade the motions to less time-consuming ones,
5. Justify new equipment,
6. Justify better tools, fixtures, and jigs.

Step 12: Select the best method. The best method is the cheapest unit cost method considering all costs. Expensive and fast equipment can be justified only if enough product is needed. High-volume production can support high-cost tooling, but low-volume production can afford only minimal tooling and machining expense.

The cost of the proposed method is subtracted from the cost of the present method, and the result is multiplied by the planned yearly production. This is the annual savings. The cost of the new proposal is then divided by the annual savings, resulting in a return on investment (ROI). An acceptable ROI varies from 25% (four-year payback) to 100% (one-year payback), depending on company goals and policy. A 100% ROI is an almost certain approval.

Step 13: Publish the approved method and time standard. The company will have a system for communicating the method and standard to all who need them. The approved method and standard will be used by all until changed officially.

Example Problems (Figures 8-7 Through 8-10)

Operation 2010 on the swing set packout line requires the operator to pack out three different parts listed on the back of the example. The work station layout shows the operator bending down into tubs to retrieve parts. The motion pattern is for both hands. This is a poor method, and an improved method follows (see Figures 8-7 and 8-8 for old method and 8-9 and 8-10 for new method).

Study these two examples closely. This plant needs to produce 300,000 swing sets per year. How much will the proposed change save the company?

The present method (Figures 8-7 and 8-8) requires 30 operators on the assembly line, and each person costs $.0458 per swing set, for a total cost of $1.375 each. The proposed method (Figures 8-9 and 8-10) requires only 15 people because they each pack out twice the number of parts than those workers in the previous method. Each person costs $.0272 per swing set, for a total cost of $.408 each. The savings are $1.375 − $.408 × 300,000 sets per year, or $290,100.00 per year. This could pay your way for a year or two.

Which method requires the worker to work harder? Even though the proposed method has each employee packing twice as many parts and more sets per day (368 verses 218), the proposed method is physically easier because the operators do not need to bend down into the tubs and arise from bends with a heavy part. The proposed method is best from everyone's point of view.

QUESTIONS

1. What will PTSS do for you?
2. What are the definitions for each motion?
3. What causes time to vary for reaches, moves, grasps, and alignment/positions?
4. What is a motion pattern construction table?
5. How do we eliminate AP3?

(Questions continue on page 158.)

FRED MEYERS & ASSOCIATES PREDETERMINED TIME STANDARDS ANALYSIS

OPERATION NO. 2010	PART NO.				OPERATION DESCRIPTION:	
DATE: 10/19/xx	TIME:				Packout Parts #1. 3. & 6.	

DESCRIPTION-LEFT HAND	FREQ.	LH	TIME	RH	FREQ.	DESCRIPTION-RIGHT HAND	ELEMENT TIME
Packout Part #1--2each							
			10	T1		Turn Left	
			40	W10F		Walk 10 feet	
			15	B		Bend	
			6	G2		Grasp #1	
Grasp #1		G2	6				
			15	AB		Arise	
Move to Box		M36	21	M36		Move to Box	
in Box		RL	--	RL		in Box	
			113				.113
Packout Parts #3 & 6 = 2 #3 & 1 #6							
			20	T2		Turn Around	
			15	B		Into Tub #3	
			3	G1		Grasp #3	
Grasp (1) #3		G1	2				
			15	AB		Arise	
			20	W2P		Side Step Twice	
			15	B		Into Tub #6	
			3	G1		Grasp Part	
			15	AB		Arise	
			20	T2		Turn Around	
		M12	9	M12		Move Parts to Box	
		RL	----	RL		Into Box	
			137				.137

TIME	STUDY	CYCLE	COST:			
.30	1.55	2.70				
.58	1.80	3.00	HOURS PER UNIT	.00458		
.91	2.10					
1.21	2.40		DOLLARS PER HOUR	10		
TOTAL		3.00				
OCC		10				
AVG. OCC		.300	DOLLARS PER UNIT	.0458		
LEV FACT		85	# People on line	30		
NORM. TIME		.255		$1.375		

TOTAL NORMAL TIME IN MINUTES PER UNIT	250
+ 10 % ALLOWANCE	.025
STANDARD TIME	.275
HOURS PER UNIT	0 0 4 5 8
PIECES PER HOUR	218

FIGURE 8-7 PTSS: present method—three tubs of parts (methods and time analysis).

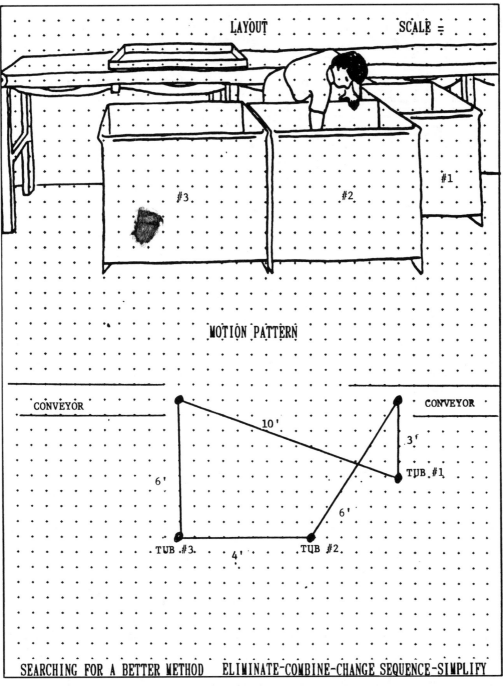

FIGURE 8-8 PTSS back page: present method—work station layout and motion pattern.

FRED MEYERS & ASSOCIATES PREDETERMINED TIME STANDARDS ANALYSIS

OPERATION NO. 2010	PART NO.				OPERATION DESCRIPTION:	
DATE: 10/19/xx	TIME:				Packout 2 each #1, 2, 3, 4, and 1 each of	
BY I.E.: Meyers					5 & 6.	

DESCRIPTION-LEFT HAND	FREQ.	LH	TIME	RH	FREQ.	DESCRIPTION-RIGHT HAND	ELEMENT TIME
Get Braces and Pack Part #1 and 2							
To part #1		R48	27	R48		To Part #2	
Grasp 2	2	G2	12				
			8	R1	2		
			12	G2	2	Grasp 2	
Regrasp 1st		G4	4	G4		Regrasp 1st	
To box		M24	15	M24		To box	
In box		RL	---	RL		In box	
			78				.078
Packout part #3 and 4							
To bag		R12	9	R12		To cap	
Grasp 2 bags	2	G1	6		2		
			6	G1		Grasp 2 caps	
Regrasp 1st		G4	4	G4		Regrasp 1st	
To box		M12	9	M12		To Box	
In box		RL	--	RL		In box	
			34				.034
Packout part #5 and 6 1 each							
To #5		R24	15	R24		To #6	
		G1	3				
			3	G1		Grasp #6	
To Box		M24	15	M24		To Box	
		RL	--36	RL			.036

TIME	STUDY	CYCLE	COST:		TOTAL NORMAL TIME	
15	.75	1.35			IN MINUTES PER UNIT	.148
30	.90	1.50	HOURS PER UNIT ____.00272			
45	1.05				+ 10 % ALLOWANCE	.015
60	1.20		DOLLARS PER HOUR $10.00			
TOTAL		1.50			STANDARD TIME	.163
OCC		10	DOLLARS PER UNIT $ .0272		HOURS PER UNIT	0 0 2 7 2
AVG. OCC		.150	#People on line = x15			
LEV FACT		95%		$.408	PIECES PER HOUR	368
NORM. TIME		.143				

FIGURE 8-9 PTSS form: proposed packout method and time analysis.

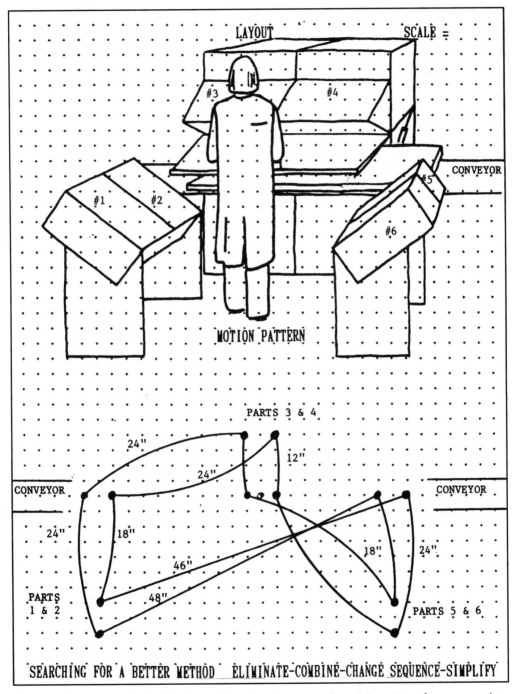

FIGURE 8-10 PTSS form: back side work station layout and motion pattern for proposed packout method.

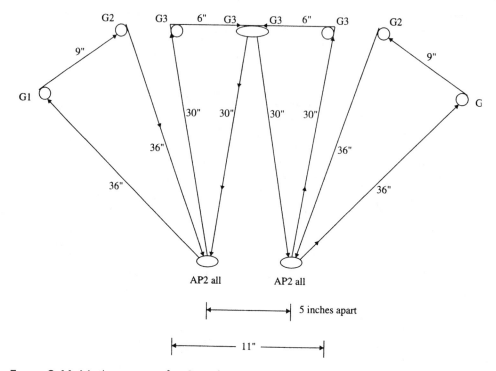

FIGURE 8-11 Motion pattern for Question 12.

6. What is the attitude toward body motions expressed in this chapter?

7. How do we use the frequency column on the PTSS form?

8. What are the decimal rules?

9. Why do we divide the job into elements, and what is an element?

10. What are the 13 steps of the PTSS procedure?

11. Complete a PTSS study for your semester project. At least four elements and .250 minutes are required. Subtotal each element.

12. Write a PTSS for the motion pattern in Figure 8-11. Pick up one of each part.

Stopwatch Time Study

INTRODUCTION

Stopwatch time study is the most common technique for setting time standards in the manufacturing area. The time standard is the most important piece of manufacturing information, and stopwatch time study is often the only method acceptable to both management and labor. Stopwatch time study was developed by Frederick W. Taylor in 1880 and was the first technique used to set engineered time standards. Because of its long history, many companies have negotiated stopwatch time study into their labor contracts. Stopwatch time study may not be the best technique for setting a particular time standard, but it may be the agreed-on method to be used. Most unions know management's strong feelings and need for time standards, and they have accepted it as a fact of life. Unions have also developed time study programs and trained people in time study. There is no real conflict between a skilled union time study person and a skilled company time study person. Both use the same scientific base. Differences do occur, but they are resolved easily between people who know what they are doing and whose goal is to be fair. The 100-plus years of work with stopwatch time study has deeply entrenched it as the technique for setting time standards.

Stopwatch time study is a difficult job only because of some employees' negative attitudes. The time study technician is under pressure from both labor and management. There is humor in the following statement, but a little truth too:

If time study technicians set time standards too tight, labor is mad at them. If time study technicians set time standards too loose, management is mad at them. If time standards are perfect, everyone is mad at them.

The time study technician's defense is that he or she knows only one way to set a time standard—the right way. The right way is fair and equitable to all. What everyone really wants is a *good* time standard.

New industrial engineering graduates have not looked favorably at time study as a career step because of the negative attitudes displayed by many production people toward it. To the detriment of all consumers, time study has had a poor public image. Nevertheless, time standards are a necessity, and we cannot operate an industrial organization successfully without them. The lean manufacturing concept has created a new life for time study. Lean manufacturing is recruiting everyone in an operation to remove waste, and one of the best tools to measure the costs and benefits of a new idea or improvement is the time study technique. The modern practitioner of time study has often been an alert person promoted out of the shop and trained in the time study technique. Now, that person will be the teacher of time study to more of the company's work force. It takes about two days to train a time study person; on-the-job experience makes a skilled time study person. The time study person who can teach others will multiply his or her effectiveness. New industrial engineers should know how to conduct time studies to be able to train people, because the need will arise frequently.

Figure 9-1 shows a stopwatch face with both seconds and decimal minutes. Decimal minutes are used in time study because the math is easier. The minute is divided up like a dollar instead of into seconds. Base 100 and base 60, respectively, are the mathematical descriptions of the two methods, and 15 seconds is equal to .25 minutes (or 25 cents). New time study people understand this concept and its advantages very quickly. The mechanical stopwatch shown in Figure 9-1 is read in .01 minute (hundredths), where digital watches keep time in .001 minute (thousandths). But once we start to extend the data, we pick up the third decimal place. Digital watches are quickly replacing mechanical watches, so this point and the problems of learning to read the watch while it is running will be eliminated with the digital watch. More will be discussed on this subject when we talk about each of the stopwatch types.

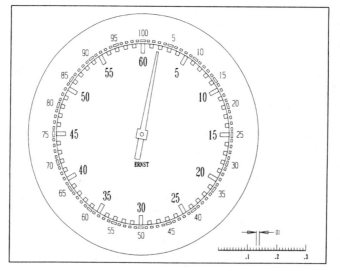

FIGURE 9-1 Stopwatch face: decimal minutes and seconds.

No matter what type of watch you use, you need practice reading and recording the end times. In a classroom environment, you can click a metal object against a hard surface and have the trainees read the watch. Giving 10 prescheduled clicks and having the trainees subtract the previous reading from the current reading will give students their first time study experience. Comparing the results among students in the class will show how accurate they are already in reading stopwatches. It really doesn't take long to learn to read any type of stopwatch, even when it is running.

This chapter is organized as follows:

Tools of stopwatch time study

Step-by-step time study procedures

Rating, leveling, or normalizing

Allowances and foreign elements

Time study practices and employee relations

Long cycle time study

Vertical time study

To eliminate redundancy, review the sections "What Is a Time Standard?" and "Stopwatch Time Study" in Chapter 4. These are two important subjects and must be understood before you proceed.

The time study procedure in this chapter starts with a standardized work station design and a skilled, well-trained operator. Both of these conditions have been discussed in previous chapters, and a time study without either is useless.

TOOLS OF STOPWATCH TIME STUDY

Because they play such an important part, the tools of stopwatch time study are important to know before we get into the technique itself. The tools discussed in this chapter are as follows:

1. Stopwatches
   a. Continuous
   b. Snapback
   c. Three-watch
   d. Methods time measurement
   e. Digital
   f. Computer
2. Boards for holding watches and paper
3. Videotape recorders
4. Tachometers

5. Calculators

6. Forms

    a. Continuous

    b. Snapback

    c. Long cycle

Stopwatches

The six types of stopwatches discussed below are still in use today, but new time study people will select a digital watch if they are given a choice. The discussions of the mechanical watches are important because your company may choose the watch you use. We feel that study of the first four watches is like a history lesson. The time study techniques can use mechanical watches, digital watches, or computers as stopwatches, but the accuracy increases with digital watches and even more so with computers.

Continuous Mechanical Stopwatch The continuous stopwatch (shown in Figure 9-2) is used in the application of the continuous time study technique. This technique can be accomplished using any type of stopwatch, but the continuous stopwatch is specifically designed for this purpose. The continuous time study technique calls for starting the stopwatch when the operator finishes a part and allowing the stopwatch to run continually until the study is complete. While the watch is running, the time study person reads the watch at the end of every element and records this time on the continuous time study form. This technique records everything that happens during the time study period. We can't hide anything, because all the time must be accounted for.

FIGURE 9-2 Continuous mechanical stopwatch. (Courtesy of Meylan Corporation.)

The continuous mechanical stopwatch has two dials (see Figure 9-2). The large dial is divided into hundredths of a minute (.01), with one revolution of the sweep hand being one minute. The small dial records minutes up to 30. The crown of the watch, when depressed, stops the watch. The crown of the watch can be depressed a second time to reset the watch to zero, but a third depression of the crown is needed to restart the watch. Three depressions of the crown is time-consuming and should be done only once per study.

The continuous time study must be extended when the readings are complete. Because only the ending elements have been recorded, every reading must be subtracted from the previous reading to calculate the elemental time. This subtracting is time-consuming and is the most undesirable part of continuous time study.

The continuous time study is the preferred technique of unions, because everything that happens during the study is recorded and open for discussion. This technique is said to have integrity. The time and nature of all foreign elements are recorded, and a decision will be made to include or exclude these elements. (A later section of this chapter covers foreign elements.)

Snapback Mechanical Stopwatch The second of the two basic time study techniques is the snapback time study technique. The snapback mechanical stopwatch (see Figure 9-3) is designed specifically for this technique. The snapback watch can be used for the continuous time study technique, but the continuous stopwatch cannot be used for a snapback time study because it needs to be reset using three depressions of the crown. Also, digital stopwatches commonly have both continuous and snapback capabilities.

The snapback time study technique is faster and easier than continuous time study. Each time an element ends, the technologist reads the watch and immediately snaps it back to

FIGURE 9-3 Snapback stopwatch. (Courtesy of Lafayette Instrument Company.)

zero, and the watch restarts automatically timing the next element. The technologist records the reading at his or her leisure during the next element. The advantage of the snapback is that each recorded reading is the elemental time, and no subtraction is needed. This is a big advantage to the industrial technologist, but the fact that there is no way of checking the industrial technologist's work makes it unacceptable to most unions. It is said that the snapback stopwatch has no integrity.

The crown of the snapback watch resets the watch every time it is pushed. The snapback watch also has a side shifter lever that turns the watch off and on. This is a useful device when interruptions occur; the time study technician can turn off the watch until the operator goes back to work. But again, there is the chance of error or controversy. The snapback technique is still a useful way of checking time studies and for quick studies.

Three-Watch Time Study The three-watch time study technique is the best of both the continuous and the snapback techniques. Three continuous stopwatches are used on one board. Remember that the first time you push the crown stops the watch, the second time resets the watch, and the third time restarts the watch. On the three-watch board, there is a watch in each stage (see Figure 9-4). When the operator finishes an element of work, the time study technician pulls a common lever that depresses all three crowns. One watch is

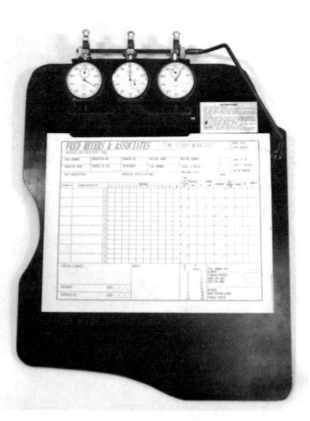

FIGURE 9-4 Three-watch time study method. (Courtesy of Meylan Corporation.)

stopped so a reading can be made, the second watch is restarted and is timing the current element, and the third watch is reset to zero and is waiting to time the next element. The technologist is able to read the watch while it is stopped. A higher-quality reading results, but the big advantage of the three-watch time study technique is the lack of subtraction to calculate elemental time. The disadvantages of the three-watch technique include the cost of the equipment and the lack of an audit trail—the technologist could leave out elements and no one would know.

Methods Time Measurement Stopwatch The methods time measurement (MTM) time study technique is used where the predetermined time standard system is MTM. MTM was developed in the 1940s and was one of the first predetermined time standards systems. Today, there are active chapters of the MTM association in every industrialized nation of the world. MTM is the grandfather system, but the association keeps it modern. MTM is a copyrighted system, and the association sponsors 80-hour courses in several of its techniques. It awards a blue card on successful completion of its programs, and many companies require that their industrial technologists have a blue card. The only way to get one is to attend the association's schools. Big companies have in-plant MTM instructors and courses.

The MTM stopwatch (see Figure 9-5) measures time in one hundred-thousandths of an hour (.00001), or one TMU (time-measured unit). The hand of the watch makes one revolution in .001 hours (3.6 seconds). This is the same unit of time as in the MTM system, and it is used to reduce the math required. It is efficient, but few people understand what is happening, so it creates more distrust of industrial engineering and technology.

FIGURE 9-5 MTM stopwatch: reads in TMUs. (Courtesy of Meylan Corporation.)

The sweep hand of the decimal hour stopwatch moves much faster than a decimal minute stopwatch. Upon first observing the MTM watch running, the technologist may think the watch is broken.

Digital Stopwatches and Electronic Timers Digital stopwatches and electronic stopwatches come in every style and type, and most of them have multiple time study technique abilities. A memory ability is available in some of the models (see Figure 9-6a). Meylan Corporation's Model 19 is a nine-memory decimal timer with the ability to time in seconds, minutes, and hours. This model can recall the last nine readings. I found this watch so useful that I now use Meylan Corporation's Model 795 MC (Figure 9-6b), which has the ability to store 500 readings. This will improve accuracy even more because of the ability to time study in three decimal places (.001). The Model 795 MC has both a continuous display and a snapback time. If the time study person takes a continuous time study using this digital watch, he or she can write down the ending times and, when back at the desk, write down the elemental times by recalling each element from the watch's memory. This eliminates the need to subtract, and the time study person can check recorded readings. The quality of time standards will improve, the time study job will be easier, and the cost of doing time studies will decrease.

Some digital watches (see Figure 9-6c) have a single readout; normally the elapsed time is displayed. When the read time button is held down, the elemental time is displayed, which allows the watch to be used as either a continuous or a snapback. If there is time between elements, both numbers can be written down, eliminating the need to subtract elemental times.

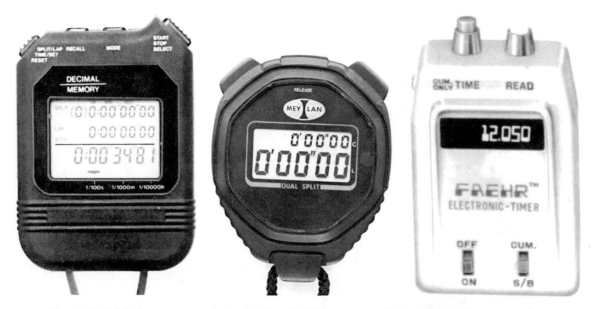

FIGURE 9-6 (a) Nine-memory decimal timer. (b) 500-memory decimal timer. (Courtesy of Meylan Corporation.) (c) Digital timer. (Courtesy of Faehr Electronic Timer Inc.)

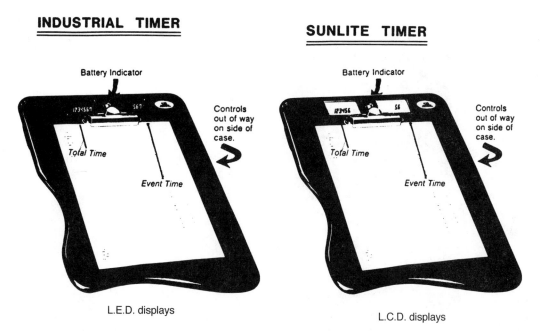

FIGURES 9-7 AND 9-8 Digital stopwatches. (Courtesy of Faehr Electronic Timers, Inc.)

Figures 9-7 and 9-8 are examples of digital watches built into time study boards. The Sunlite (L.C.D. display) can be used in bright light.

Digital watches are electronic and need to be recharged. The maintenance of the charge is important. Digital watches can be used in both continuous and snapback time studies.

Computers Computers can be programmed to do time study. Several commercial microcomputer time study programs and specially built hardware are available.

Hand-held data collectors are taken to the shop floor where the time study is performed (see Figures 9-9 and 9-10). The Datamyte 1000 data collector is connected to a computer terminal for extending the time study. The math is totally automatic. Once ending elements are defined and the time study is started, the time study technician needs only to push a button on the collector to record the time. Accuracy is much improved.

Figure 9-10 shows the RateSetter^{TM}, a computer-aided time study system by Faehr Electronic Timers, Inc. The RateSetter timer records and stores the time study observations as they are made. The data can be transmitted later to an IBM-compatible PC for the time study clerical work. The time study person is freed of the tedious, time-consuming clerical calculations for more studies and analysis. These automated systems will lower the cost of time study and increase the quality of time standards. The math of the manual time study systems can be mind numbing, so automating these tasks will improve the quality of our time standards.

Summary of Stopwatches and Time Study Techniques The continuous time study technique is the most widely used because of its integrity, and the continuous stopwatch is used the most frequently. Mechanical or digital watches are both used,

FIGURE 9-9 Computer time study: hand-held data collector taken to the shop floor. (Courtesy of Datamyte.)

but digital watches are being adopted at a fast rate. Snapback watches can be used for continuous time study, but they can be manipulated and do not have the integrity of the continuous watch. Integrity is very important for a time study person, and we should do everything within our power to maintain high ethical standards of operation. We will never get some experienced time study people to give up their mechanical watches, and some union contracts call for their use. But digital watches will improve the accuracy of time standards. Computer extension of time study data will increase this quality improvement.

Boards

Time study boards vary from cheap clipboards to multiwatch digital boards, and they have one goal—to hold equipment for ease of use. If a simple clipboard is used, the watch must be held in the same hand. This is not impossible, but it is not comfortable (see Figure 9-11).

Continuous and snapback time study boards will have one watch holder and a clip for paperwork. The watch holder is reversible for left-handed technicians. The board is also cut out for the arm and stomach, for comfort's sake.

Three-watch boards are designed to hold the three watches and a common lever for depressing all three crowns at the same time. They also have clips and cut-outs, but left-handed boards must be special ordered (see Figure 9-4).

Digital boards usually have the watches built in. Two watch displays are common. These boards are very expensive (see Figure 9-7).

Computer time study does not need a board, just a data collector keyboard.

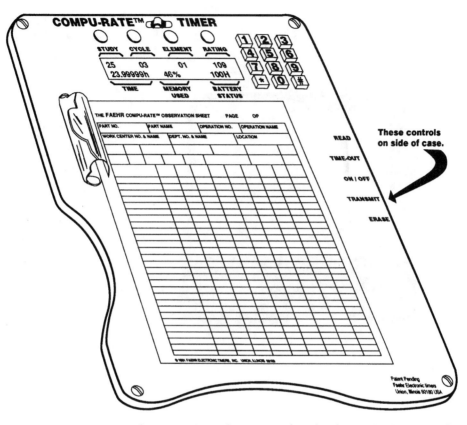

FIGURE 9-10 RateSetter electronic timer. (Courtesy of Faehr Electronic Timers, Inc.)

Videotape Recorder

One of the newest and best tools for studying and recording the method and time standard is the videotape recorder. Operation description is an important part of time study. A time standard is good for only one set of circumstances, and if anything changes, the time standard must change. When a change in time standard is made, it is often challenged by unions because the operation description was not recorded clearly enough. Most unions correctly include a statement like this:

> No time standard (rate) will be changed unless there is a change in the machine, tooling, material, method, or working conditions that creates a change of more than 5%.

Videotape recording a few minutes of operation of a work station costs only a few cents. What better technique is there to record exactly what the time study technician studied than to videotape it?

FIGURE 9-11 One-watch time study board. (Courtesy of Meylan Corporation.)

Another use of videotape is to tape an operation and review the tape for methods analysis and methods improvement. This is called *micromotion study*. The tape can be slowed, speeded, and frozen. It can be replayed to watch one hand at a time. The videotape is a great methods improvement tool.

The third use of videotape recording is to use it as a stopwatch. Many recorders have time recorders built into them. The beginning time can be subtracted from the ending time, and an element or cycle time results. In addition, the time study technician can time-study a tape in the comfort of his or her office. Consider a corporate industrial time study technician going out to a remote plant with a video recorder. One tape can hold fifty operations or more. For a few dollars of tape, the whole plant can be taped and returned to the corporate office for analysis (methods and standards). The corporation will save enough money on travel expenses to pay for the equipment in one trip, plus it will have the best possible record of existing methods. Every time study department should have a videotape recorder.

Tachometer

A tachometer is used for determining the speeds of machines and conveyers. A center point attachment is placed on the tachometer and then placed against the center of a turning shaft, chuck, or arbor. The number of revolutions per minute (RPM) is recorded on the time study form as part of the operation description. The center point attachment can be replaced with a 4-inch diameter wheel to convert RPM to feet per minute (FPM) (see Figure 9-12). When the wheel is held against a moving conveyer, FPM is indicated. This information is also an important part of the operation description. The tachometer is an important tool of motion and time study, and without it the technician is left with imperfect information.

Calculator

The importance and use of a calculator does not have to be emphasized to an industrial engineer, but a few comments are needed. Time study involves a lot of math, and the accuracy cannot be overemphasized. The calculator will speed up the process and make the results more accurate.

One special feature is recommended for time study—the $1/x$ function. The $1/x$ function is 1 divided by another number, and hours per unit and units per hour are the $1/x$ of each other. The math is much easier and faster, and the time study forms have been set up to be math efficient (i.e., the fewest steps possible). The $1/x$ button can be one of your favorite functions once you learn its importance and use. Have you ever divided something upside down (the numerator into the denominator)? If you do it again, just push the $1/x$ button and it straightens out everything. This is much easier than entering all the calculations again.

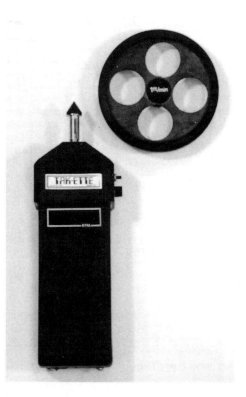

FIGURE 9-12 Tachometer: RPM and FPM record. (Courtesy of Meylan Corporation.)

Forms

Time study forms are the hardest part of learning how to do a time study. Time study forms are set up to lead the technician into the correct procedure. The next section is a step-by-step procedure for making a stopwatch time study, using a form.

TIME STUDY PROCEDURE AND THE STEP-BY-STEP FORM

The time study procedure has been reduced to 10 steps, and the time study form has been designed to help the time study technologist perform the 10 steps in the proper sequence. (Figure 9-13 shows a blank time study form with circled numbers.) This section is organized according to the following 10 sequential steps:

Step 1:   Select the job to study.

Step 2:   Collect information about the job.

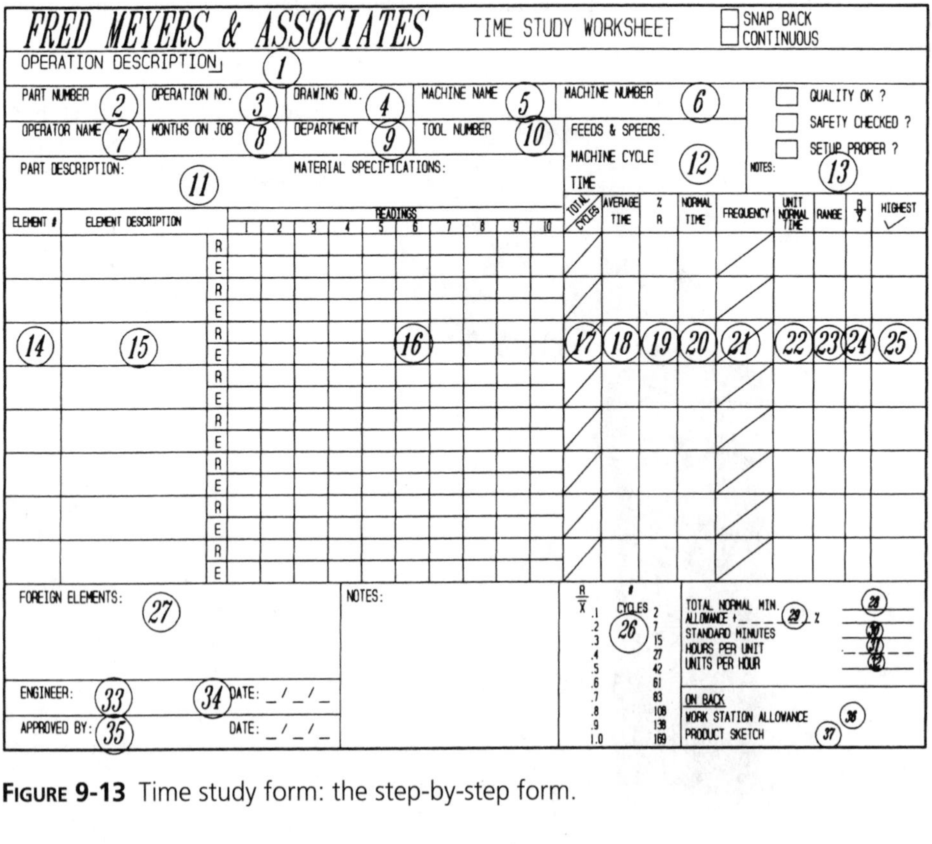

FIGURE 9-13 Time study form: the step-by-step form.

Step 3: Divide the job into elements.

Step 4: Do the actual time study.

Step 5: Extend the time study.

Step 6: Determine the number of cycles to be timed.

Step 7: Rate, level, and normalize the operator's performance.

Step 8: Apply allowances.

Step 9: Check for logic.

Step 10: Publish the time standard.

In the following discussion, within each step the blocks of the time study form involved are defined. The circled number refers to the block on the time study form. The form is designed for both continuous and snapback time study techniques. Everything except block 16 is exactly the same.

Step 1: Select the job to study. Requests for time study can come from every direction:

1. Unions can question time standards and request a restudy.
2. Supervisors, who are judged partly on the performance of their subordinates, can request a restudy.
3. The job could change, requiring a new standard.
4. New jobs may have been added to the plant.
5. New products can be added, requiring many new time standards.
6. Anyone can improve methods, requiring a new time standard.
7. Cost reduction programs can require new standards—new machinery, tools, materials, methods, etc.

Once a reason for studying a job has been determined, the time study technician may have several people doing the same job. Which person do you time-study? The best answer is two or three, but those people you do *not* want to time study are

1. The fastest person on the job. The other employees may think you are going to require them to keep up. Even though you can do a good job of setting a time standard on this person, you don't want to create employee relations problems.
2. The slowest person on the job. No matter how you rate the job and no matter how good the time standard is, the employees will wonder how you came up with that standard.
3. Employees with negative attitudes that will affect their performance while being studied. If you can sidestep a potential problem, you should.

The person or persons to be time-studied should have sufficient time on the job to be a qualified, well-trained operator. For this reason, ⑦ and ⑧ have been included on the time study form:

⑦ Operator's name

⑧ Months on the job.

The employee should have been on the job for at least two weeks.

Once the job to study has been selected, the following information has been determined:

② Part number

③ Operation number

④ Drawing number

⑤ Machine name: a generic name like press, welder, lathe, drill, etc.

⑥ Machine number: a specific machine with specific speeds and feeds

⑨ Department: where the machine is located (this can be a number or name).

Step 2: Collect information about the job.

Now that the job has been identified, the technologist must collect information for the purpose of understanding what must be accomplished. The information required is as follows:

① Operation description: a complete description of what needs to be accomplished

④ Drawing number: will lead to a blueprint to show items like the following:

    a. ⑪ and ㊲: part description and material specification (a place on the back of the time study form has been set aside for a product sketch, if needed).

    b. ⑩ Tool numbers and sizes of tools like fixtures, drill sizes, etc.

    c. ⑫ Feeds and speeds of equipment: these depend on sizes of parts and material specifications found on the blueprint; they must be recorded.

⑬ When reviewing the work station and before starting the time study, the technologist must check the following:

    ✓ Is quality okay? Quality control must confirm that the quality of the product being produced is high. Is the operator checking parts on the proper schedule? Time standards from producing scrap are worthless.

    ✓ Has safety been checked? If all the safety devices are not in place, then the technologist would be wasting time setting a standard for the wrong method.

    ✓ Is setup proper? This is the time to see that the proper method, tools, and equipment are in place. Are the materials and tools correctly positioned? Are there unnecessary moves or elements being performed?

If anything is wrong, it must be corrected before a time study can be performed. If the operator must be retrained, the time study should be postponed until retraining is complete.

㊱ A big part of collecting the information is the work station layout. The back of the time study form has been developed for a work station layout, but this may not be needed if done on another of the previous forms (multiactivity form). The work station layout is one of the best ways to describe the operation. Review Chapter 7, on work station design, for what must be included on a work station layout.

Step 3: Divide the job into elements.

Elements were discussed in Chapter 8 as being units of work that are indivisible. This is not a useful definition in time study, because the operation could be too fast to time-study. Time study elements should be as small as possible, but not less than .030 minutes. The element should be as descriptive as possible. The elements must be in the sequence that the methods call for and should be made as small as is practical.

Principles of Elemental Breakdown
---

1. It is better to have too many elements than too few.
2. Elements should be as short as possible, but not less than .030 minutes. Elements over .200 minutes should be examined for further subdivision.
3. Elements that end in sound are easier to time because the eyes can be looking at the watch while the ears are anticipating the sound.
4. Constant elements should be segregated from variable elements to show a truer time.
5. Separate the machine-controlled elements from the operator-controlled elements so work pace can be differentiated.
6. Natural breaking points are best. The beginning and ending points must be recognizable and easily described. If the element description isn't clear, the description or breakdown must be rethought.
7. The element description describes the complete job, and the ending points are clearly marked.
8. Foreign elements should be listed in the order of occurrence. Foreign elements are not listed until they occur during the study.

The reasons for breaking down a job into elements are as follows:

1. It makes the job easier to describe.
2. Different parts of the job have different tempos. The time study technician will be able to rate the operator better. Machine-controlled elements will be constant and normally 100%, whereas the operator may be more or less proficient at different parts of the job.

3. Breaking down the job into elements allows for moving a part of the job from operator to operator. This is called *line balancing*.

4. Standard data can be more accurate and more universally applied with smaller elements. All work is made up of common elements. After a number of time studies, the technologist can develop formulas or graphs to eliminate the need for time study. Standard data is the goal of all time study departments.

On our time study form, two columns have been assigned to elements:

⑭ Element #: The element number is just a sequential number and is useful when more than 10 cycles are timed. Instead of describing each element over and over again, we just reference the element number.

⑮ Element description: Be as complete as possible. The ending points should be clear.

㉗ Foreign elements: These foreign elements will be eliminated from the study, but we don't want to hide anything. Therefore, a reason for throwing out the time is required. Foreign elements marked with an asterisk (*) in the body of the study are referred to this box.

Step 4: Do the actual time study: ⑯.

This is the guts of the stopwatch time study. Block 16 on the step-by-step form is for recording the time for each element. The form has room for 8 elements (8 lines) and 10 cycles (columns) for 80 readings. Most studies will have only 3 or 4 elements, so there is room on one sheet for 20 cycles. This form can be used for either snapback or continuous time study.

Continuous time study is the most desirable time study technique. The stopwatch remains running through the duration of the study, and element ending times are recorded behind the "R" for reading.

Continuous Example

		1	2	3	4	5
	R	.16	.83	1.50	2.17	2.83
Load and clamp	E					
	R	.56	1.23	1.90	2.57	3.23
Run machine	E					
	R	.66	1.33	2.01	2.67	3.32
Unload and place aside	E					

Note that each time is getting larger and that five parts were run in a total time of 3.32 minutes. In Step 5, we calculate the elemental times, but at this time we are still out in the plant collecting data.

Snapback studies allow the technician to read the watch and reset it immediately to time the next element. The exact same study is shown next using the snapback technique.

Snapback Example

		1	2	3	4	5
	R					
Load and clamp	E	.16	.17	.17	.16	.16
	R					
Run machine	E	.40	.40	.40	.40	.40
	R					
Unload and put aside	E	.10	.10	.11	.10	.09

Note that the elemental time (E) is already calculated. Look at the load and clamp time; the times look consistent—.16, .17, .17, .16, and .16. The time for loading and clamping is immediately obvious. This same information will be available in a continuous time study, but a lot of arithmetic is required first. In the snapback time study technique, the "R" row can be used for rating the operator on each element of work. (We discuss this in more detail later when we discuss the rating, leveling, and normalizing sections.)

Step 5: Extend the time study.

Now that the time study has been taken, the bigger job comes. The continuous method has one more step than the snapback method, so we will concentrate on the continuous method. After a brief description of the steps, an example problem is given for you to extend (see Figure 9-17 on page 000).

⑯ Subtract the previous reading from each reading. The previous element reading was its ending time and the beginning of this element. Subtracting the beginning time from the ending time gives elemental time.

⑰ Total/cycles: The total refers to the total time of the appropriate cycles timed. Some cycles may be eliminated because they include something that doesn't reflect the elemental time. We circle these elements. Circled elements are eliminated from further consideration. Cycles are the number of applicable elemental times included in the total time.

⑱ Average time: Average time is the result of dividing total time by the number of cycles. On the average, there was .40 minute of machine time in our last example.

⑲ % R: Percent rating refers to our opinion of how fast the operator was performing. The rating divided by 100, multiplied by the average time, equals normal time.

$$\text{Average time} \times \frac{\text{rating \%}}{100} = \text{normal time}$$

Later in this chapter we discuss rating in detail.

⑳ Normal time: Normal time is defined as the amount of time a normal operator working at a comfortable pace would take to produce a part. Normal time is calculated above and is explained further in Step ㉒.

㉑ Frequency: Frequency indicates how often a task is performed. For example, moving 1,000 parts out of the work station, moving the empty tub to the other side of the work station, and bringing in a full tub of 1,000 new parts to the work station would occur only once in 1,000 cycles (1/1,000). If Quality Control asked the operator to inspect one part out of every 10, $\frac{1}{10}$ would be placed in this column. The biggest use of this column is when the operator is doing two parts at a time; then $\frac{1}{2}$ is placed in this column. If $\frac{1}{1}$ would be entered in the column, it can be left blank.

㉒ Unit normal time: Unit normal time is calculated by multiplying the frequency by the normal time.

EXAMPLES:

NORMAL TIME		FREQUENCY		UNIT NORMAL TIME
1.160	×	1/1,000	=	.001 minute
0.400	×	1/10	=	.040 minute
0.100	×	1/2	=	.050 minute
0.050	×	1/1	=	.050 minute

Every element must reflect the time to produce one unit of production. No one wants a standard for pairs, and mixing frequency of units leads to bad time standards. Be very careful here.

Step 6: Determine the number of cycles to be timed.

The accuracy of time study depends on the number of cycles timed: the more cycles studied, the more accurate the study. Almost all time study work is aimed at an accuracy of ±5% with a 95% confidence level, so the question is, How many cycles should be studied to achieve this accuracy?

Graphs and tables are easier and more cost effective than formulas. The graphs and tables (see Figure 9-14 and Table 9-1, respectively) used in this book are based on the following formula:

$$\sqrt{N} = \frac{2R}{Ad_2\bar{x}} \quad \text{or} \quad N = \frac{4R^2}{(A)^2(d_2)^2(\bar{x})^2}$$

N = Number of cycles to time study.

R = Range of the sample of observations (highest value of elemental data minus lowest value).

A = Required precision (±5% or ±10%) (must be extended as decimal ±.05 or ±.10, etc.).

d_2 = A constant used to estimate the standard deviation of a sample. It is a function of the sample size. Must be obtained from a statistical table.

$\bar{x}$ = Arithmetical average; sum of the observations divided by the number of observations.

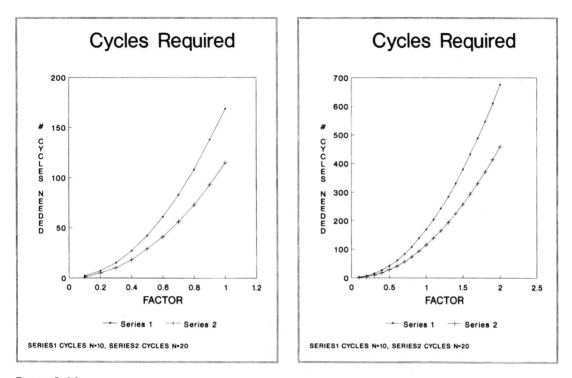

FIGURE 9-14

Table 9-1 The Number of Cycles to Time Based on 95% ± 5% Accuracy.

RANGE = R	NO. OF CYCLES TO TIME
0.1	2
0.2	7
0.3	15
0.4	27
0.5	42
0.6	61
0.7	83
0.8	108
0.9	138
1.0	169

Notes:

1. If ±10% accuracy is okay, divide the number of cycles by 4.

2. A result over 169 means that you have timed an unskilled operator, your readings are off, or your elements are too small.

3. This chart is printed at the bottom of the time study forms in this book.

EXAMPLE: Confidence 95% precision ±5%

Readings .08, .07, .09, .10, .07, .11, .08, .07
Range = .11 − .07 = .04
Required precision = .05
d_2 = 3.078
$\bar{x}$ = .82/10 = .082

$$N = \frac{4(.04)^2}{(.05)^2(3.078)^2(.082)^2} = 40.18, \text{ or } 40$$

EXAMPLE: All data is the same except the precision $(A) = ±10\%$ (.1).

$$N = \frac{4(.04)^2}{(.1)^2(3.078)^2(.082)^2} = 10.01, \text{ or } 10$$

If ±10% accuracy is an acceptable quality standard for our time study data, one-fourth the number of cycles are required (40 versus 10).

Figure 9-14 shows two graphs as examples of how many cycles to time. These two graphs are based on an initial 10 or 20 timed cycles. Calculate the range (R) and the arithmetic average $\bar{x}$. Divide R by $\bar{x}$ and the factor results. Look up the factor on either chart. The chart on the left is used when the factor is under 1.0, and the chart on the right is used when the factor is over 1.0. The charts tell you how many cycles to time to achieve ±5% accuracy. Divide this number by 4 if ±10% accuracy is adequate.

The number of cycles to time-study is a measure of how confident we are in the time standard we are setting. The time standard is much like a batting average or a bowler's average. In the beginning of a season, a good or bad game can change the average considerably. At the end of a season, however, a good or bad game will have very little effect on the average. We want to take enough cycle readings to ensure that a good or bad cycle will not affect the average. We can then have the confidence that our time standard is achievable and that it is a fair reflection of what the operators can do.

The following blocks on the time study form help calculate the number of cycles needed:

㉓ The range

㉔ $R/\bar{x}$ or the factor

㉕ Highest

㉖ Factor table.

The procedure for determining the number of cycles to be timed is as follows:

1. Time-study 10 cycles for jobs less than 2 minutes long and 5 cycles for jobs longer than 2 minutes.

2. Determine the range (R) ㉓ of the elemental times for each element of the job. The range is the highest elemental time less the lowest elemental time. The smaller this range, the fewer cycles are needed for the accuracy level.

3. The average time has already been determined in column ⑱ of the time study; $\bar{x}$ is the mathematical symbol for arithmetic average.

An example of one element of a job is: .08, .09, .08, .08, .07, .08, .08, .10, .08, .09. The total of these elemental times is .83; the number of elemental times is 10; therefore, the average time equals .083 minute. The range (R) is $.10 - .07 = .03$.

4. Determine the factor $R/\bar{x}$ ㉔; $R/\bar{x}$ is the range divided by the average time. $R = .03$, $\bar{x} = .083$, and $R/\bar{x} = .36$.

5. Determine the number of cycles needed. ㉖ on the time study form is a list of factors most commonly used. The number .36 on the table is 60% of the way between .3 and .4. Sixty percent of the difference between 27 cycles and 15 cycles is 7.2 cycles; therefore, 23 cycles must be timed ($7.2 + 15 = 22.2$). The time study technologist must go back to the job and collect 13 more cycles to have a 95% ± 5% accurate time study (see Table 9-1). Factors over 1.0 are possible, as the graphs in Figure 9-14 show, but the cost would be very high. Something is wrong if you have a factor over 1.0—probably an inexperienced operator or time study person.

Normal Distribution

The number of cycles to time-study is based on the laws of probability as best explained by the normal distribution curve. If we were to collect data on any specific characteristic of a large group of individuals, or if we were to measure a particular dimension on a large number of parts, and then plot the frequency of each observation, the resulting distribution would resemble a symmetrical bell-shaped curve. Such distributions are called *normal* or *Gaussian distributions* and are of significant importance in any statistical study. Although there are many other forms of distributions, most phenomena in nature and industry have a normal distribution, or most often these distributions approach normality to such an extent that the normal laws of probability are applicable.

Understanding normal distributions is simple. In such cases, most values are bunched or clustered around the central value or the average. As the values differ from the average value, and as the size of this difference increases, the frequency or occurrence of these values decreases. Consider an employee who, on the average, takes 15 minutes to perform a task. We know from experience that sometimes he or she will take a little more, and sometimes a little less, time to do the same task. However, seldom does the employee take too much time under or over the average time. Over a long period of time, the employee will have an approximately equal number of extremely low and high readings, although not very many of each.

Standard Deviation

In any normal distribution, the average indicates the center value around which most of the data are clustered, but it does not indicate how variable or spread out the data may be. Imagine two groups of employees performing the same task (see Figure 9-15). One group consists of individuals with an equal amount of training and experience. The employees in the second group, however, are varied in training and experience. While the *average* time per employee for both groups may be the same (say, 30 minutes), the individual times in the first group may range from 25 to 35 minutes, and in the second group, from 10 to 50 minutes. Would you say that both distributions are the same? Of course not. Even though both may be normal and have the same average, they do not have the same spread or variability. The quantitative value that states the degree of variability or spread of a population is called *standard deviation* and is denoted by *s*. The larger the variability or the spread of the data, the larger the standard deviation.

There is a definite and interesting relationship between the variability of the data as measured by the standard deviation and the frequency or occurrence of the values for any distribution. Given the average and the standard deviation, *s*, approximately 68.26% of all the measurements for any normal distribution will be located between the average ±1*s* (see Figure 9-16). If the average is extended to ±2*s* on both sides (i.e., average ±2*s*), then 95.45% of all the data are included. Finally, the area enclosed by average ±3*s* will include approximately 99.73% of the data.

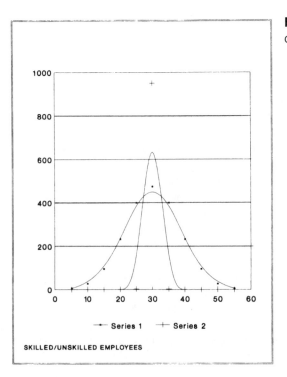

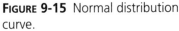

FIGURE 9-15 Normal distribution curve.

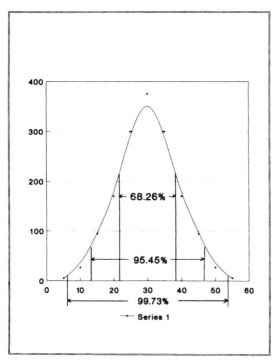

FIGURE 9-16 Standard deviation curve.

Step 7:   Rate, level, and normalize the operator's performance.

⑲% rating is the technologist's opinion of the operator's performance. Rating, leveling, and normalizing all mean the same thing, and the term *rating* is used from this point on. Rating is the most challenged aspect of motion and time study, and for that reason it is the most important subject of this chapter (it is discussed in detail later in this chapter).

$$\text{Average time} \; \times \; \frac{\text{rating}}{100} = \text{normal time}$$

⑱   ⑲   ⑳

Step 8:   Apply allowances ㉙.

Allowances are added to a time study to make the time standard practical.

$$\text{Total normal time} \; + \; \text{allowances} = \text{standard time}$$

㉘   ㉙   ㉚

There are several methods of applying allowances, and there are several types of allowances. We discuss allowances in detail later in this chapter.

Step 9:   Check for logic.

   Once the time study has been extended, the test for logic should be applied in two ways:

1. The average time ⑱ should look like the elemental times. If an error in adding was made, a test for logic will save a mistake. The easiest mistake to make is decimal error. Be careful not to make decimal errors, because they look bad— 1,000% of errors result from misplacing a decimal only one place, which is why being consistent with decimal placement is so important:

   a. Read stopwatches in two places: .01.

   b. From average time on, use three places: .001.

   c. Hours per unit are five places: .00001.

2. The second test for logic is the total normal time for one unit. During your study, you timed a specific number of parts in a certain amount of time. For example, 10 cycles were timed in 7.5 minutes (7.5 was the last reading in the tenth column). The average time should be somewhere around .75 minute each. Are you close with the total normal time? If not, there is a major error. Warning: Don't forget that if the operator is producing two at a time, twice as many parts are being produced.

Step 10:   Publish the time standard.

   Three numbers are required to communicate a time standard:

1. Decimal minutes: ㉚

2. Hours per unit: ㉛

3. Pieces per hour: ㉜.

Starting with standard minutes, dividing ㉚ by 60 minutes per hour equals hours per unit ㉛, and pieces per hour ㉜ is $1/x$ of ㉛ (or divide hours per unit into 1 hour).

   Every company has a method of recording time standard information. In Chapter 4, an operation sheet for a water valve factory was shown. The time standards could be placed on that operations sheet. (Please review Figure 4-1.) The production route sheet is another common tool for communicating the time standard. The computer is the most common method of storing and communicating to everyone what the time standard is for each job.

   A few more pieces of information remain to be discussed on the step-by-step time study form:

㉝ Engineer: The time study technologist puts his or her name here.

㉞ Date: A time study with an incomplete date is worthless.

㉟ Approved by: This is where the chief engineer or manager signs, approving your work. You never fill this in.

FRED MEYERS & ASSOCIATES   TIME STUDY WORKSHEET   ☐ SNAP BACK ☐ CONTINUOUS

OPERATION DESCRIPTION⌐   ASSEMBLE PARTS 2 & 4, MACHINE SCREW & STAKE. INSPECT

PART NUMBER 4650-0950	OPERATION NO. 1515	DRAWING NO. 4650-0950	MACHINE NAME	MACHINE NUMBER 21	☐ QUALITY OK ?
OPERATOR NAME MEYERS	MONTHS ON JOB 5	DEPARTMENT ASSEMBLY	TOOL NUMBER M61	FEEDS & SPEEDS. MACHINE CYCLE TIME	☐ SAFETY CHECKED ? ☐ SETUP PROPER ?
PART DESCRIPTION:		MATERIAL SPECIFICATIONS:			NOTES:

ELEMENT #	ELEMENT DESCRIPTION		READINGS 1 2 3 4 5 6 7 8 9 10	TOTAL CYCLES	AVERAGE TIME	% R	NORMAL TIME	FREQUENCY	UNIT NORMAL TIME	RANGE	R/X	HIGHEST ✓
•	ASSEMBLY	R E	9 41 71 1.07 38 77 2.08 48 77 3.07			90		1	1			
	DRIVE SCREW	R E	15 46 79 13 43 82 14 53 82 93			100		1	1			
	PRESS	R E	28 59 94 27 66 95 28 66 96 4.06			110		1	1			
	INSPECT	R E	32 62 92 30 69 98 41 69 99 4.09			100		1	1			
	LOAD SCREWS	R E	3.83			125		1	10			
		R E										
		R E										
		R E										

FOREIGN ELEMENTS:	NOTES:	R/X CYCLES TOTAL NORMAL MIN. .1 2 ALLOWANCE + ____ .10 % .2 7 STANDARD MINUTES .3 15 HOURS PER UNIT .4 27 UNITS PER HOUR .5 42 .6 61 .7 83 ON BACK .8 108 WORK STATION LAYOUT .9 138 PRODUCT SKETCH 1.0 169
ENGINEER: DATE: _ / _ / _	
APPROVED BY: DATE: _ / _ / _	

FIGURE 9-17 Time study problem: continuous technique. (Answer on page 60)

An example problem has been included in Figure 9-17. The data has been collected, the job has been broken down into elements, and the time study has been made. You need to extend the study and develop a time standard. This was a continuous time study, and that should be obvious because the times are always getting larger. The extension will start with the subtraction of element readings to find elemental time.

RATING, LEVELING, AND NORMALIZING

Rating is the process of adjusting the time taken by an individual operator to what could be expected from a normal operator. The industrial technologist must understand the industry standards of *normal*. Rating an operator includes four factors:

1. Skill
2. Consistency
3. Working conditions
4. Effort (most important).

Three of these four factors are accounted for in other ways and have little effect on rating. Effort will be our primary concern.

1. *Skill:* The effect of skill is minimized by timing only people who are skilled. Operators must be fully trained in their work classification before being time-studied. A welder must be a qualified welder before being considered a subject for time study. Two years of training may be required to become a welder, and in addition, this welder must be on this job for at least two weeks before performing the job sufficiently. Habits of motion patterns must be routine enough that the operator doesn't have to think about what comes next and where everything is located. Very skilled operators make a job look easy, and the industrial technologist must let this skill affect the rating. On the other hand, if an operator shows lack of skill, such as dropping, fumbling, inconsistent timing, stopping/starting, etc., the technologist should postpone the study or find someone else to time-study.

2. *Consistency:* Consistency is the greatest indication of skill. The operator is consistent when he or she runs the elements of the job in the same time, cycle after cycle. The time study technician begins to anticipate the ending point while looking at the watch and listening for the ending point. The operator is said to be like a machine. Consistency is used to determine the number of cycles. A consistent operator needs to run only a few parts before the cycle time is known with accuracy. The skill of the operator should be evident to the time study technician, and the technician's rating of the operator should be high. When inconsistency is present, the technologist must take many more cycles to be acceptably accurate in the time study. This inconsistency tends to affect the technologist's attitude and rating of the operator in a negative way, and the best thing to do is to find someone else to study. It is more fun rating and working with operators with great skill.

3. *Working conditions:* Working conditions can affect the performance of an operator. In the early twentieth century, this was much more of a problem than it is today. But if employees are asked to work in hot, cold, dusty, dirty, noisy environments, their performance will suffer. These poor working conditions can be eliminated if the true cost is shown. The way we account for poor working conditions today is to increase the allowances (discussed later in this chapter). If operators are required to lift heavy materials in the performance of their duties, 25% more time can be added to the time standard as an allowance. Working conditions are not part of modern rating.

4. *Effort:* Effort is the most important factor in rating. Effort is the operator's speed and/or tempo and is measured based on the normal operator working at 100%. A 100% performance rating is defined as

    a. Walking 264 feet in 1.000 minute or 3 miles per hour,

    b. Dealing 52 cards into four hands around a $30'' \times 30''$ card table in .500 minute,

    c. Assembling thirty $\frac{3}{8}'' \times 2''$ pins into a pinboard in .435 minute.

Effort can be seen easily in walking. Walking at speeds less than 100% is uncomfortable for most people, and walking at 120% requires a sense of urgency that indicates increased effort.

Psychology has been good to the time study technician. The normal tendency of people being watched is to speed up. Being watched makes people nervous, and nervous energy is converted by the body into a faster tempo. The time study technician then gets a frequent chance to rate over 100%. When an operator works at 120%, the technologist has the pleasant experience of telling the operator, "You are fast. I'm going to have to give you 20% more time so that an average person can do the job." That is fun to say, and it happens often.

When rating, you must keep tuned into normal pace. This requires continued practice on your part, forever. The experiments in the next section will help keep your rating accurate.

All PTS systems have been developed based on the concept of normality according to industry standards, and a synthetic rating is developed by time-studying a job that has been proven by PTS. A good learning technique used at many companies is to have new technologists time-study known jobs and compare their time standards to the known time standards. Another good learning experience is to time several people on the same job. Effort and skill are the only differences in time, so proper rating should make all the normal times the same.

Many companies use time study rating tapes developed by industrial associations and professional organizations:

1. Society for the Advancement of Management (SAM)
2. Tampa Manufacturing Institute
3. Ralph Barnes and Associates
4. Faehr Electronic Timers, Inc.

All of these groups produce time study rating films. *Industrial Engineering Solutions* magazine would also be a good source.

100% STANDARDS AND EXPERIMENTS

Industrial technologists can teach themselves to be good raters by setting up some simple experiments.

Walk a 50-Foot Course

Set up a 50-foot course with a starting line and an ending line. With a stopwatch in hand, start about 10 feet in front of the starting line and start the walk. When crossing the starting line (already up to speed), start the stopwatch. Maintain the same pace until the finish line is crossed, then stop the watch. Slow down and stop only after the finish line is crossed. Now read the watch. Let's use an example of .18 minute.

How fast was the walk? The time should have been .19 minute (the standard). The time standard for walking has been universally accepted as 3 miles per hour.

$$\frac{3 \text{ mph} \times 5280 \text{ ft/mile}}{60 \text{ min/hr}} = 264 \text{ ft/min}$$

$$\frac{50 \text{ ft}}{264 \text{ ft/min}} = .19 \text{ min} \quad \text{(time standard for 50 ft)}$$

Notice that only two decimal places are used when reading a stopwatch.

$$\% \text{ performance} = \frac{\text{time standard}}{\text{actual time}} = \frac{.19}{.18} = 106\%$$

What is the percent performance for the following?

STOPWATCH TIME (ACTUAL)	STANDARD TIME	% PERFORMANCE
.24	.19	
.16	.19	
.28	.19	
.11	.19	
.20	.19	

A test of logic is, did you walk faster (over 100%) or slower (under 100%) than the standard? Practice walking the 50-foot course at 100% first. Then try 120%, 80%, and 60%. Notice how difficult walking below 100% is. Notice the sense of urgency at 120%.

Walking is a good first study of rating. Most everyone is a skilled walker, so effort is all that is being rated. Some people are tall and have long legs, while other people are short and have short legs. Is it fair to have one standard for all? The answer doesn't speak to the fairness, but only to the fact that one standard is all that can be used. Would a customer be willing to pay more because a short person worked on this job? No. Only one cost can be charged, and only one time standard can be used. The "average" person must be used in all time study. Table 9-2 shows time and percents for walking a 50-foot course.

Dealing 52 Cards into Four Equal Stacks on a Card Table in .500 Minute

Set up an area to deal cards. A 30″ square card table was used for the following data, but any area of at least that size can be used. A good way to begin would be to start a stopwatch and place it in front of the dealer, who then deals a card every .01 minute. When 52 cards are dealt, .52 minute has been used. What is that performance?

$$\text{Performance} = \frac{\text{time standard}}{\text{actual time}} = \frac{.50}{.52} = 96\%$$

Table 9-2 Rating: Percents and Time for Walking a 50-Foot Course.

TIME	%	TIME	%	TIME	%
.09	211	.18	106	.27	70
.10	190	.19	100	.28	68
.11	173	.20	95	.29	66
.12	158	.21	90	.30	63
.13	146	.22	86	.31	61
.14	136	.23	83	.32	59
.15	127	.24	79	.33	58
.16	119	.25	76	.34	56
.17	112	.26	73	.35	54

Ninety-six percent is close to 100%, and if a small speed-up were to be made, 100% would result. Dealing cards is something that we all do very well. Again, give it a try. Deal 100% a few times, then 120%, 80%, etc. Watch other people and estimate their rating without the use of the stopwatch. Check yourself with the watch, but don't let it affect your rating estimate.

What are the percent performances for the following?

DEALING TIME	STANDARD	% PERFORMANCE
.40	.50	
.45	.50	
.55	.50	
.60	.50	

Try dealing the cards as fast as you can without losing pile quality. Try dealing the cards as slowly as you can without stopping between cards. The range will probably be a ratio of 1:2; the slowest time takes twice as long as the fastest.

Dealing cards is a good rater training experiment because it is much like simple assembly work. Reaches and moves account for half of all work and are the only actions a technologist can rate. Dealing cards consists of reaches and moves. Table 9-3 shows times and percentages for dealing 52 cards into four equal stacks.

Assemble 30 Pins into a Pinboard in .435 Minute Using the Two-Hand Method

The pinboard assembly experiment has been around for almost 100 years. The time calculated by PTSS (chapter 8) is .435 minute. The two hands are working at the same time; therefore, 15 sets of motions are required. The pinboard experiment is realistic and similar

Table 9-3 Ratings: Percents and Times for 52 Dealing Cards (Four Equal Stacks).

TIME	%	TIME	%	TIME	%
.25	200	.46	109	.67	75
.26	192	.47	106	.68	74
.27	185	.48	104	.69	72
.28	179	.49	102	.70	71
.29	172	.50	100	.71	70
.30	167	.51	98	.72	69
.31	161	.52	96	.73	68
.32	156	.53	94	.74	68
.33	152	.54	93	.75	67
.34	147	.55	91	.76	66
.35	143	.56	89	.77	65
.36	139	.57	88	.78	64
.37	135	.58	86	.79	63
.38	132	.59	85	.80	63
.39	128	.60	83	.81	62
.40	125	.61	82	.82	61
.41	122	.62	81	.83	60
.42	119	.63	79	.84	60
.43	116	.64	78	.85	59
.44	114	.65	77	.86	58
.45	111	.66	76	.87	57

to production work. It has been used in the past to screen new employees. Figure 9-18 is a picture of the 30-pin pinboard assembly job, and Table 9-4 shows the times and percents for assembling the pinboard. What would be the percent performance of the following pinboard times?

	STOPWATCH TIME IN MINUTES	STANDARD	% PERFORMANCE
1.	.32		
2.	.40		
3.	.50		
4.	.45		

Rating is not an exact science. The industrial technologist should aim for ±5% accuracy, and ±10% accuracy is of value. When rating, always round off to the nearest 5% (80, 85, 90, 95, etc.). Anything else is a false claim to more ability than we possess.

Figure 9-18 1. The pinboard is 7″ × 8″ with five rows of six pins, $\frac{3}{8}$″ in diameter. The holes are countersunk on one side. 2. The 30 pins are $\frac{3}{8}$″ in diameter, $2\frac{3}{4}$″ long, beveled on one end. 3. Assemble 30 pins using two hands with the countersink facing up in .435 minute.

Table 9-4 Ratings: Percents and Times for Assembling Pinboard (30 Pins, Two Hands).

TIME	%	TIME	%	TIME	%
.27	161	.43	101	.59	74
.28	155	.44	99	.60	73
.29	150	.45	97	.61	71
.30	145	.46	95	.62	70
.31	140	.47	93	.63	69
.32	136	.48	91	.64	68
.33	132	.49	89	.65	67
.34	128	.50	87	.66	66
.35	124	.51	85	.67	65
.36	121	.52	84	.68	64
.37	118	.53	82	.69	63
.38	114	.54	81	.70	62
.39	112	.55	79	.71	61
.40	109	.56	78	.72	60
.41	106	.57	76	.73	60
.42	104	.58	75	.74	59

TIME STUDY RATER TRAINER FORM

The time study rater trainer form is an aid to help the time study technician improve his or her rating ability. The form has been developed to point out the technician's problems and to give direction to improvement. The step-by-step procedure that follows follows the circled numbers on the form (see Figure 9-19). The step-by-step procedure will tell the technician exactly how to use the form. The experiments just discussed (walking, dealing cards, and pinboard) or rating films can be used to test the technologist's rating ability, but this form is used to test the technologist's ability to rate the performance of any operation. Ten rating samples are needed and are compared to the actual time.

Step-by-Step Procedure for Completing the Time Study Rater Trainer Form

① Name: Your name—the person doing the rating—goes here. This form is a measure of rating ability, so the person doing the rating is the focus of this form.

② Date: The full date of the test.

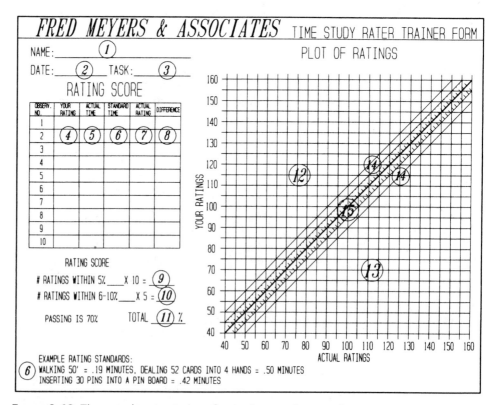

FIGURE 9-19 Time study rater trainer form: the step-by-step form.

③ Task: Walking, dealing cards, pinboard, film name or number, or job title. What job did you watch?

④ Your rating: There are 10 rows in this column, and one rating goes in each column. Look at the completed example for walking in Figure 9-20. When 10 people were observed (one at a time) walking through a 50-foot course, the rating for each of these people was placed in this column. This column is completed before any further work is done.

⑤ Actual time: While observing people walking through the 50-foot course, the technologist should time each one with a stopwatch. This time is recorded in this column. This is the actual time it took the person to walk through the 50-foot course. Don't let the watch reading affect your rating.

⑥ Standard time: The standard time is

   a. Walking = .19 minute

   b. Dealing cards = .500 minute

   c. Pinboards = .435 minute

   d. Films = actual ratings ⑦.

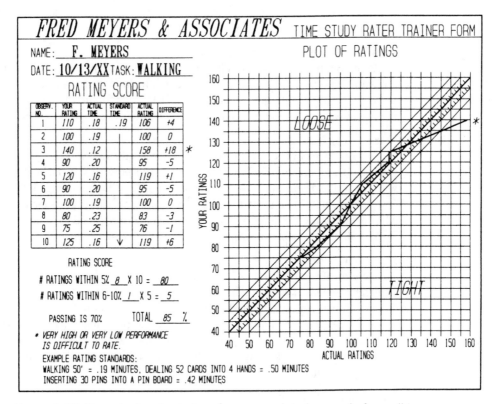

FIGURE 9-20 Time study rater trainer form: completed example for walking.

Standard time is the time a skilled person would take if working at a normal pace. The standard time will be the same for each observation, as long as the task is the same.

⑦ Actual rating: The actual rating is

$$\frac{⑥ \text{ standard time}}{⑤ \text{ actual time}} = \text{actual rating } ⑦$$

The actual rating is the correct answer that the technologist was shooting for.

⑧ Difference:

Actual rating ⑦ − your rating ④ = difference ⑧
or
Your rating ④ − Actual rating ⑦ = difference ⑧

(being over or under makes no difference here).

⑨ Ratings within 5% _____ × 10 = _____.
    Ratings with a difference of 5% or less are perfect ratings and, for the purposes of grading, earn 10 points each. If a technologist rated all 10 observations within ±5%, a 100% grade would result (10 × 10 = 100).

⑩ Ratings within 6% to 10% _____ × 5 = _____.
    Ratings with a difference ⑧ of less than 10%, but more than 5%, are good ratings and are worth 5 points each on a grading scale. If a technologist rated all 10 observations with 6% to 10%, a 50% grade would result (10 × 5 = 50). Fifty percent isn't great, but 0% is worse. As can be seen, ratings of over 10% difference ⑧ earn no grade points.

⑪ Total _____%.
    The total percent is calculated by adding the results of ⑨ + ⑩ = ⑪. Just like class grades, 60% is passing, 70% = C, 80% = B, and 90% = A. A technologist working with motion and time study must maintain an 80% or higher rating skill grade.
    A comment must be made for the ± aspect of rating. The ± gives the technologist an edge on rating; a +10% error will be offset by a −10% error, and the total will be correct. Errors will have a chance of washing out if the technologist is zeroed in at the beginning.

⑫ Plot of rating (loose) above the line.
    The plot of rating allows the technologist a chance to see his or her rating ability. The technologist takes each pair of ratings (his or her rating ④ and the actual rating ⑦) and plots them on a graph. His or her rating ④ goes on the vertical axis, and the actual rating ⑦ on the horizontal axis. Where the two points intercept on the graph, place an X. The ⑫ area on the graph indicates that the technologist is too loose or that the technologist rates too high. If many of the ratings are too high, the technologist should consciously lower the rating.

⑬ Plot of rating (tight) under the line.
    Plotted points under the line indicate that the technologist is rating too low or is rating tight. Many ratings in this area indicate that the ratings should be increased.

⑭ Between the ±10% line.

This is good rating—not great, but good.

⑮ Between the ±5% line.

This is the goal line. In rating, the line is ±5%, and a rating within ±5% is great rating. This is the goal of all our ratings.

Once all the points are plotted, connect the X's by drawing a straight line between them. (See the example in Figure 9-20.) Adjust your thinking, if needed, and try again. Your rating score needs to be 80% or higher.

Figure 9-21 is a list of 10 fundamentals of pace rating provided by the Tampa Manufacturing Institute. Figure 9-21 is the back side of the time study rater trainer form in Figure 9-20.

---

1. Healthy people in the right frame of mind easily turn in 100% performance on correctly standardized jobs. For incentive pay, good performers usually work at paces from 115% to 135%, depending on jobs and individuals.

2. For most individuals, *it is uncomfortable to work at a tempo much below 100% and extremely tiring to operate for sustained periods at paces lower than 75%;* our reflexes are naturally geared to move faster.

3. Poor efficiency on a correctly standardized job usually results from stopping work frequently—"goofing off" for a variety of reasons. Specifically, *substandard production seldom results from inability to work at a normal pace.*

4. Some standards of 100%   1. Walking 3 mph or 264 ft/min
                              2. Dealing cards into 4 stacks in .5 min
                              3. Filling the pinboard in .435 min

5. Very seldom can *true* performance of over 140% be found in industry.

6. When an operator consistently comes up with extremely high efficiencies, it is usually a sign that the method has been changed or the standard was wrong in the first place.

7. An operator's work pace during a time study does not affect the final standard. His or her *actual* time is multiplied by the performance rule to give a job standard that is fair for all employees.

8. Inasmuch as healthy employees can *easily* vary work pace from approximately 80% to around 130%—through a range of 50%—reasonable inaccuracies in the setting of standards should be sensibly accepted.

9. Ineffective foremen usually fight job standards. *Good supervisors, however, sincerely help in the standard-setting effort, clearly realizing that such information is their best planning and control tool.*

10. *Methods usually influence production more than work pace.* Don't ever get so absorbed in how quickly or how slowly an operator "seems" to be moving that you fail to consider whether or not he or she is using the right method.

FIGURE 9-21 Ten fundamentals of pace rating. (Courtesy of Tampa Manufacturing Institute.)

ALLOWANCES

Allowances are extra time added to the normal time to make the time standard practical and attainable. No manager or supervisor expects employees to work every minute of the hour. What should be expected of the employee? This was the question asked by Frederick W. Taylor over 100 years ago. Would you expect the employee to work 30 minutes per hour? How about 40 minutes? 50 minutes? This section will assist the technologist in answering Taylor's question.

Types of Allowances

Allowances fall into three categories:

1. Personal
2. Fatigue
3. Delay.

Personal Allowance Personal allowance is that time an employee is allowed for personal things such as

a. Talking to friends about nonwork subjects,
b. Going to the bathroom,
c. Getting a drink,
d. Any other operator-controlled reason for not working.

People need personal time, and no manager would begrudge an appropriate amount of time spent on these activities. An appropriate amount of time has been defined as about 5% of the work day, or 24 minutes per day.

Fatigue Allowance Fatigue allowance is the time an employee is allowed for recuperation from fatigue. Fatigue allowance time is given to employees in the form of work breaks, more commonly known as coffee breaks. Breaks occur at varying intervals and are of varying duration, but all breaks are designed to allow employees to recuperate from on-the-job fatigue. Most employees today have very little physical drudgery involved with their jobs, but mental fatigue is just as tiring. If an employee uses less than 10 pounds of effort during the operation of his or her job, then 5% fatigue allowance is normal. A 5% increase in fatigue allowance is given for every 10-pound increase in exertion required of the employee (see Figure 9-22).

The breaks are calculated into the fatigue allowance because the reporting practice is to not report the time spent on breaks separately. Because lunch time is punched out and not reported, it does not enter into the allowance calculation. Remember, allowances are for times when the employee is expected to perform but can't.

EXAMPLE:

An employee must pick up a 50-pound part.
Fatigue allowance is $50 - 10 \div 10 = 4.0$ units of ten pounds.

$$5\% + (4 \times 5) = 25\% \text{ allowance}$$

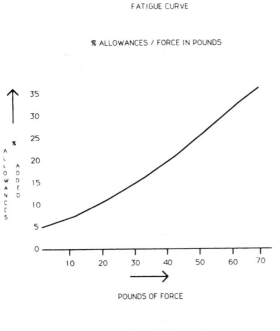

FIGURE 9-22 Fatigue allowance curve: % allowances per pound of force.

FATIGUE CURVE

% ALLOWANCES / FORCE IN POUNDS

1. 5% IS MINIMUN FATIGUE ALLOWANCE
2. 5% OF INCREASE FATIGUE ALLOWANCES PER 10 POUNDS
   OF INCREASE FORCE OVER 10 POUNDS
3. FORCE IS WEIGHT OF PART IF LIFTED

EXPLANATION OF EXAMPLE:

The basic fatigue allowance is 5% and an additional 5% fatigue allowance is added for each 10 pounds of force required over 10 pounds. 50 pounds is 40 pounds more than basic. 40 pounds is 4 units of excess weight (10 pounds is 1 unit). 4 units × 5% = 20% of excess fatigue. 20% + 5% basic equals 25% fatigue allowance.

The weight has to be picked up every minute. If the frequency were once every 5 minutes, the 50 pounds would be divided by 5.

$$5\% + (\tfrac{4}{5} \times 5) = 9.0\%$$

The basic fatigue allowance is still 5%. When lifting only one 50-pound object every five minutes, only one-fifth of the excess weight is considered. 40 pounds is 4 units of weight, so:

$$\frac{4 \text{ units} \times 5\%}{5} = 4\% \text{ excess fatigue allowance}$$

5% basic + 4% excess equals 9% fatigue allowance.

The duration of breaks must now be calculated. The normal 5% fatigue allowance is commonly interpreted as two 12-minute breaks, one in midmorning and one in midafternoon, or a combination of the two, adding up to 24 minutes. Five percent of the 480 minutes in an 8-hour day is 24 minutes.

Seventeen percent allowances would equal 82 minutes per day. How will this 82 minutes be split for frequency and duration of breaks? I suggest that 11 minutes be given every hour except the hour before lunch. Seven 11-minute breaks equals 77 minutes, plus a 5-minute clean-up at the end of the shift. Note that a heavy job such as the one we are discussing here will tire out the employee faster than light or mental work, and the increased breaks are not only justified but will increase production. Breaks from work allow employees to recuperate, so when they return to work their production rate is higher than it would have been without a work break. The break more than pays for itself.

1. 5% is the minimum fatigue allowance.
2. 5% increased fatigue allowances per 10 pounds of increased force over 10 pounds. Force is weight of part if lifted.

Delay Allowances Delay allowances are called unavoidable because they are out of the operator's control. Something happens to prevent the operator from working. The reason must be known and the cost accounted for to develop the cost justification. Examples of unavoidable delays include:

1. Waiting for instructions or assignments.
2. Waiting for material or material-handling equipment.
3. Machine breakdown or maintenance.
4. Instructing others (training new employees).
5. Attending meetings, if authorized.
6. Waiting for setup. Operators should be encouraged to set up their own machines. A setup is complete when quality control approves.
7. Injury or assisting with first aid.
8. Union work.
9. Reworking quality problems (not operator's fault).
10. Nonstandard work—wrong machine or other problem.
11. Sharpening tools.
12. New jobs that have not been time-studied yet.

The operator's performance must not be penalized for problems out of his or her control. (Delays that are controlled by the operator are called personal time and are not considered here.)

Three methods are available to account for and to control unavoidable delays:

1. Add delay allowances to the standard.
2. Time-study them and add them to the time standard.
3. Charge the time to an indirect charge (see Chapter 13 for a more detailed discussion).

The goal of time study is to eliminate delay allowances. This is best done by time-studying the delay and adding that time to the time standard. However, some delays are so complicated that negotiating an allowance with the operator will save time and money for the company. For example, suppose you ask the question, "How much time do you spend a day cleaning the machine?" The operator will always say, "Well, it depends," and the technologist must ask something like

What is the longest time?

What is the shortest time?

Do you think 15 minutes is a good average?

If the operator agrees that 15 minutes per day is a good figure, the technologist will calculate a delay allowance as follows:

$$\frac{15 \text{ min clean-up}}{480 \text{ min/shift}} = 3\%$$

A 3% allowance will be added to the personal allowance of 5% plus a fatigue allowance of 5% to produce a 13% total allowance.

Generally, unavoidable delays can be eliminated or anticipated. Time standards in the form of standard data can be established and added to the time study to compensate the operator. An unavoidable delay is a foreign element, discussed later in this chapter. Those unavoidable delays that cannot be anticipated will require the operator to charge his or her time to an indirect account—meeting, injury, machine breakdown, and rework are examples. Supervisors will be required to approve all indirect charges, and the time should be more than 6 minutes to be statistically significant. The employee must not be penalized for management's lack of planning, but the supervisor must be given as much advance notice as possible. Reassignment may be in order.

One last warning about delay allowances: Don't put anything in the time standard that you cannot live with. It is difficult to get it out of the standard once it is included. Most companies have eliminated delay allowances but have allowed operators to punch out for anything not covered by the time standard.

Personal, fatigue, and delay allowances are added together, and the total allowance is added to the normal time:

$$\text{Normal time} + \text{allowance} = \text{standard time}$$

Methods of Applying Allowances

Allowances are added in four different ways. The forms in this text use just one of these methods, but there are good reasons for using the other methods. Each company has its own time study form and procedure. The form tells you which method of applying allowances to use. The four methods are presented here in order of ease of application.

Method 1: 18.5 Hours per 1,000 This method is the simplest of all and reduces the mathematical steps. It is also based on a constant allowance—in this case, 10%.

If a job takes 1.000 minute normal time, how many pieces per hour could be produced? At the rate of one per minute, 60 could be produced per hour, but we want to be practical and add 10% allowances. Ten percent of 60 is 6, so 54 pieces per hour would be an appropriate time standard. How many hours would it take to produce 1,000 units at the rate of 54 per hour? One thousand divided by 54 equals 18.5 hours per 1,000—the name of this method. Three numbers are required to communicate a time standard:

$$\text{Decimal minute} = 1.000$$
$$\text{Hours per } 1,000 = 18.5$$
$$\text{Pieces per hour} = 54$$

All time standards start with a decimal minute, so if our next standard is .5 minute, the hours per 1,000 equals $.5 \times 18.5 = 9.25$ hours per 1,000, and the pieces per hour is $1/x$, or 108 pieces per hour. Try these examples:

NORMAL MINUTES	HOURS/1,000 18.5	1/X PIECES PER HOUR
0.250	4.625	216
0.333		
0.750		
1.459		
2.015		

Notice that no calculations are made to add allowances; it is all in the 18.5 hours. What would the hours per 1,000 be with 15% allowances?

Method 2: Constant Allowance Added to Total Normal Time Method 2 is used in this text and is the most common technique used in industry. Each department or plant has only one allowance rate. The average allowance is between 10% and 15%. An explanation of what makes up the allowance, like the one below, must be included:

$$\text{Personal time} = 24 \text{ min}$$
$$\text{2 breaks at 10 min} = 20 \text{ min}$$
$$\text{Clean-up time} = \underline{\ 4 \text{ min}}$$
$$\text{Total allowances} = 48 \text{ min}$$

$$\frac{48 \text{ min}}{480 - 48 \text{ min}} = 11\%$$

Eleven percent is added to normal time to get standard time, or 111% times normal time equals standard time.

$$1.000 + .11 \quad = 1.110 \text{ min}$$
$$1.000 \times 111\% = 1.110 \text{ min}$$

The time study form will tell you which calculation to make.

Method 3: Elemental Allowances Technique

The theory behind this technique is that each element of a job could have different allowances, as in the following example:

ELEMENT DESCRIPTION	UNIT NORMAL TIME	ALLOWANCE	STANDARD TIME
1. Load machine	.250	15%	.288
2. Machine time	.400	5	.420
3. Unload machine	.175	10	.193

Note that each element allowance is different. Element 1 is operator controlled, and a heavy part is involved. Therefore, more allowances were included. Element 2 is a machine element, and the operator just stands there—no fatigue was given. Element 3 is a normal 5% fatigue, plus 5% personal allowance.

The obvious advantage of this method is improved elemental time standards. The disadvantage is the increased math effort required. The time study form would have to be redesigned to accommodate this method and, as with all allowances, the form would show you which technique to use.

Method 4: The PF&D Elemental Allowance Technique

As in Method 3, the allowance is placed on each element, and this method shows everyone exactly how the allowance was developed. This technique is the most complete of all the techniques.

EXAMPLE:

ELEMENT DESCRIPTION	UNIT NORMAL TIME	ALLOWANCES %				STANDARD TIME
		P	F	D	TOTAL	
1. Load machine	.250	5	10	0	15	.288
2. Run machine	.400	0	0	5	5	.420
3. Unload machine	.175	5	5	0	10	.193

This allowance technique takes a lot of time and effort. It is very descriptive, but the cost is too high for most companies.

Allowances are an important part of the time standard, and properly established allowances will assist in the continued improvement of the quality of work life. If a job has undesirable aspects that do not reflect on the individual cycle, the allowances must reflect this undesirability. In that way, the money exists to justify a needed change. A plantwide base rate of 10% is still very desirable, but additional allowances can be added as needed. The forms used in this text allow for a range of allowances.

Allowances are negotiated, while foreign elements are time standards for the same thing. Time study is best, but allowances are quicker to set.

FOREIGN ELEMENTS

Foreign elements are any elements of work not planned for by the time study technologist. Foreign elements may be absolutely necessary, but they don't occur every cycle and they may not be known when the time standard is set. There are two basic types of foreign elements:

1. Productive
2. Nonproductive.

Productive Foreign Elements

Productive foreign elements are necessary jobs that must be performed or the operation halts. Some examples are

1. Cleaning the chips or slugs out of a machine,
2. Loading parts into a feeder,
3. Moving finished material out and new material into the work station,
4. Changing tools,
5. Loading welding rod coil into welder.

These examples can be considered unavoidable delays, and we calculate their effect by time-studying them. Table 9-5 shows an example of productive foreign elements.

Table 9-5

ELEMENT DESCRIPTION	DESCRIPTIONS				
	1	2	3	4	5
1. Load machine	.16	.83	1.50	3.17	4.15
2. Machine time	.56	1.23	2.90	3.57	4.55
3. Unload	.66	1.33	3.01	3.99	4.64
4. Clean machine			2.40		
5. Inspect parts				3.89	

Extend this study. Subtract the previous time from each time as always, but be careful. When did we clean the machine? Between 1.50 and 2.40, or .90 minute cleaning time. When did we inspect a part? On the fourth cycle, between 3.57 and 3.89, or .32 minute inspecting time. Once the element cycle time is calculated, the frequency must be determined. How many parts can we run before the machine needs cleaning? This is discussed with the operator. How many parts are run before we check a part? Quality control procedures tell us this.

$$\text{Clean machine} = \frac{.90 \text{ min}}{300 \text{ parts}} = .003 \text{ min/part}$$

$$\text{Inspect} = \frac{.32 \text{ min}}{10 \text{ parts}} = .032 \text{ min/part}$$

These times are added to the total normal time just like any other element of work.

One time study may not give the technologist enough information to set a good quality time standard for these foreign elements. With enough time studies, however, these foreign element times will eventually start making sense.

Nonproductive Foreign Elements

Nonproductive foreign elements are eliminated from the time study. A nonproductive foreign element is a goof that should not be part of the operation, such as

1. Dropping a part or fumbling,
2. Stopping to talk to the time study technician,
3. Tying a shoe,
4. Time study technologist not getting a good reading.

The real question is, Should this be a part of the standard? The continuous time study requires the technician to record everything that happens during the study, but everything does not have to be included in the time standard. When something unusual happens during a time study, the technician places an asterisk (*) next to the ending point and describes what happened in the foreign element block ㉗. Once the time study is extended, the foreign element being discarded is circled. By circling the bad element time, it is highlighted but not obliterated.

One more point regarding foreign elements: If the operator performs a productive foreign element but the time study technician doesn't realize it in time to record the ending point of the previous element, it is not a problem. The technician enters the next normal ending point, marks the reading with an asterisk, and discards that reading because it contains work time and the foreign element time. Our objective is to set a good time standard for a job, so throw out any element of work that detracts from this goal. An explanation in the foreign element block ㉗ is important.

LONG CYCLE TIME STUDY

The long cycle time study worksheet is used for the following:

1. Long cycle time—15 minutes and longer
2. Inconsistent element sequence
3. 8-hour performance studies.

On long cycle jobs, many foreign elements tend to be a part of the study, and the sequence is not the same every time. These problems can create poor results from the continuous and snapback form.

Figures 9-23 and 9-24 show two parts of an 8-hour time study. The data was collected on eight pages of the long cycle time study worksheet and summarized on the graph. This example study was taken on a quart oil canning line. The purpose of the study was to determine and eliminate the reasons for stopping the line. In one hour, 24,000 quart cans could be filled and packed—24 per case. One thousand cases per hour was the potential of the line, but 3,500 cases per 8-hour shift was the average output. The 8-hour time study showed 72 work stoppages in one 8-hour shift. Each work stoppage was recorded and timed, and the

FRED MEYERS & ASSOCIATES LONG CYCLE TIME STUDY WORK SHEET					
PART NO. Quart Line	OPERATION DESCRIPTION: 4 people (loader, Machine,				
OPERATION NO. Line	Cartons, Unloader) Automatic Quart Line Cann				
DATE/TIME 10/10/xx	MACHINE; TOOLS, JIGS: #1--300 CANS/Minute				
BY I.E. Meyers	MATERIAL: Motor oil any weight				

ELEMENT I	ELEMENT DESCRIPTION Started 7:00 AM Ended 3:30 PM	ENDING WATCH READING	ELEMENT TIME	% R	NORM. TIME
1	Shift Start up--No production	7:05	5.0	100	5.0
2	Run	7:06	1.0'	100	1.0
3	Stopped--operator forgot something	7:07½	1½	0	----
4	Run	7:14	6½	100	6.5
5	No Lids	7:16½	2½	0	----
6	Run	7:19	2½	100	2.5
7	Check Temperature	7:20½	1½	70	1.05
8	Run	24	3½	100	3.5
9	Box Jammed in Former	25	1	110	1.1
10	Run	28½	3½	100	3.25
11	Bad Can	29	3/4	120	.9
12	Palletizer Jam-Bad Pallet	31	2	110	2.2
13	Run	33	2	100	2
14	Bad Box in Former	7:33½	½	130	.65
15	Run	7:41	7½	100	7.5
16	Bad Box in Former	7:41½	½	140	.7
17	Run	7:51	9½	100	9.5
18	Bad Box in Former	7:54	3	120	3.6
19	Run	7:56	2	100	2.0
20	No Lids in Machine	8:00½	4½	0	----

FIGURE 9-23 Long cycle time study: page one of eight.

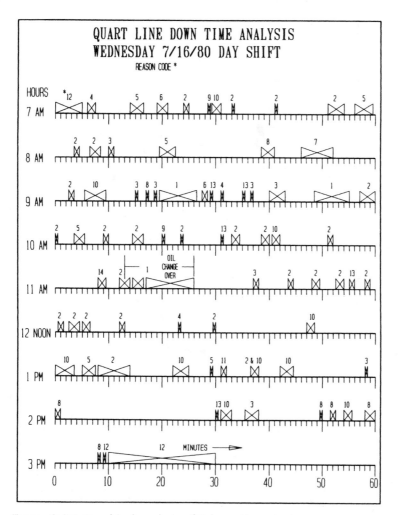

FIGURE 9-24 Graphical analysis of 8-hour time study.

reason for stoppage was recorded. Once the problems were identified and costed, economical solutions could be found and implemented. The resulting improvements are immediate and financially significant.

Step-by-Step Instructions for Preparing the Long Cycle Time Study Worksheet

See Figure 9-25 for an example of the worksheet.

① Part no.: The part number identifies the part being studied.

② Operation no.: The operation number identifies the specific operation on the part being studied.

| FRED MEYERS & ASSOCIATES | LONG CYCLE TIME STUDY WORK SHEET |

PART NO. ①	OPERATION DESCRIPTION: ⑤
OPERATION NO. ②	
DATE/TIME ③	MACHINE; TOOLS, JIGS: ⑥
BY I.E. ④	MATERIAL: ⑦

ELEMENT #	ELEMENT DESCRIPTION	ENDING WATCH READING	ELEMENT TIME	% R	NORM. TIME
⑧	⑨	⑩	⑪	⑫	⑬
					⑭
					⑮
					⑯

FIGURE 9-25 Long cycle time study worksheet: the step-by-step form.

③ Date/time: The date of the study and the time of day the study was started is written here.

④ By I.E.: The name of the time study technician doing the study goes here.

⑤ Operation description: A detailed description of the operation being performed is needed to communicate to future generations of operators, managers, and technicians what was being done when the standard was set.

⑥ Machine: Tools/jigs. Any machine, tool, or jig is written here.

⑦ Material: The material specification may have an effect on machine speeds and feeds, so this information is important. Blueprints of parts are desirable and can be attached to the study.

⑧ Element #: The element number is just a sequential number and is used for reference.

⑨ Elemental description: The elemental description in long cycle time study may be lengthy, but it is important to know what is included in the time standard. Be as descriptive as possible and include sizes, number of parts, weight, and any unit of measure that may be the reason for time to vary. A good rule is, "You can't have too much information."

⑩ Ending watch reading: As in all time study, the ending time is the only time to be recorded. The continuous time study technique is used.

⑪ Element time: Subtract the previous time from each watch reading to calculate element time. This is the time it took to perform this element of work.

⑫ % R: Percent rating is the technologist's opinion of the speed and/or tempo of the operator. The speed of the operator can change the time required significantly. We must always rate the operator on each and every element of work.

⑬ Normal time:

$$\frac{\text{Element time} \times \% \text{ R}}{100} = \text{normal time}$$

This is the time it should take an average person working at a comfortable pace.

⑭ Total normal time: Even though this block on the form is not labeled, this is where it should go. The reason that total normal time does not appear at the bottom of the form is that many pages of long cycle time study worksheets are involved, and the last page is the only one that needs the total normal time block. This is the same location and math used on the previous time study and PTS forms.

⑮ Standard time: Normal time plus allowances equals standard time.

⑯ Hours per unit: Standard time divided by 60 minutes equals hours per unit.

⑰ Pieces per hour: Pieces per hour is the $1/x$ of hours per unit, or divide the hours per unit into 1.

VERTICAL TIME STUDY FORM AND PROBLEM

The procedure and results obtained from the use of the vertical time study form are the same as for the horizontal time study form. The big difference in the vertical form is that the elements are listed across the top and the cycles are listed down the page, whereas the elements were listed down the page and the cycles were listed across the top of the horizontal form. You will notice that the extensions are across the bottom instead of on the right side of the

Element \ Cycles	Load machine	Run	Unload	Inspect	Material Handling		
1	.15	.55	.80				
2	.94	1.34	1.61				
3	1.77	2.17	3.52		3.27		
4	3.67	4.07	5.24	5.00			
5	5.39	5.79	6.50*				
Total time							
Number				10	1000		
Average time							
Rating %	120	100	80	130	60		
Normal time							
Allowances (add 10%)							
Standard time							
					Total standard time =		

*Eliminate.

FIGURE 9-26 Vertical time study form and problem.

page. Those who use this form say that it is much more logical, but your company will decide which form you will use, and you should know how to use both the vertical and horizontal forms. This text has concentrated on the horizontal form because it is the most widely used. The vertical time study form and problem are included here to give you the confidence that you can use any form your company throws at you.

The example form shown in Figure 9-26 shows a time study where the data have been collected but the extension needs to be completed. Inspection is to be done every tenth piece, and there are 1,000 pieces in each tub of parts. Please extend the study and make the total standard time equal to .994 minute. Standard times by element are 1—.198, 2—.440, 3—.222, 4—.133, and 5—.001. The pieces per hour will equal 60. A completed study can be found at the back of the book in the answers to Chapter 9 questions.

TIME STUDY PRACTICES AND EMPLOYEE RELATIONS

Time study practices and employee relations can be summarized in a list of practices and attitudes that have been developed over the years to promote and improve the time study technician's image and improve results.

1. Always subtract the previous watch reading from each watch reading using a red pen. The elemental time is more important than the watch reading, and extending the study in red will highlight the elemental time. Only the subtraction needs to be in red.

2. Always stand up while taking a time study. Sitting or slouching presents a lazy, nonproductive attitude. The time study technician is a leader in productivity improvement, and presenting the right attitude is important.

3. Talk to the operator. The operator is the expert on the job you are time-studying and therefore is the source of much information you need. Asking questions shows respect, and respect is what we all want. There is no room in our profession for time study technicians who do not talk with people. They may be shy, but employees see them as being stuck up.

4. Be positive about your time standards. If you don't believe in your time standards, who will? If you allow doubt about your time standards, they will not be as useful. You need to take every opportunity to sell the accuracy of your standards and defend them quickly. If needed, restudy the challenged job immediately. There can be no excuse for not achieving the time standard.

5. Get the supervisor's permission to enter his or her area. The supervisor's area is like a home, and you wouldn't enter someone's home without permission. The employee works for the supervisor, not the time study technician. If anything needs changing, have the supervisor give instructions.

6. Try to put the operator at ease. Tell the operator what you are going to do and why. The operator may have a bad attitude about time study, and it is your job to sell yourself as an honest person.

7. *Be* an honest person. The objective of a good time study technician is to set fair and equitable time standards. Nothing should interfere with this goal. Tell operators that you are honest and wouldn't knowingly do anything to hurt them.

8. Be a friendly and happy person. The negative attitude of an employee cannot affect your attitude. Make it a challenge to win over even the most obnoxious employee. This person was probably hurt by a time study in the past, but you can correct those errors.

9. Stand in a position where you can see what the operator is doing, but out of the way of moving machines or flying parts. Never hide while doing a time study. Hiding is not professional.

10. Never change a time standard without good reason, and if a change is made,

   a. It must be over 5%.

   b. It must be communicated to the employees—the reason for the change, and the amount of change.

This is an ever-growing list, and if you think of something that should be here, please share it.

QUESTIONS

1. Stopwatch time study can be learned only by doing. Identify and time-study two production operations using everything learned in this chapter.
2. Why is stopwatch time study the most popular method of setting time standards?
3. Why is time study a good entry-level job for an industrial engineer?
4. What are the six different stopwatches, and what are the benefits of each? Which one will we use and why?
5. What are the uses of a video camera in motion and time study?
6. What are the 10 steps of the time study procedure?
7. Who should we time-study, and who should we not time-study?
8. What should the technologist check before starting a time study and why?
9. What are the eight principles of elemental breakdown?
10. What are the four reasons for breaking down a job into elements?
11. How is the frequency column ㉑ used? (Refer to Figure 9-13.)
12. Determine the number of cycles needed for every element of your time studies in Question 1.
13. What are rating, leveling, and normalizing?
14. What are allowances?
15. What do we mean by "check for logic," and why is it important to do so?

16. What are the four factors of rating, and how important is each?

17. What are some of the standards used to define 100% performance?

18. Check your performance calculations on page 191 with Table 9-4.

19. What is the rater trainer form?

20. What are the three categories of allowances? Give an example of each, and a typical percentage allowance for each.

21. How does excessive weight affect the allowance?

22. How do we apply allowances?

23. What is a foreign element?

24. How do we handle foreign elements?

25. Time-study two projects. Show everything that is required in a proper time study. Use the continuous time study method.

26. Complete the time study in Figures 9-17 and 9-26.

Standard Data and Its Uses in Balancing Work

INTRODUCTION

Standard data is a common term given to any collection of time values. There are a number of methods of making the data easily available. Graphs, tables, formulas, and worksheets are explained in this chapter. The formatted data can then be utilized in conjunction with calculated cycle times. Examples are given here for the development of standard data for the special case of machine speeds and feeds for various shop machines. The formatted data is used for calculating performance and may be applied in the comparison of one task against others for the equitable distribution of work. Such calculations are commonly called *balancing* and contain many of the concepts required for operations analysis. Formatted data is also usable for determining the operating parameters for a plant providing a lean manufacturing environment. It is equally important that the technologist be able to apply optimization techniques to the overall performance of the system as well as the optimization of processes and operations.

Formatted standard data not only reduces the amount of effort the time study analyst must spend collecting data but it also provides a data set for the calculation of alternative optimization approaches. This could not be done on the strength of individual work measurement studies.

DEFINITION

Standard data is a catalog of elemental time standards developed from a database collected over years of motion and time study. Machine names or numbers and job descriptions organize the catalog of time standards. When a new part is designed and the fabrication steps have been identified, the time study person looks up the machine in the catalog. That machine page tells the time study person what causes the time to vary, so he or she can take measurements from the blueprint of the new part and find the time for the new job.

Each job may have several elements, so several elemental times are developed for each job. One of the main reasons for dividing a job into elements when studying it, as stated in Chapter 8, is to develop standard data. Each element has a different reason for its time to vary. Some elements are constant and their times do not vary at all; some are variable and their times will vary with some measure of size or weight. Some elements are machine controlled and do not require ratings; some are operator controlled, which require rating. Therefore, standard data times will be more accurate when they are broken into elements, and the more elements, the more accurate the standard data will be.

Standard data times are for full elements, whereas PTS consists of time data for smaller divisions of work—basic motions. It can take up to 30 minutes to develop an elemental time using PTS or time study, while the same standard can be set in 2 minutes using standard data. PTS and time study are used to develop standard data.

A distinction must be made between setting standard data and applying standard data. Setting standard data is the investigation of what causes time to vary on a job and developing a communications technique for all future generations of time study people to understand. This is very high-level time study work, whereas any trained clerk can apply the data. Our goal is to set time standards as accurately and quickly as possible.

ADVANTAGES

The advantages of standard data time standards include the following:

1. Standard data standards are chosen from a book of standards. It takes about 2 minutes to select a standard, compared to about 30 minutes per standard using PTS or stopwatch time standards techniques.

2. Standard data time standards are more accurate than other time standard techniques because individual time studies are compared to all other time studies of the same machine or work center, and differences are averaged to make all standards uniform.

3. Consistency is another way of saying that the standard is fair. Standard data time standards are more consistent than other time standard techniques because individual differences between jobs are smoothed out in curves, formulas, or graphs. During time studies, rating and watch readings may vary, creating small errors ($\pm 5\%$), but these small errors create what the employees call good jobs or bad jobs because the time standards are easy to achieve or difficult to achieve, respectively. Rating and watch reading are eliminated. Standard data smoothes these differences.

4. Time standards can be set before production starts. When engineering and marketing departments develop a new model, manufacturing is asked to cost the product before production can be approved. This costing requires a detailed manufacturing plan as a basis for costing. It includes work station design, material flow, and time standards. How else would we know how many machines to buy or people to hire?

5. Time standards for short-duration jobs can be set economically. Because we can set time standards so fast, we can afford to set time standards for jobs that were previously too small to be covered economically by time standards. A job that occurs only once a month and lasts only an hour is an example of this. One hundred percent time standards coverage is also a measure of industrial engineering performance, so we want to cover as many jobs as possible.

6. Standard data reduces the need for time study. Stopwatch time study is costly and can create some employee relations problems. Standard data can minimize both these disadvantages.

7. Time standards are easier to explain and adjust if needed. Employees probably know what makes the time vary between jobs better than an industrial engineer does. When standard data time standards are set, the new standard can be explained easily to the operator, because it is similar to some other job the station has run.

8. The cost of time standard application is greatly reduced. Standard data time standards are easy, fast, and accurate, which results in a lower cost for time standards.

9. Adjustments can be made quickly, if needed. A graph can be raised or lowered a percentage and change every job on that machine. A standard data element may be tight or loose, but all jobs will be tight or loose and can be corrected by a minor change in the slope of a line or an adjustment in a formula.

METHODS OF COMMUNICATING STANDARD DATA TIME STANDARDS

Time standards can be presented in many forms, and this is the main subject of this chapter. The development of standard data is the ultimate job in motion and time study, and standard data development must be the goal of every industrial engineering department. Standard data development is much like detective work. The task is to find what causes time to vary. The better the job the technologist did in breaking the job into elements during the time study phase, the easier will be the standard data development.

The following methods of communicating standard data are discussed in this chapter:

Graphs

Tables

Formulas

Worksheets

Machine speeds and feeds.

Graphs

Graphs and a worksheet are shown in Chapter 4, Figure 4-8. Please review this example before continuing here.

The time study shown in Figure 10-1 is a classroom example of how a standard data graph is created. The first step is to time study (or PTS) enough different jobs at a work station to determine what causes time to vary. The time study in Figure 10-1 has three elements of walking (of different distances) and three elements of counting cards (of different amounts). Figure 10-2 shows how the three data points look on a graph. Notice the perfect straight line. This straight line is not normal because of small errors in rating or watch readings. For this reason, many time studies (data points) should be made: the more time studies, the better the accuracy.

Once the graphs are produced, the technologist can go to the graphs and look up the number of paces or cards the next time a time standard is needed for a new walking element or counting cards element. He or she moves the eye up the graph to the standard line, and then across horizontally to the time line. That number is the normal time for the new job. Never again will we have to time study counting cards or walking, unless the new job exceeds the limits of the graphs.

FRED MEYERS & ASSOCIATES TIME STUDY WORKSHEET ☐ SNAP BACK ☒ CONTINUOUS

OPERATION DESCRIPTION: Walking (w) in paces (p) & Dealing Cards (c)

PART NUMBER	OPERATION NO.	DRAWING NO.	MACHINE NAME	MACHINE NUMBER	
0052	05	0052-005	None	None	☒ QUALITY OK ?
OPERATOR NAME	**MONTHS ON JOB**	**DEPARTMENT**	**TOOL NUMBER**	**FEEDS & SPEEDS.**	☒ SAFETY CHECKED ?
Fred	60	Technology	None	MACHINE CYCLE TIME: NONE	☒ SETUP PROPER ?
PART DESCRIPTION: Standard play cards		**MATERIAL SPECIFICATIONS:** Plastic coated			NOTES:

ELEMENT #	ELEMENT DESCRIPTION		1	2	3	4	5	6	7	8	9	10	TOTAL CYCLES	AVERAGE TIME	% R	NORMAL TIME	FREQUENCY	UNIT NORMAL TIME	RANGE	R/X̄	HIGHEST ✓
1	W5P	R	05	64	19	99	54	12	89	47	5.03	60	.41				1				
		E	.05	.04	.04	.05	04	05	25	05	04	05	9	.046	105	.048	1	.048	.01	.217	
2	C15C	R	18	77	33	2.13	68	26	4.03	61	17	74	1.38				1				
		E	.13	.13	14	.14	.14	.14	.14	.14	.14	.14	10	.138	115	.159	1	.159	.02	.145	
3	W10P	R	27	87	41	22	78	34	12	70	26	82	.89				1				
		E	09	.10	.08	.09	.10	.08	.09	.09	.09	.08	10	.089	105	.093	1	.093	.02	.225	
4	C10C	R	38	96	74	32	88	44	21	80	36	93	.90				1				
		E	.11	.09	.33	.10	.10	.10	.09	.10	.10	.11	9	.100	115	.115	1	.115	.02	.200	
5	W15P	R	52	1.09	89	46	302	59	35	93	50	6.07	1.40				1				
		E	.14	.13	.15	.14	.14	.15	.14	.13	.14	.14	10	.140	100	.140	1	.140	.02	.143	
6	C5C	R	58	15	94	50	7	64	42	99	55	14	.58				1				
		E	.06	.06	.05	.06	.05	.05	.07	.06	.05	.07	10	.058	120	.070	1	.070	.02	.286	✓
		R																			
		E																			
		R																			
		E																			

FOREIGN ELEMENTS:
*Stop to talk
*Dropped deck

NOTES: This time study to be used for standard data.

R/X̄	# CYCLES
.1	2
.2	7
.3	15
.4	27
.5	42
.6	61
.7	83
.8	108
.9	138
1.0	169

(circled: .2–.3, 14 cycles)

TOTAL NORMAL MIN.	.625
ALLOWANCE + 11 %	.069
STANDARD MINUTES	.694
HOURS PER UNIT	0 11 5 7
UNITS PER HOUR	86

ON BACK
WORK STATION LAYOUT
PRODUCT SKETCH

ENGINEER: Meyers DATE: 2 / 8 / xx
APPROVED BY: DATE: 2 / 9 / xx

FIGURE 10-1 Standard data time study: three walking elements and three of counting cards.

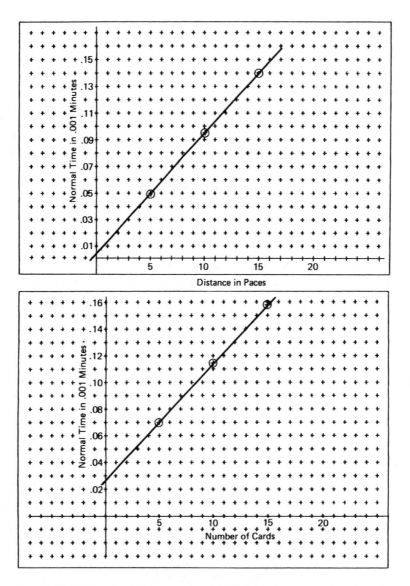

FIGURE 10-2 Standard data graphs: (a) Walking in paces, developed from Figure 10-1. (b) Counting cards, developed from Figure 10-1.

Tables

Whatever can be communicated with graphs can be converted into tables. When untrained people are required to use standard data, the table is the method of choice because it is easy to use. Look at the graphs in Figure 10-2 and create a table for dealing cards and walking, as follows:

NUMBER OF PACES OR CARDS	WALK IN PACES	DEAL CARDS
5	.050	.070
10	.095	.115
15	.140	.160

It is important not to expand the table beyond the data collected, but walking and dealing cards lend themselves to some logic. We can expand the table for walking to include every pace because, once we get started, the time per pace will be constant except for the effects of fatigue, which is accounted for by providing allowances. (If you remember from the PTS system in Chapter 8, the predetermined time for walking one pace was .010 minute.) A handful of cards, or at least a deck of cards, can be dealt before we need to pick up a new deck of cards, so we can expand the card section of the table to at least 52 cards.

The table of card counting and walking in paces is used in the next section (Formulas). At a minimum, three elemental times are needed to develop formulas, and the two elements of counting cards and walking in paces are used every day.

Tables should be used when numerous jobs are included under one heading but have nothing in common. Material handling is an example of this. Table 10-1 is an example of material-handling standard data. Tables are used in the automobile dealership service departments for estimating work on cars. This time standards book is called the flat-rate manual. *Means Construction Guide* is another catalog of facts, this one used by the construction industry. You can look up each part listed on the building blueprints and go to the *Means Construction Guide* to learn how many hours per unit each part takes. Figure 10-3 is a page from the *Means Construction Guide*. When you put all these times together, you have the labor hours by craft. Cost estimating for construction is much more accurate because of this catalog of time standards. Every industrial engineering department should be striving for the same time standards results as the construction industry.

Formulas

Formulas are the most efficient way of communicating standard data, especially if you use a computer. Many types of formulas are used in standard data: straight-line, curvilinear, and special formulas that we create. The straight-line formula is used most frequently. See Figure 10-4. The formula for a straight line is

$$y = a + bx$$

y = vertical axis, measures time. In the card-dealing example, this is measured in thousandths of a minute (.001 minute).

a = y intercept (where the line crosses the y axis). a is the time for getting started and finishing up. This time is required, even though no work is done. In our card-dealing example, it is the time for picking up the deck and placing the remaining cards back on the table.

b = slope of the line, or the time per unit of work. In our card-dealing example, this is the time per card dealt.

x = horizontal axis, or the units-of-work axis. In the card-dealing example, this is the number of cards.

Table 10-1 Standard Data.

SETUP AND MATERIAL HANDLING

TIME IN .001 NORMAL MINUTES

(THE FOLLOWING DATA ARE ONLY A FEW OF THE THOUSANDS AVAILABLE)

1. Punch in and out of job, get blueprint, process sheet, reset counter, receive instructions	2.700
2. Plus walk to and from crib, supervisor, and time clock	0.004/foot
3. Adjust adjustable gauge	2.340
4. Set up punch press	
15 min	
30 min	
60 min	
5. Pick up and put aside hand tool	
Small .060	
Large .078	
6. Loosen or tighten nuts or bolts	
With allen wrench	0.084
With open or box wrench 1/2–	0.132
1/2+	0.192
Finger tighten	0.078
7. Pick up and hand-assemble nut or bolt	
From setup table to machine	0.480
From machine bed	0.258
8. Remove and put aside nut and bolt	
To setup table	0.318
To machine bed	0.204
9. Pick up or put aside punches, dies, gauges, etc.	
To or from setup table	0.192
To or from rack on machine	0.234
To or from storage rack	1.320
10. Move tub load of material out of a work station, move empty tub to other side,	
move new tub of material in (fork truck, hand truck, or hand jack)	1.000
11. Move box of parts into a work station by hand	0.500
12. Pack parts in box	
1 at a time—large part	0.150
2 at a time—medium parts	0.050/part
Per additional part from same tub	0.010/part
13. Form carton and	0.150
Tape—manual	0.100
Staple—4 staples	0.050
Stitch—3 places	0.075
14. Pick up loaded pallet (lifting time only)	
Manual—mechanical	0.126
Manual—hydraulic	0.402
Electric	0.162
15. Position truck to pick up pallets	
Manual	0.258
Electric	0.090
16. Move load	
Manual	0.003/foot
Electric	0.004/foot
17. Set down loaded skid	
Manual—mechanical	0.084
Manual—hydraulic	0.192
Electric	0.066

EXTERIOR FINISH CARPENTRY	
TYPE	LABOR HOURS
Siding, narrow lap on bevel	3.0 per 100 square feet
Siding, wide lap on bevel	2.5 per 100 square feet
Siding, drop	2.25 per 100 square feet
Shingles, on walls	5.0 per 100 square feet
Building paper on walls	0.8 per 100 square feet
Cornice	6.0 to 12.0 per 100 linear feet
Mold	6.0 per 100 linear feet
Water table	5.0 per 100 linear feet
Corner board	4.0 per 100 linear feet
Shutters	1.5 per pair
Porch columns	1.5 each
Porch rail	50 per 100 linear feet
Porch steps, set of 4	10 per set
Door, single exterior	6.0 complete
Door, single exterior, frame	2.0 each
Door, single exterior, trim	1.0 each
Door, single exterior, hang	1.5 each
Door, single exterior, hardware	1.5 each
Door, double exterior	9.5 complete
Door, double exterior, frame	3.0 each
Door, double exterior, trim	1.5 each
Door, double exterior, hang	2.5 each
Door, double exterior, hardware	2.5 each
Window, single	5.0 complete
Frame	1.0 each
Trim	1.0 each
Hang, 2 sash	1.5 each
Hardware	1.5 each
Window, double	7.5 complete
Frame	1.5 each
Trim	1.5 each
Hang, 4 sash	2.5 each
Hardware	2.0 each
Window, triple	10.0 complete
Frame	2.5 each
Trim	2.0 each
Hang, 6 sash	3.0 each
Hardware	2.5 each
Screen, doors	2.5 each
Windows	1.0 each
Allow for Scaffolds Where Required.	

FIGURE 10-3 *Means Construction Guide* (Carpentry Section).

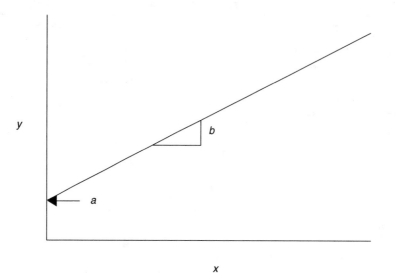

y

b

a

x

Figure 10-4 Straight-line graph components.

EXAMPLE:

How much time will it take to count out 25 cards?

$$x = 25 \text{ cards}$$
$$y = \text{unknown time}$$
$$a = .025 \text{ min}$$
$$b = .009 \text{ min}$$

Therefore,

$$y = .025 + .009(25)$$
$$y = .250 \text{ min}$$

The computer is loaded with an element number for counting cards, and the a and b time standards are assigned to that element. When a new standard is needed, you call up the element number and key in the quantity x, and the answer would appear immediately. Time standards for a whole new product can be set in a fraction of the time that any other system would require. The computer listing of standards would look something like this:

JOB NUMBER	JOB DESCRIPTION	UNIT OF WORK	a	b
0025	Walk	Pace	.002	.009
0067	Deal card	Card	.026	.009
0100	Shrink pack	Length (in inches)	.125	.080
etc.				

To use this data, a clerk keys in the job number and the quantity of work, and a time standard results. All other calculations are within the computer program.

Calculating the variables (a and b) is now necessary. Both a and b can be estimated from a graph, but formulas are much more accurate. Let's use the counting cards and walking examples. The graphs of counting cards and walking can be used to estimate a and b. (When we finish calculating them using formulas, please compare your estimate with the math answer.) Many good time standard data for a and b have been estimated from graphs.

The formula for fitting a line to a set of data is called *regression analysis*. The x and y coordinates are used to calculate the variables a and b. The x and y data come from time studies, Figure 10-1 in this example.

A table of counting cards and walking would look like this (again based on the time study):

WALKING (IN PACES)	NORMAL TIME (IN MINUTES)	COUNT CARDS	NORMAL TIME (IN MINUTES)
x	y	x	y
5	.048	5	.070
10	.093	10	.115
15	.140	15	.159

The regression line equations to solve for a and b are

$$a = \frac{(\Sigma x^2)(\Sigma y) - \Sigma x(\Sigma xy)}{N(\Sigma x^2) - (\Sigma x)^2}$$

$$b = \frac{N(\Sigma xy) - \Sigma x(\Sigma y)}{N(\Sigma x^2) - (\Sigma x)^2}$$

From the preceding table data, an extension of the data is needed.

Walking

n	x	y	x^2	xy
1	5	.048	25	.24
2	10	.093	100	.93
3	15	.140	225	2.10
Σ	30	.281	350	3.27

Substitution gives us the following:

$$a = \frac{350(.281) - 30(3.27)}{3(350) - (30)^2} = .002 \text{ min}$$

$$b = \frac{3(3.27) - 30(.281)}{3(350) - (30)^2} = .009 \text{ min}$$

Therefore, the y intercept is .002 minute, and the unit time per pace is .009 minute. The time required to walk 20 paces is $y = a + bx$. The a and b are now constants of .002 and .009, respectively. The a, or y intercept, is the acceleration/deceleration time, and b is the time per pace. So $y = .002 + .009(20) = .182$ minute to walk twenty paces.

Counting Cards

n	x	y	x^2	xy
1	5	.070	25	.350
2	10	.115	100	1.150
3	15	.159	225	2.385
Σ	30	.344	350	3.885

$$a = \frac{350(.344) - 30(3.885)}{3(350) - (30)^2} = \frac{120.4 - 116.55}{1050 - 900} = \frac{3.85}{150} = .026 \text{ min}$$

$$b = \frac{3(3.885) - 30(.344)}{3(350) - (30)^2} = \frac{11.655 - 10.32}{1050 - 900} = \frac{1.335}{150} = .009 \text{ min}$$

Therefore, the counting of cards starts with picking up the deck, which takes .026 minute. Putting down the deck of cards after dealing is included in this .026 minute. Dealing cards takes .009 minute per card. How much time would it take to count 25 cards?

$$y = a + bx$$
$$y = .026 + .009(25)$$
$$y = .251 \text{ minute to count 25 cards}$$

Isn't this easier than time study? Once the a and b have been calculated, the answer is yes.

The interchangeability of graphs, tables, and formulas should be evident. Any of the techniques can be used, but one technique will be better than another for a particular set of circumstances.

Worksheets

Even though worksheets come in many different forms, they are generally like lengthy formulas. The worksheet for one job may have many elements of work, both constant and variable. The worksheet has blanks to be filled in for the variables; the constants as well as the element standards are preprinted on the worksheet. The bottom of the worksheet is just like the PTS system form or time study form, because total normal time, allowances, hours per unit, and pieces per hour are calculated. The advantages of a worksheet are

1. Any clerk can set standards. We just fill in the blanks from blueprints or bills of materials.
2. Standards can be set before production starts.
3. Many variables can be handled.

4. Constant elements are preprinted on the form.

5. All the elements are listed, so we leave nothing out.

6. Each worksheet can be set up on a computer spreadsheet. With only a few changes to an old time standard, a new time standard is set. We can also place many similar job standards on the same page. (This will help factory people understand what causes time to vary.)

The worksheet standard data system is easy to use and is a valuable technique.

Figure 10-5 is a standard data worksheet used by a swing set manufacturer to set time standards for hardware bags. Every swing set required a different collection of hardware, but there were many similarities. An experienced clerk filled out this form in less than five minutes, and the company knew the labor cost of the new hardware bag. This worksheet can be completed as a homework assignment. The bold numbers are from the parts list. This plant produces swing sets and gas grills. Over 100 different bags of parts are designed every year. One of their customers, Sears, requires the company to set its cost for the entire year before production starts. The price is included in the catalog, and changes in price are not allowed. How important is it to have a good time standard? Making a $1.00 error in cost estimating

---

Model number **1660**   Set name **GAS GRILL**

No. of different parts **15**   Date **2-16-xx**

No. of individual parts **105**   Name of technician _____

No. of chains ___—___   Instruction book no. **1660-1200**

Bag part number **1660-1550**   **1660-1250**

Element #—Operation description   Normal time in minutes

1. Get bag, open bag, place on funnel   0.52

2. Pack parts

    A. # of different parts   **15** × .018 = _____

    B. # of individual parts:

        Small parts ⅛" or less thick   **50** × .011 = _____

        Medium parts ⅛" to ¾" thick   **30** × .008 = _____

        <u>Large parts ¾" and over</u>   **25** × .005 =

    C. # chains   ___ × .075 = _____

    D. # pages or booklets   **2** × .025 = _____

3. Remove bag, fold, staple, and set aside   .110

4. Total normal time   _____

5. Plus 10% allowance (NT × 1.10)   _____

6. Hours per bag = line 5 ÷ 60 minutes/hour   _____

7. Bags per hour 1/x of 6   * _____

*Make the answer come out to 39 bags per hour.

FIGURE 10-5 Standard data worksheet—parts-bagging operation.

is just like including a dollar bill in the package. Making a $1.00 error in labor is like packing $3.00 in the box because of overhead costs.

Machine Speeds and Feeds

Constant Cycle Time Some machines (for example, mechanical punch presses) have constant cycle times. A time study person holds down the cycle trip button and times 10 cycles to get an average time. This time is recorded on a machine cycle time sheet for future reference. The cycle time can also be a constant part of a worksheet. It is used in PTS. Constant cycle time is the easiest type of time standard to set and use.

Constant Feed Rates Other machines (for example, those used in welding) have constant feed rates. Table 10-2, provided by a manufacturer of welding equipment, tells the standard setter the feed rate.

If we need a time standard for an 18″ long weld of the type shown in Figure 10-6, we go to this page in the standard data book (see Table 10-2) and look up the weld size, which is specified by product engineering. Let's look up a $\frac{1}{4}$″ fillet weld. The middle of the chart shows 24″ per minute, or .042 minute per inch, so .042 times 18″ = .756 minute (18″ ÷ 24″ per minute = .750 minute). The small difference in times is due to rounding, but both are correct because both are just for welding time. Time spent moving to and from the weld and any material-handling time must be developed by the PTS technique.

Variable Feeds and Speeds The speeds and feeds of chip cutting machines, such as lathes, drills, and mills, depend on

1. The material being cut,
2. The type of tool being used.

Table 10-2 Standard Data for a Horizontal $\frac{3}{32}$″ Electrode Fillet Weld Using Carbon Dioxide Gas Metal Arc Welding Process and a Composite Electrode of Steel (One of 75 Pages).

SIZE WELD	NUMBER OF PASSES	ELECTRODE SIZE	FEED (IN./MIN)	DEPOSITION (FT)	WIRE #/ FOOT OF WELD	VOLTS	AMPS	TRAVEL (IN./MIN)	MINUTES PER INCH OF WELD
$\frac{1}{8}$	1	$\frac{3}{32}$	120	.031	.034	24–36	350	60	.017
$\frac{3}{16}$	1	$\frac{3}{32}$	120	.060	.075	24–36	400	36	.028
$\frac{1}{4}$	1	$\frac{3}{32}$	120	.088	.103	24–36	400	24	.042
$\frac{3}{8}$	1	$\frac{3}{32}$	120	.204	.225	24–36	475	15	.067
$\frac{1}{2}$	3	$\frac{3}{32}$	120	.345	.399	24–36	400	20	.15
$\frac{5}{8}$	3	$\frac{3}{32}$	120	.562	.593	24–36	450	14	.213
$\frac{3}{4}$	6	$\frac{3}{32}$	120	.834	.966	24–36	400	16	.375

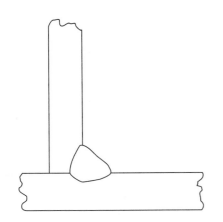 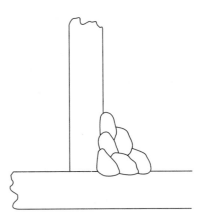

FIGURE 10-6 Examples of fillet welds.

Speeds and feeds for all of the aforementioned machines are scientifically determined to minimize cost, and the source of speeds and feeds information is the *Machinery's Handbook,* published by Industrial Press (New York, NY). Other sources for speeds and feeds information are tool and machinery manufacturers. Some manufacturers provide technologists with slide rules and plastic-laminated speeds and feeds tables. Examples are provided in Tables 10-3 and 10-4.

Speeds and feeds are the basis of machine run time. The time standard includes the machine run time and the load/unload time, which is controlled by the operator. A time study would be needed to determine the load/unload time, and after enough time studies have been made, standard data tables could be developed.

Table 10-3 Cutting Speeds and Feeds for Common Metals.

WORK MATERIAL	HARDNESS BHN	HIGH SPEED STEEL (FEET PER MIN)	CEMENTED CARBIDE (FEET PER MIN)	FEED/INSERT (INCHES/REV.)
Alumium	60–100	300–800	1000–2000	.010–.030
Brass	120–220	200–400	500–800	.010–.040
Bronze				
Hard drawn	220–	65–130	200–400	.010–.050
Gray cast iron				
ASTM 20	110–	50–80	250–350	.015–.050
Low carbon steel	220–	60–100	300–600	.010–.025
Medium carbon				
Steel alloy 4140	229–	50–80	225–400	.005–.012
High carbon steel	240–	40–70	150–250	.003–.010
Aisil 8620 steel	200–250	40–70	150–350	.010–.030
Stainless steel	120–200	30–80	100–300	.002–.004

Table 10-4 Feed Chart—High Speed Steel Drills (Average Conditions; Feed in Inches Per Revolution).

	DIAMETER OF DRILL							
	$\frac{1}{8}$	$\frac{1}{4}$	$\frac{3}{8}$	$\frac{1}{2}$	$\frac{3}{4}$	1	$1\frac{1}{4}$	$1\frac{1}{2}$
Aluminum and alloys	.003	.005	.006	.008	.010	.013	.015	.016
Brass and bronze (soft)	.004	.007	.010	.014	.018	.022	.026	.030
Cast iron	.004	.006	.009	.012	.016	.020	.025	.028
Steel (low carbon)	.003	.005	.007	.010	.014	.016	.022	.025
Steel (medium carbon)	.003	.005	.006	.010	.014	.016	.022	.025
Steel (alloy)	.003	.004	.005	.006	.010	.012	.014	.016
Steel (tool)	.003	.004	.006	.009	.012	.014	.016	.020

The general rule for using drill feed is:

DIAMETER OF DRILL	FEED PER REVOLUTION
$\frac{1}{8}$ and less	.001 to .002
$\frac{1}{8}$ to $\frac{1}{4}$	.002 to .004
$\frac{1}{4}$ to $\frac{1}{2}$	.004 to .008
$\frac{1}{2}$ to 1	.008 to .015
1 and larger	.015 to .025

Speed rates are specified in feet per minute. If a speed rate of 500 feet per minute is called for, a point on the diameter of the tool or part must travel at that speed. For example, if a 2-inch-diameter bar was placed in a lathe, a spot on the diameter would need to be moved at a speed of 500 feet per minute. A 2-inch drill would also have to move a point on its outside diameter 500 feet per minute. A 2-inch-diameter part or tool is a little over 6 inches around (πD), and the machine would have to turn this 6 inches (half foot) 1,000 times per minute to produce 500 feet per minute. This is the logic of the RPM (revolutions per minute). The formula is as follows:

$$(1)\ RPM = \frac{\text{speed rate}}{\pi D}$$

Speed rate is in feet per minute. Multiply by 12 for inches per minute. Three formulas are required to calculate a time standard from feeds and speeds:

$$(1)\ RPM = \frac{\text{speed rate}}{\pi D}$$

$$(2)\ \text{Number of revolutions required} = \frac{\text{length or depth of cut}}{\text{feed rate}}$$

$$(3)\ \text{Time} = \frac{(2)\ \text{Number of revolutions}}{(1)\ \text{RPM}}$$

EXAMPLE: Two-inch part with a speed of 500 feet/min. Two inches comes from the blueprint, and 500 feet per minute comes from the *Machinery's Handbook*.

$$\text{RPM} = \frac{500 \times 12}{\pi(2)} = 955\ \text{RPM}$$

This is close to our logic example.

Feed rates are given in inches of advancement per revolution of the tool or part. A .002-inch feed rate means advance the tool two thousandths of an inch every time the tool turns one revolution. In our answer to the foregoing RPM formula, that tool would move .002 × 955 = 1.91 inches per minute. If we were to turn a part 2 inches back, or drill a hole 2 inches deep, about one minute would be required. Again, this is a logic approach, but for more accuracy and speed, the formula for number of revolutions is needed.

$$(2)\ \text{Number of revolutions needed} = \#\ \text{rev.}$$

$$(2)\ \#\ \text{rev.} = \frac{\text{length or depth of cut}}{\text{feed rate}}$$

$$(2)\ \#\ \text{rev.} = \frac{2''}{.002} = 1{,}000\ \text{revolutions}$$

The time required to turn 1,000 revolutions when the speed of the tool is 955 RPM is a little over one minute. The formula for this is

$$(3)\ \text{Time} = \frac{\#\ \text{rev.}(2)}{\text{RPM}(1)}$$

$$(3)\ \text{Time} = \frac{1{,}000\ \text{rev.}}{955\ \text{rev./min}} = 1.047\ \text{min}$$

Let's look at some specific examples.

LATHE:

How much time would be required to turn the part shown in Figure 10-7 on a lathe?

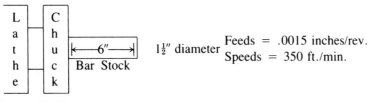

Feeds = .0015 inches/rev.
Speeds = 350 ft./min.

FIGURE 10-7

$$(1)\ \text{RPM} = \frac{350 \times 12}{\pi\ (1\frac{1}{2})} = 892\ \text{RPM}$$

Let's be realistic. This machine can run 800 or 900 RPM, but not 892, so we will use 900 RPM.

$$(2)\ \#\ \text{rev.} = \frac{6''}{.0015} = 4,000\ \text{rev.}$$

$$(3)\ \text{Time} = \frac{4,000}{900} = 4.444\ \text{min}$$

Load/unload time is still required, but 4.444 minutes is the machine cutting time. Depth of cut may require two passes.

DRILL:

Drills are a little more complicated because of the effect of the drill tip on the length of the cut. A drill tip is very close to .4 times the diameter of the drill. A $\frac{1}{2}$-inch drill would have a .2-inch drill tip (.4 × .5 = .2). How much time will it take to drill a $\frac{3}{8}$-inch hole through a 2-inch part with a speed rate of 500 ft/min and a .0025 feed rate? (See Figure 10-8.)

$$(1)\ \text{RPM} = \frac{500 \times 12}{\pi\ (\frac{3}{8})} = 5,096\ \text{RPM}$$

Rounding off for actual machine capability = 5,000 RPM.

$$(2)\ \#\ \text{rev.} = \frac{2 + .15}{.0025} = 860\ \text{rev.}$$

$$(3)\ \text{Time} = \frac{860}{5,000} = .172\ \text{min}$$

Again, the load/unload time must be determined by some other technique, because machine times are only cutting time, and the machine time is .172 minute.

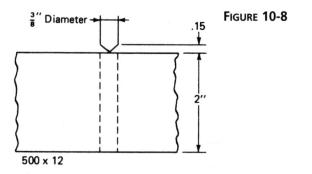

FIGURE 10-8

MILLS AND BROACHES:

Mills are the most difficult of the three basic chip-producing machines because of

1. Overtravel,
2. The number of teeth.

Regarding overtravel, the mill diameter can be quite large, and the diameter must be added to the length of the cut if the machine is cutting through the part. One half of the diameter would account for cutting through the part, but a tail cut often leaves an uneven surface under the trailing half of the cutter. If surface finish is important, the whole cutter must pass through the part.

A cutter could have 1, 4, 8, 16, or 32 teeth, and some cutters have 64. Each tooth can take a feed rate of its own.

EXAMPLE: How much time would it take to mill off the surface of an engine head made of cast iron with a speed rate of 225 and a feed rate of .0025? (See Figure 10-9.)

$$(1)\ \text{RPM} = \frac{225 \times 12}{\pi(6)} = 143\ \text{RPM, but 150 RPM is available}$$

$$(2)\ \#\ \text{rev.} = \frac{24 + 6}{.0025 \times 8} = 1{,}500$$

$$(3)\ \text{Time} = \frac{1{,}500}{150} = 10.000\ \text{min}$$

Time standards resulting from standard data are in normal time. Allowances must be added afterward.

Standard data is the most efficient method of setting time standards. Every company should be setting standard data, and a good place to start is with a simple machine that is popular in that company. A toy company had a shrink-wrap operation on every packout line. It had been setting standards for years on the shrink-wrap job using PTS and time study. The resulting standard data was developed in less than one day and reduced the standard-setting time from a 45-minute job to under 1 minute. This reduction in time-setting cost is typical. The industrial engineer has an unlimited number of problems to solve in any manufacturing plant, so don't make a career of time study. Standard data will put you out of the time study business.

FIGURE 10-9

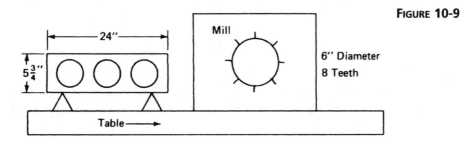

WORK CELL, ASSEMBLY LINE, AND PLANT BALANCING

Purpose

The balancing technique is a use of elemental time standards for the purposes of

1. Equalizing the workload among the people, cells, and departments. It doesn't help if one person, cell, or department can do one more unit of work if those departments feeding them with work or departments they are sending work to cannot keep up. All people, cells, and departments need to be balanced with each other. To make it fairer for all, we can take work away from a busy station and give it to someone who doesn't have enough work.

2. Identifying the bottleneck operation. The person, cell, or department with the most work is the bottleneck station, and we need to work on bringing this station into balance with the rest of the plant. This station needs more industrial engineering and supervisory help than any other station. If we have one person with 10% more work than the other 20 people on an assembly line, we can save the equivalent of one-fifth of a person with every 1% reduction in the bottleneck station time until we reduce the bottleneck station 10%. We can justify 20 times the normal tooling cost because of this multiplier. The balancing technique is also a good cost reduction tool.

3. Establishing the speed of the assembly line. Conveyor speeds need to be adjusted to the rate of the plant. Even if we don't have conveyors, move schedules are needed.

4. Determining the number of work stations. When one job has more work than one person can handle to achieve the quantity goals set by the customers, additional work stations must be added. The question is, How many? The time standard divided by the plant rate gives us the number of stations.

5. Helping determine the labor cost. Adding all operations' hour-per-piece time standards will give us total hours. Total hours multiplied by the average hourly wage rate gives us labor cost.

6. Establishing the percent workload of each operator. This determines how busy each person is compared to the bottleneck station, *Takt* time, or plant rate.

Information Necessary for Balancing an Operation or Plant

Balancing techniques must build on previously determined facts:

1. Blueprints and bills of materials from product engineering tell us what needs to be done.
2. Output required (schedule) from marketing or production control gives us the quantity. We develop the rate of the plant (*R* value) and *Takt* time of the plant from their data.

3. Elemental time standards from industrial engineering tell us how long each task takes.

Plant Rate and *Takt* Time   The plant rate (R value) and *Takt* time tell the industrial engineer how fast the industrial plant has to run to meet customer demand. Every machine and operation in the plant is keyed to this rate, and parts must be supplied at the same rate as the assembly line uses them. *Takt* time is a German expression meaning the available production time divided by the rate of customer demand. If the customers demand 120 units per day and we work 480 minutes per day, then the *Takt* time would be 4 minutes. We need to produce 1 part every 4 minutes, and every work station needs to supply parts at this rate. The R value is similar, but with factoring to consider time standards, performance percentages, and allowances. An R value of .250 minute of cycle time means that 1 finished product must come off the assembly line every .250 minute or the plant will not produce enough product. Every other machine and operation in the plant must produce a part every .250 minute (4 parts per minute), or it will be behind schedule. If two parts are needed per assembly (like wheels on a bike), the R value for that part would be .125 minute. If a work station or machine takes .4 minute, and the R value is .250, how many machines would you need?

$$\frac{\text{Time standard}}{R \text{ value}} = \frac{.400 \text{ min}}{.250 \text{ min}} = 1.6 \text{ stations}$$

If this were a fabrication machine, we would add all these fractions of work stations together, then round up and buy or build that many work stations. However, on an assembly line, we would have to round up immediately and provide two work stations. Each work station would be only 80% loaded (busy), but because the people at the work station in front of and behind this work station are producing 1 part every $\frac{1}{4}$ minute, these two 80% loaded people will need to stay at their work stations all the time.

Plant Rate Calculation   Assembly line balancing starts with the plant rate calculation. Some information is required from other sources before beginning:

1. The production volume (for example, 1,500 per shift) comes from sales or marketing management, and it determines how many units the company can sell. The production inventory control department calculates how fast to produce. Factors like season, warehousing costs, training costs, and manufacturing costs enter the production volume determination. The industrial engineer cannot do anything about plant layout or balancing without a production volume estimate, and the industrial engineers are not a good source of sales information.

2. The allowances in the average plant are 10%. We have been using this allowance throughout the book and it will suffice for our example. The plant will be down 48 minutes per day, or 10% of the time. We cannot expect to produce products every minute of every day. And when the assembly line is stopped for one operator, all operators are stopped.

3. The efficiency must be anticipated. Experience shows what our efficiency rate averages, and we will use that experience. First-year production plants produce at 70% of standard; normally 85% could be expected on continuing operations. If the industrial engineer designs a plant to produce at the rate of 100%, what is the chance of meeting this goal? The chances are very low, and the engineer who has not prepared management for the reasonable expectation had better find a new job. Standards are set at 100%, but first-year production averages only 70%, and continuing production (production the second year and beyond) can expect 85% performance. If we need 2,000 units per day, we need 2,000, not 85% of 2,000. Mathematically, the plant rate calculations are as follows:

$$
\begin{aligned}
&480 \text{ min/shift (8 hr} \times 60 \text{ min per hour)} \\
&\underline{- 48} \text{ min downtime (10\% allowances in this plant)} \\
&= 432 \text{ min available} \\
&\underline{\times 75} \text{ \% (anticipated first-year performance based on our experience)} \\
&= 324 \text{ effective min/shift} \\
&\underline{\div 1,500} \text{ units/shift (needed production)} \\
&\quad = R \text{ value} = .216 \text{ min/unit (plant rate, or } Takt \text{ time)} \\
&\quad 4.63 \text{ units/min (1 min} \div .216 \text{ min/unit)}
\end{aligned}
$$

PROOF:

$$4.63 \text{ units/min} \times 432 \text{ min/shift at } 75\% = 1,500 \text{ units/shift}$$

In this example, 4.63 parts each minute must come off the end of the assembly line. Every other cell and machine in the plant needs to do the same, or we will not make our goal of 1,500 units per day. The R value is our starting point for assembly line balancing.

This same concept can be used in any business. Think of a restaurant. How many customers can be seated at one time? How many waiters or waitresses, cooks, dishwashers, etc. must be hired? Every enterprise must be balanced or waste occurs.

Standard Elemental Time Time standards for each part must be calculated before parts (elements) can be combined into jobs. When designing a new production line, these times are calculated using PTS or standard data. With an R value of .216, the industrial engineer will combine these elements of work into jobs. These jobs will have times as close as possible to the multiples of .216 minutes (.216, .432, .648, and the highest, .864.) Four people doing the same job is normally the largest group on an assembly line because of line layout problems. More than four operators trying to receive parts from one source and then sending the completed parts to another source makes for very complicated movement of inventory.

A step-by-step procedure using the line balance example shown in Table 10-5 and the form in Figure 10-12 with the circled numbers will help you understand the logic and math for solving assembly line balancing problems. This assembly line balance solution can be greatly improved, and later in this chapter we will discuss a better solution.

Table 10-5 Assembly Line Balance Example.

⑦ OPERATION NUMBER	⑧ OPERATION DESCRIPTION	⑨ R VALUE	⑩ CYCLE TIME	⑪ NUMBER OF STATIONS	⑫ AVG. CYCLE TIME	⑬ % LOAD	⑭ HOURS/ 1,000	⑮ PIECES/ HOUR
5	Assemble	.216	.357	2	.179	85	7.770	129
10	Assemble	.216	.441	3	.147	70	1.655	86
15	Cement	.216	.210	1	.210	100	3.885	257
20	Rivet	.216	.344	2	.172	82	7.770	129
25	Form carton	.216	.166	1	.166	79	3.885	257
30	Label	.216	.126	1	.126	60	3.885	257
35	Packout	.216	.336	2	.168	80	7.770	129
Total			1.980	12 people			46.620	

Assembly Line Balancing

The purpose of the assembly line balancing technique is to

1. Equalize workload among the assemblers,
2. Identify the bottleneck operation,
3. Establish the speed of the assembly line,
4. Determine the number of work stations,
5. Determine the labor cost of assembly and packout,
6. Establish the percent workload of each operator,
7. Assist in plant layout,
8. Reduce production cost.

The assembly line balancing technique builds on the operations chart (Figure 5-2) and the plant rate (R value) calculated in Chapter 4. See footnote 2 on page 00. The objective of assembly line balancing is to give each operator as close to the same amount of work as possible. This can only be accomplished by breaking the tasks into the basic motions required to do every single piece of work and reassembling the tasks into jobs of near equal time value. The work station or stations with the largest time requirement is designated as the 100% station and limits the output of the assembly line. If an industrial engineer wants to improve the assembly line (reduce costs), he or she would concentrate on the 100% station. Reduce the 100% station in our example by 1% and save the equivalent of 0.25 people, a multiplying factor of 25 to 1.

An example assembly balance problem for a tool box example appears in Table 10-6. SA3 could be taken off the assembly line and handled completely separate from the main line, and we can save money. SA3 .250 = 240 pieces per hour and .00417 hour each. If balanced, the standard would be 180 pieces per hour and .00557 hour each.

Table 10-6 Initial Assembly Line Balance.

TASK	TIME STANDARD	NUMBER STATIONS	ROUNDED OFF	AVERAGE TIME	PERCENT LOADED	HOURS PER UNIT	PIECES PER HOUR
SSSA1	.306	1.77	2	.153	92	.00557	180
SSA1	.291	1.68	2	.146	87	.00557	180
SSA2	.260	1.50	2	.130	78	.00557	180
SA1	.356	2.06	3	.119	71	.00834	120
A1	.310	1.79	2	.155	93	.00557	180
A2	.555	3.20	4	.139	83	.01112	90
A3	.250	1.44	2	.125	75	.00557	180
SA2	.415	2.40	3	.138	83	.00834	120
SA3	.250	1.44	2	.125	75	.00557	180
P.O.	.501	2.90	3	.167	100	.00834	120
Total	3.494		25			.06950	

R value = .168 minutes = line rate or *Takt* time

*Number of stations = time standard + R value. This plant wants to make 2,000 units per day at 70%.

$$
\begin{array}{rll}
 & .00557 & \text{balanced cost} \\
- & .00417 & \text{by itself cost} \\
\hline
 & .00140 & \text{savings hours/unit} \\
\times & 500,000 & \text{units/year} \\
\hline
 & 700 & \text{hours per year} \\
@ & \$\ \ 15.00 & \text{per hour} \\
= & \$10,500.00 & \text{per year savings}
\end{array}
$$

This is called the *cost of balancing*. In this case it is too high.

Subassemblies that can be taken off the line must be

1. Poorly loaded. The less percent loaded, the more desirable to be subassembled. For example, a 60% load on the assembly line balance would indicate 40% lost time. If we take this job off the assembly line (not tied to the other operators), we could save 40% of the cost.

2. Small parts easily stacked and stored.

3. Easily moved. The cost of transportation and the inventory cost will go up, but because of better labor utilization, total cost must go down.

A subassembly on the assembly line balance form would look like this:

	TIME STANDARD	NUMBER STATIONS	ROUNDED UP	AVERAGE TIME	PERCENT LOADED	HOURS PER UNIT	UNITS PER HOUR
SA3	.250	1.44	2	2.50	Sub	.00417	240

Look back at Table 10-6 and SA3. We have saved plenty, but can we do this to SA1? No, because it's a large part not easily stacked, stored, or moved.

The assembly line balance in Table 10-6 is not a good balance because of the low percent loads. An improvement is possible (look at the 100% station). If we add a fourth packer, we will eliminate the 100% station at P.O. Now the new 100% (bottleneck station) is A1 (93%). By adding this person, we will save 7% of 25 people, or 1.75 people, and increase the percent load of everyone on the assembly line (except P.O.). We might now combine A1 and A2 and further reduce the 100%. The best answer to an assembly line balance problem is the lowest total number of hours per unit. If we add an additional person, his or her time is in the total hours. Try to improve the tool box assembly line balance, and then see how that affects the assembly lines in Figures 10-10 and 10-11.

Notes on the Assembly Line Balance (Table 10-6):

1. The busiest work station is P.O. It has .167 minute of work to do per packer. The next closest station is A1, with .155 minute of work. As soon as we identify the busiest work station, we identify it as the 100% station, and communicate that this time standard is the only time standard used on this line from now on. Every other work station is limited to 360 pieces per hour. Even though other work stations could work faster, the 100% station limits the output of the whole assembly line.

2. .06950 hour is the total time required to assemble one finished tool box. The average hourly wage rate times .06950 hour per unit gives us the assembly and packout labor cost. Again, the lower this cost, the better the line balance.

Line balancing is an important tool for many aspects of industrial engineering, and one of the most important techniques used in the assembly line layout. The back of the assembly line balancing form is designed for an assembly line layout sketch. Look at the examples in Figures 10-10 and 10-11. See how they agree with Table 10-6.

Packout work is considered to be the same as assembly work as far as assembly line balancing is concerned. Many other jobs may be performed on or near the assembly line but are considered subassemblies and are not directly balanced to the line because subassemblies can be stockpiled. Time standards for subassemblies stand on their own merit.

Step-by-Step Procedure for Completing the Assembly Line Balancing Form

Refer to the assembly line balancing form shown in Figure 10-12.

① Product no.: The product drawing or product part number goes here.

② Date: The complete date of development of this solution goes here.

③ By I.E.: The name of the technician doing the assembly line balance—your name—goes here.

④ Product description: The name of the product being assembled goes here.

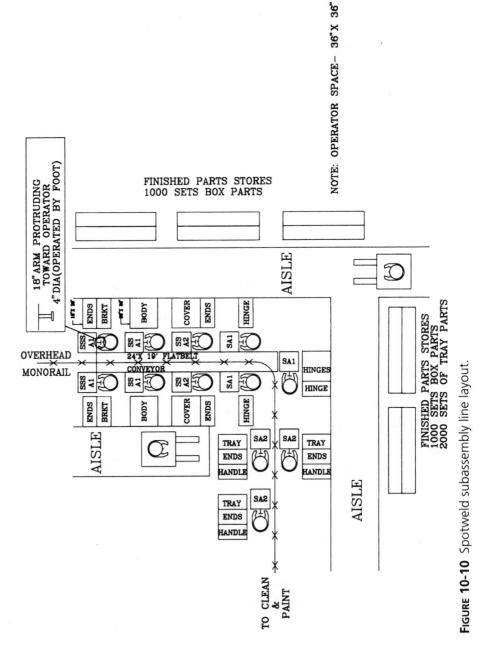

Figure 10-10 Spotweld subassembly line layout.

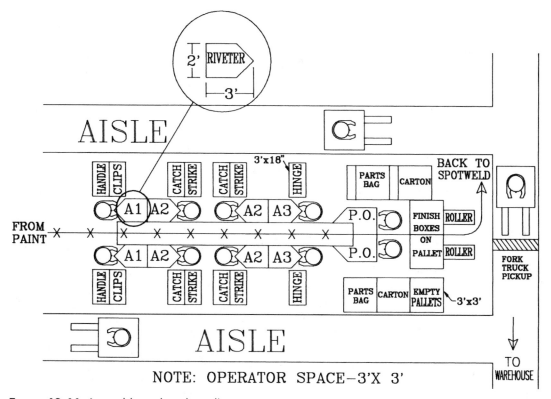

FIGURE 10-11 Assembly and packout line.

⑤ Number units required per shift: This is the quantity of production required per shift—given to the industrial engineer by the sales department. The industrial engineer's objective is to get as close to this quantity as possible without going below.

⑥ R value: The plant rate has been discussed previously in this chapter, but this block is designed for a specific plant with the following past experience:

  a. Existing products have run at 85% efficiency.

  b. New products average 70% efficiency during the first year.

  c. Eleven percent allowances are added to each standard. The R value in this plant is calculated by dividing 300 or 365 minutes by the number of units per shift (Step ⑤). The result is the plant rate—the R value.

⑦ No.: This is a sequential operation number. The use of operation numbers is to give a simple, useful method of referring to a specific job.

⑧ Operation description: A few well-chosen words can communicate what is being done at this work station. Part names and job functions are the key words.

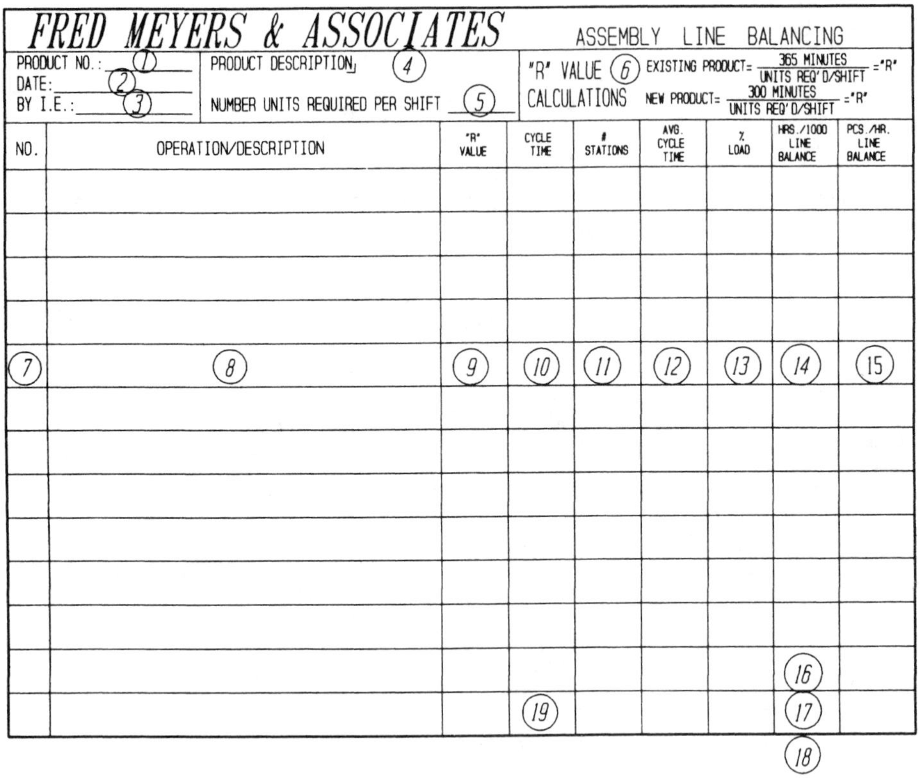

FIGURE 10-12 Assembly line step-by-step balancing form.

⑨ R value: The R value calculated in block ⑥ goes to the right of each operation. The plant rate is the goal of each work station, and putting the R value on each line keeps that goal in focus.

⑩ Cycle time: The cycle time is the normal time standard set by combining elements of work together into jobs. Our goal is the R value, but that specific number can seldom be achieved. Cycle time can be changed by moving an element of work from one job to another, but elements of work are a large proportion of most jobs. Faster equipment or smarter methods may reduce the cycle time, and this is a good cost reduction tool.

⑪ No. stations: The number of stations is calculated by dividing the R value ⑨ into the cycle time ⑩ and rounding up. If the number of stations is rounded down, the goal (number of units per shift, ⑤) will not be achieved. Management may round down the number of work stations because of cost, but if they do, they know the goal will not be achieved without overtime, etc. But that is management's decision, not the technologist's. If the number of work stations is rounded down, that work station will be the bottleneck, the restriction, the slowest station, or the 100% station.

⑫ Avg. cycle time: The average cycle time is calculated by dividing the cycle time ⑩ by the number of work stations ⑪. This is the speed at which this work station produces parts. If the cycle time of a job is 1 minute and 4 machines are required, the average cycle time is .250 minute (1.000 ÷ 4 = .250) or a part will be produced by those 4 machines every .250 minute. The best line balance would be for every station to have the same average cycle time, but this never happens. A more realistic goal is to work at getting them as close as possible. The average cycle time will be used to determine the percent work load of each work station, the next step.

⑬ % load: The percent load tells how busy each work station is compared to the busiest work station. The highest number in the average cycle time column ⑫ is the busiest work station and is therefore called the 100% station. One hundred percent is written in the % load column. Every other station is compared to this 100% station by dividing the 100% average station time into every other average station time and multiplying the result by 100, thus determining the % load of each station. The % load is an indication of where more work is needed or where cost reduction efforts will be most fruitful. If the 100% station can be reduced by 1%, then we will save 1% for every work station on the line.

EXAMPLE: % load calculation.

Look again at the example in Table 10-5. The average cycle times were .179, 147, .210, .172, .166, .126, and .168. Reviewing these average cycle times reveals that .210 is the largest number and is designated the 100% work station. A good practice is to circle the .210 and the 100% to remind ourselves that this is the most important work station on the line and that no other time standard has any further meaning. Now that the 100% station is determined, the percent load of every other work station is determined by dividing .210 into every other average cycle time:

$$\text{Operation } 05 = .179 \div .210 = 85\%$$
$$10 = .147 \div .210 = 70\%$$
$$20 = .172 \div .210 = 82\%$$
$$\text{etc.}$$

Where will the supervisor put the fastest person? Operation 15. Where will the industrial engineer look for improvement or cost reduction? Operation 15, the 100% loaded station.

A good line balance would have all work stations in the 90% to 100% range. One work station below 90% can be used for absenteeism. A new person can be put on this station without holding up the whole line.

⑭ Hours/1,000: The hours per 1,000 units produced can be calculated most easily by multiplying the 100% average cycle time (which is circled on the line balance) by 18.5 (the standard developed in Chapter 9 under allowances—see page 000). 18.5 is the hours per 1,000 for a 1-minute job. 18.5 adds in a constant 10% allowance, and because all normal time standards are in minutes, the number of minutes times 18.5 hours/1,000/minute equals the hours per 1,000 for that job.

From our example in Table 10-5, .210 was the 100% station:

$$.210 \times 18.5 = 3.885 \text{ hr}/1,000$$

If more than one person is working at a station, the hours per 1,000 is multiplied by the number of people:

> 2 people = 7.770 hours (twice as much time)
> 3 people = 11.655 hours (three times as much time)

Hours per 1,000 equals the 100% station average time times 18.5 hours/1,000 times the number of operators at that station for every station. Every work station will be 3.885 hours per 1,000 or multiples of that number. This is what line balance means—everyone works at the same pace. The logic for a work station with three people taking three times as many hours per 1,000 should be obvious. Another piece of logic is that everyone on an assembly line must work at the same rate. The person with the least work to do still cannot do one more than comes to the operator and cannot do one more than the following operator can do.

⑮ Pieces/hour: Pieces per hour is $1/x$ of hours/1,000 times 1,000 (or divide the hours per 1,000 into 1,000). Notice in our example, Table 10-5, that all the stations produce 257 pieces. Station 05 has two operators, each producing 129 pieces per hour for 258 pieces per hour total.

⑯ Total hours/1,000: Total hours per 1,000 is the number of hours for all the operations together. The hours per 1,000 for one operator times the total number of operators on the line also equals the total hours per 1,000. The total of column ⑪ is the total operators.

⑰ This entry is an average hourly wage rate. This would come from the payroll department, but let's say $7.50/hour is the average hourly wage rate.

⑱ This entry is the labor cost per 1,000 units. In our example, 46.62 hours times $15/hour = $699.30 per 1,000 units of $.70 each labor cost. The lower the cost, the better the line balance.

⑲ This entry is the total cycle time. The total cycle time tells us the exact work content of the whole assembly and, if treated like any other time standard, can show us what a perfect line balance would be.

In our example, 1.980 minutes × 18.5 hours/1,000 equals 36.63 hours per 1,000. Our line balance came out to 46.62—10 hours more. This 10 hours is potential cost reduction, and what cannot be removed by cost reduction is called the *cost* of line balance.

Improving Assembly Line Balancing The end of an initial line balance will lead to improvements. To improve line balance, we look at

1. Reducing the 100% station. This can be done by
   a. Adding an operator
   b. Reducing cost
2. Combining the 100% station with an operation in front or behind, but the sequence of operations must be maintained.
3. Combining other operations to eliminate one of them.

Let's do our example again with the addition of one more operator at station 15, the 100% station. Refer to Table 10-6 (page 234). The original line balance was 46.62 hours. Subtracting the new line balance of 43.05 shows a savings of 3.57 hours, and at the rate of $15/hour, the savings are $53.55 per 1,000, or about $100 per day or $25,000 per year. This is still not a good balance. Can you do better?

Speed of the Conveyor Line

On the first day of production, someone will ask you how fast to run the conveyor belt. Conveyor speed is a combination of product size and R value. Conveyor speed is measured in feet per minute, and when an R value is determined, a finished part must roll off the assembly line at that speed. R values are normal time, and the line speed is calculated or based on normal time. The R value example of .250 minute is a good one: .250 minute is 4 parts per minute, and that is the number of finished units to cross the finish line every minute. Since a conveyor moves at the same rate from the first operator to the last operator, 4 parts per minute will be the speed at every work station. The only other piece of information needed is the length of a finished product. For example, a swing set is 10 feet long. A swing set coming down the line takes up 10 feet of belt. So we need four swing sets per minute at 10 feet each, or 40 feet per minute belt speed. A 2-foot part at 4 per minute equals 8 feet per minute. Remembering our time standard for walking (264 feet/minute) from Chapter 9 helps us to develop the proper perspective about these belt speeds.

No allowances are included in belt speed, because the line will be off during breaks and delays. If allowances were added to belt speed, we would never achieve 100% performance.

$$\text{At the rate of 4 pieces/minute} = 240 \text{ possible per hour}$$
$$.250 \text{ minute} \times 18.5 = 4.625 \text{ hours}/1,000 = 216 \text{ pieces/hour}$$

Our standard is 216 pieces per hour, but our belt speed is for 240 per hour. The extra parts will be lost when the belt is off. Two hundred and sixteen parts per hour $\times$ 8 hours per shift equals 1,728, but we wanted only 1,200 units per shift. No plant operates at 100% performance. If this is a new product, the average performance during the first year will be closer to 70%.

$$70\% \text{ of } 1,728 \text{ units} = 1,209 \text{ units/shift}$$

Is that close enough?

EXAMPLES:

Figure 10-13 is an initial line balance, and Figure 10-14 is an improvement. Both of these examples have allowances included in the cycle time column, so the hours per 1,000 column is calculated by dividing 60 minutes into the 100% average cycle time $(.250/60 \times 1,000 = 4.17)$.

Figure 10-13 shows a cost of $.54 each. Figure 10-14 shows a cost of $.42 each. The quantity requested was 1,200 units per shift, so the savings through improved balance is $144.00 per shift, or $36,000 per year ($.54 − $.42 × 1,200 units per day × 250 days per year = $36,000.00). Notice that we took two people out of the assembly line, which is now producing at a faster rate. We could have just as easily added people to get a better balance, and a better line balance is one with lower unit cost, so the cost of adding employees is in the new cost. In this example, we are now producing at the rate of 1,310 units per shift, which is 110 units per day faster than the sales department wanted. But to save $.12 cents each or to achieve a 22% reduction in labor cost, management feels (and it is a management decision, not industrial engineering's) that this is a good decision. As a company, we have several choices:

FRED MEYERS & ASSOCIATES — ASSEMBLY LINE BALANCING

PRODUCT NO.: 1670
DATE: 10-10-xx
BY I.E.: F. Meyers

PRODUCT DESCRIPTION,
New plastic charger
NUMBER UNITS REQUIRED PER SHIFT 1200

"R" VALUE CALCULATIONS

EXISTING PRODUCT = $\frac{365 \text{ MINUTES}}{\text{UNITS REQ'D/SHIFT}}$ = 'R'

NEW PRODUCT = $\frac{300 \text{ MINUTES}}{\text{UNITS REQ'D/SHIFT}}$ = 'R'

NO.	OPERATION/DESCRIPTION	'R' VALUE	CYCLE TIME	# STATIONS	AVG. CYCLE TIME	% LOAD	HRS./1000 LINE BALANCE	PCS./HR. LINE BALANCE
1	Place bottom housing on line and lub	.250	.200	1	200	80	4.17	240
2	Assemble parts 3, 4, & 5	.250	.250	1	250	100	4.17	240
3	Assemble parts 6 & 7 together and Place sub-assembly in housing	.250	.305	2	153	61	8.34	120
4	Drive 6 bolts holding sub-assembly in bottom housing	.250	.600	3	200	80	1250	80
5	Get vent cover & cement in place in bottom housing	.250	.198	1	.198	79	4.17	240
6	Get top housing, apply cement & assemble to bottom housing	.250	.290	2	.145	58	8.34	120
7	Place in carton & in master carton (6 per) and aside to pallet	.250	.625	3	.208	83	12.50	80
							54.19	
	EXAMPLE AND ASSIGNMENT:						× $10.00/hr.	
	Improve this line balance – Reduce cost.						$541.90/1000	
	Your goal is to make this product as cheaply as possible.						$ .54 each	

FIGURE 10-13 Assembly line balance: initial try.

		FRED MEYERS & ASSOCIATES	ASSEMBLY LINE BALANCING

PRODUCT NO.: __1670__
DATE: __10-11-XX__
BY I.E.: __MEYERS__

PRODUCT DESCRIPTION
NEW PLASTIC CHARGER
NUMBER UNITS REQUIRED PER SHIFT __1200__

"R" VALUE CALCULATIONS

EXISTING PRODUCT= $\frac{365\ MINUTES}{UNITS\ REQ'D/SHIFT}$ = "R"

NEW PRODUCT= $\frac{300\ MINUTES}{UNITS\ REQ'D/SHIFT}$ = "R"

NO.	OPERATION/DESCRIPTION	"R" VALUE	CYCLE TIME	# STATIONS	AVG. CYCLE TIME	% LOAD	HRS./1000 LINE BALANCE	PCS./HR. LINE BALANCE
1 & 2	Place Bottom Housing on line, Lubricate & assemble parts # 3, 4 & 5.	.250	.450	2	.225	98	7.63	131
3 & 4	Assemble Parts # 6 & 7 Together, Place in Housing & Drive 6 Bolts Holding sub to Housing.	.250	.905	4	.226	99	15.26	65
5	Get Vent Cover & Cement in Place in Bottom Housing.	.250	.198	1	.198	86	3.82	262
6 & 7	Get Top Housing, apply cement & assemble to Bottom Housing, Pack Out.	.250	.915	4	.229	100	15.26	65
				11			41.97	
							x 10.00/HR	
							419.70/1000 or	
							$ .42 EACH	

FIGURE 10-14 Assembly line balance: improved solution.

1. Hold fast to the 1,200 units per day.
2. Lower the selling price to motivate the sales department.
3. Spend more in advertising to sell more.
4. Quit production earlier in the year because we are producing 110 units faster than we planned.

To make this subject even more complicated (and real), warehousing costs will increase if we don't move the finished product 9% faster. What management wants is lower total cost, and it would not be wise to lower labor cost and increase warehousing cost by a substantial amount.

Summary of Assembly Line Balance

Line balancing is an important tool for many aspects of industrial management, and one of the most important tools of motion and time study. The assembly line balance is the starting place for assembly line layout. The back of the assembly line balancing form is designed for an assembly line layout sketch. Review the example in Figures 10-13 through 10-15. Figure 10-15 is developed from the data in Figure 10-14.

CHAPTER 10

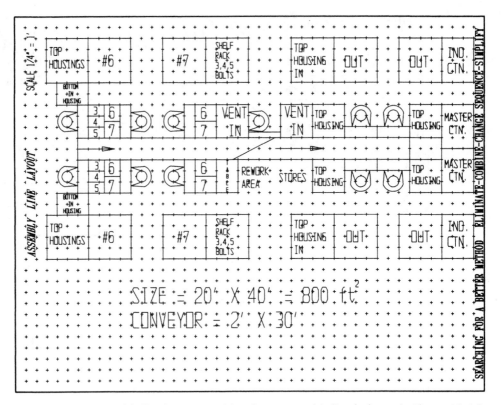

FIGURE 10-15 Assembly line layout resulting from assembly line balance in Figure 10-14.

Packout work is considered to be the same as assembly work as far as assembly line balancing is concerned. Many other jobs may be performed on or near the assembly line, but they are considered subassemblies and are not balanced directly to the line (because subassemblies can be stockpiled). Their time standards stand on their own merit. They can be part of the assembly line form, but they do not enter the balancing procedure, their load is 100%, and their time standards are their own. Assembly line balancing and assembly line industrial engineering is fun and interesting work. I hope you get a chance to experience it.

Work Cell Plant Balancing

The cell concept is not new; we have had single cells in manufacturing for generations. But the new concept is to combine cells into strings that can be a conveyor. More often, the material is moved manually from cell to cell as demanded by the next cell. No material is moved until the next cell demands it; then the cell providing the part or parts builds to replace that inventory. This inventory waiting to be moved can be in either a designated finished storage area, kanban, or waiting in the cell manufacturing work area. In this case, no additional work can be performed in the cell until the material waiting is moved. Material inventory control is a big part of this cell concept. Minimum inventory is desired, and no inventory between cells is the most desirable. When one cell needs to change over from one part to another, some inventory is required. In these situations kanban (inventory control cards, location, and inventory) are created as a buffer between the two cells. When one kanban is pulled away from the first cell, that cell must set up to fill the empty space.

Some Definitions

1. Cell—The layout of machines required to make a unit of production in a specific sequence around an operator or operators. The machines are usually arranged in a U-shaped layout to permit single piece flow.

2. Kanban—Inventory control pull system where parts are placed in containers waiting for the next cell. When material is pulled away from a cell, it must be replaced at this cell so there are always parts available for the next cell. Press shops supporting a spotweld subassembly is a good example. One punch press may run many parts and needs to be set up many times. We cannot set up the machine for every part, so we stockpile some parts, which forces us to speed up changeover.

3. Changeover—The removing of one tool or set of tools from a machine and replacing it with a new tool or set of tools to produce a new part.

4. Just-in-time—A system of producing and delivering the right item at the right time in the right amounts. Just-in-time is achieved when the upstream producers finish their job just moments before the downstream cell needs it. Single piece flow is promoted.

5. Multimachine working—Training employees to operate and maintain different types of production machines. It is an important part of work cells where one worker may operate many different types of machines.

6. *Takt* time—The production pace of the plant set by the customer. If we need 10 units per day and we have 480 minutes per shift, then we have a *Takt* time of 48 minutes.

7. Single piece flow—A layout where one unit of production flows through the operation one at a time, without interruptions, backtracking, or setting aside. An example follows.

EXAMPLE: A manufacturer of flatbed, over-the-road trailers pulled one of his most popular trailers out of his original production plant and set up a new plant designed as a single-piece-flow, cell-type plant. The plant operated during its first year averaging only 5 trailers per day. An industrial engineering consultant was hired to develop a plan to achieve the designed goal of 8 trailers per day. The results of that project follow.

The first step of the project was to time study each work cell to determine its work content. Table 10-7 shows the cell names, the crew size, their work time standards, and the number of hours of work required from each cell to produce 8 trailers. Figure 10-16 is a bar chart showing the percent load of each cell. Study Table 10-7 and Figure 10-16 for improvement. Four cells are properly loaded, 1 cell is overloaded, and 8 cells are very poorly loaded. Before time study, these conditions were unknown. Table 10-8 and Figure 10-17 are the line balance and percent loads, respectively, required for 9 trailers (without adding additional people).

Table 10-7 Assembly Line Balance—8 Trailers.

	A	B	C	D	E	F	M	O
1	Machine center	Number in crew	Hours per day available	Mins needed per trailer	Hours needed for 8 trailers	% load (column E ÷ column C)		
3	Press	1	8	21.86	3.09	38.62		
4	Parts assembly	2	16	74.56	10.54	65.86		
5	Plasma	2	16	139.54	19.72	123.26		
6	Beam weld	2	16	80.08	11.32	70.74		
7	Beam finish	8	64	294.00	41.55	64.92		
8	Jig assembly	4	32	148.28	20.96	65.49		
9	Weld out	2	16	79.50	11.24	70.22		
10	Axle subassembly	1	8	48.84	6.90	86.28		
11	Axle and grind	4	32	156.40	22.10	69.08		
12	Wash	2	16	109.96	15.54	97.13		
13	Paint	2	16	79.84	11.28	70.53		
14	Air	4	32	219.76	31.06	97.06		
15	Sill leader	1	8	39.93	5.64	70.54		
16	Sill	4	32	216.40	30.58	95.58		
17	Longeron fabrication	4	32	228.28	32.26	100.82		
18	Longeron installation	4	32	181.12	25.60	79.99		
19	Electric	3	24	147.06	20.78	86.60		
20	Z brace weld	1	8	35.42	5.01	62.58		
21	Floor layout	3	24	145.95	20.63	85.95		
22	Floor installation	3	24	132.24	18.69	77.87		
23	Tires	2	16	73.16	10.34	64.62		
24	Total	59	472	2652.18	374.84	84.18		
25								

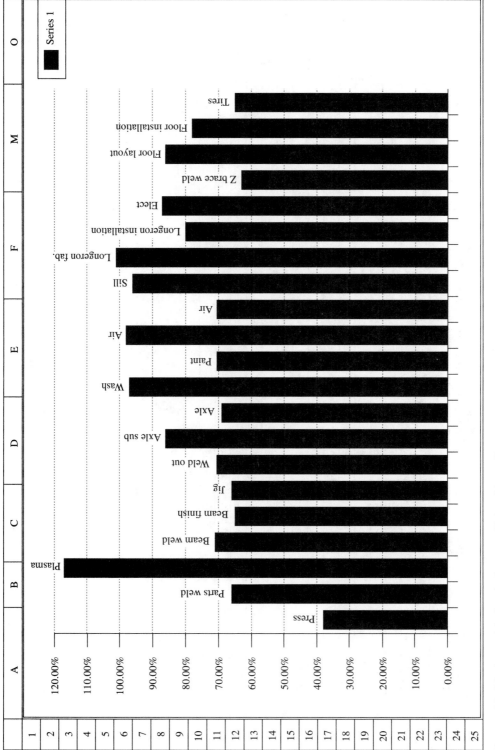

FIGURE 10-16 Assembly line balance—% load by cell.

Table 10-8 Assembly Line Balance—9 Trailers Per Day Schedule.

	A	B	C	D	E	J	K	L	M
1	Machine center	Number in crew	Hours per day available	Minutes needed per trailer	Hours needed for 8 trailers	Number in crew for 9	Hours per day available	Hours needed for 9 trailers	% load (column G ÷ column C × 108%)
2									
3	Press	1	8	21.86	3.09	With parts assembly and eliminate one operator			
4	Parts assembly	2	16	74.56	10.54	2.0	16	15.33	95.81
5	Plasma	2	16	139.54	19.72	3.0	24	22.19	92.45
6	Beam weld	2	16	80.08	11.32	2.0	16	12.73	79.58
7	Beam finish	8	64	294.00	41.55	6.0	48	46.75	97.39
8	Jig assembly	4	32	148.28	20.96	3.0	24	23.58	98.24
9	Weld out	2	16	79.50	11.24	2.0	16	12.64	79.00
10	Axle sub	1	8	48.84	6.90	Combine with axle and grind			
11	Axle and grind	4	32	156.40	22.10	5.0	40	32.57	81.42
12	Wash	2	16	109.96	15.54	2.0	16	14.00	87.50
13	Paint	2	16	79.84	11.28	2.0	16	12.69	79.34
14	Air	4	32	219.76	31.06	4.5	36	34.94	97.06
15	Sill leader	1	8	39.93	5.64	1.0	8	6.35	79.36
16	Sill	4	32	216.40	30.58	4.5	36	34.41	95.58
17	Longeron fab.	4	32	228.28	32.26	5.0	40	36.30	90.74
18	Longeron inst	4	32	181.12	25.60	4.0	32	28.80	89.99
19	Electric	3	24	147.06	20.78	3.0	24	23.38	97.43
20	Z brace weld	1	8	35.42	5.01	1.0	8	5.63	70.40
21	Floor layout	3	24	145.95	20.63	3.5	28	23.21	82.88
22	Floor inst	3	24	132.24	18.69	3.5	28	21.03	75.09
23	Tires	2	16	73.16	10.34	2.0	16	13.64	85.25
24	Total	59	472	2652.18	374.84	59.0	472	420.16	89.02
25									

Figure 10-17 Assembly line balance, % load—9 trailers per day schedule.

Takt time or plant rate: We have 450 minutes in which to make 9 trailers. Therefore, we must move a trailer out every 50 minutes. The following is a schedule that must be followed or we will not make 9 trailers:

7:00 A.M. _____ Shift starts—supervisors fill in absences.
7:10 A.M. _____ Start work on trailer #1.
8:00 A.M. _____ Move trailer #1 out and trailer #2 in.
8:50 A.M. _____ Move trailer #2 out and trailer #3 in.
9:00 A.M. _____ Take 10-minute break.
9:50 A.M. _____ Move trailer #3 out and trailer #4 in.
10:40 A.M. _____ Move trailer #4 out and trailer #5 in.
11:30 A.M. _____ Move trailer #5 out and trailer #6 in, and go to lunch.
12:50 P.M. _____ Move trailer #6 out and trailer #7 in.
1:40 P.M. _____ Move trailer #7 out and trailer #8 in, and take a 10-minute break.
2:40 P.M. _____ Move trailer #8 out and trailer #9 in.
3:30 P.M. _____ Move trailer #9 out and go home.

We could be ahead all day long.

If a work station is not ready to move when the time arrives, the lead person fills out a late movement report telling why they were late. If they are not late, no report is needed. The supervisor must be especially vigilant when it is time to move.

FIGURE 10-18 Production schedule based on producing 9 trailers per day.

Figure 10-18 is a schedule of move times developed on the *Takt* time and plant rate of 50 minutes (480 minutes − 30-minute breaks ÷ 9 trailers/day = 50 minutes). In the finishing end of the plant, one work cell could not move until the next cell was empty. If a cell was open, the previous cell held up the plant and needed help. Supervisors and cell leaders did everything possible to ensure that holdups were minimized because the plant was on an incentive system and each employee would lose $4.00 per day per trailer less than 9. It took the plant two days to reach the goal of 9 trailers, and in the first five days of the new system, they produced 43 trailers. Figure 10-19 is a plant layout sketch of the trailer plant.

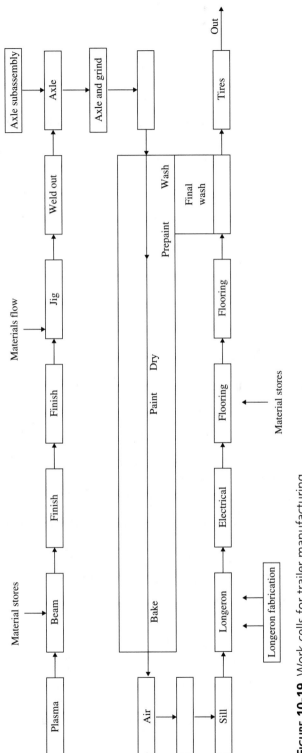

Figure 10-19 Work cells for trailer manufacturing.

QUESTIONS

1. Define standard data.

2. What are the advantages of standard data?

3. What are the five methods of communicating standard data? What is the advantage of each?

4. Calculate a and b for the following:

SIZE	TIME
49″	.98
55″	1.10
75″	1.50
104″	1.80
115″	2.30

Note: One of these data points is in error. Plot a graph first and eliminate the error before calculating a and b.

5. What would be the time standard for a 100″ part? Use the figures that you calculated in Question 4 for a and b.

6. What is the time required to turn (lathe) a ½-inch diameter part 3 inches back? The speeds and feeds are 250 ft/min and .001 in./rev.

7. What would be different if the part in Question 6 were a drill or mill?

8. Complete the worksheet example in Figure 10-5.

9. What are the purposes of assembly line balance?

10. What information is required before an assembly line balance can be tried?

11. What is an R value?

12. What is the average cycle time of ⑫ in Figure 10-12?

13. What does % load mean in ⑬ of Figure 10-12?

14. What does 18.5 mean?

15. Why is only the 100% station standard used?

16. What is the significance of total hours per 1,000 in ⑯ of Figure 10-12?

17. What is the significance of total cycle time in ⑲ of Figure 10-12?

18. Improve the line balance that appears in Table 10-6.

19. When should subassemblies be placed on the assembly and packout line?

20. What should the speed of the conveyor be with the following?

R	LENGTH	CONVEYOR SPEED
.162	96"	
.440	24"	
1.100	6"	

How would incentives change the above?

21. Improve the line balance of the example in Figure 10-13. How much can you save?

22. Review the layout in Figure 10-15.

23. How does the assembly line balance relate to the line layout?

Work Sampling

Work sampling, TV sampling (Nielsen ratings), political poll sampling (Gallup polls), and unemployment statistics are all random sampling, and as such are scientifically based on the same theories or laws of probability. Work sampling is the process of randomly observing people working to determine how they spend their time. Everyone who has ever worked with others has work sampled. The attitudes developed about fellow employees' work ethics or productivity are based on random observations. Conclusions are drawn that this person is a "workhorse" or a "goof-off" because of random observations. Supervisors are work sampling their employees all the time. These informal work samples can be much more scientific and fair if done properly. The science and technique of work sampling is the subject of this chapter.

Work sampling is subdivided into four subtechniques:

1. Elemental ratio studies
2. Performance sampling studies
3. Time standard development studies
4. Process effectiveness studies.

Each technique gets more complicated and must be learned in this sequence. Each technique is also a usable tool in itself. A time standard development study uses elemental ratio and performance sampling studies, and a process effectiveness study includes all of the others.

ELEMENTAL RATIO STUDIES

The primary task (what the operator does most of the time) performed by a person defines his or her job title, but many other activities (productive and nonproductive) take time, too. Each activity must be measured and compared to the total time. This is the element ratio. An elemental ratio study will determine the percent of time that each element of work takes.

Elemental Breakdown and Ratio Estimates

When starting an elemental ratio study, the elements of work must be listed, and the ratios must be estimated. The elemental ratios are what we want to determine, so this early estimate is nothing more than an educated guess. A few early observations are made before the study starts to estimate these ratios, but the estimates are to determine the total observations needed for a specific confidence level and accuracy. An example of a list of elements for a machinist's job and the estimated elemental ratios follows:

ELEMENT NUMBER	ELEMENTAL DESCRIPTION		ELEMENTAL RATIO
1	Load and unload	20%	Productive
2	Machine time	35	Productive
3	Setup	15	Productive
4	Tool change	7	Productive (in delay allowance)
5	Inspection	5	Productive (in delay allowance)
6	Material handling	4	Productive (in delay allowance)
7	Away and idle	14	Nonproductive (10% in allowances)
	Total	100	

Total allowances will be 26% of the total time: 10% of away and idle, plus 4%, 5%, and 7%, which are the three delay occurrences. This is adjusted to the percent of productive time so that the allowance can be applied directly to the elemental times. It is calculated as follows:

$$\frac{\text{Percent allowances}}{\text{Percent total} - \text{percent allowances}} = \frac{\text{percent allowances}}{\text{percent productive}}$$

$$\frac{26\%}{100\% - 26\%} = 35\%$$

This elemental ratio study may be for a person, a department of many people, or a whole plant. It makes no difference how many people are in the study; only the number of total observations is important. The smallest percentage (4% in this case) will determine the total number of observations required for a confidence level and accuracy, because larger percentages require fewer samples. Before discussing how many observations are required, accuracy and confidence level must be defined.

Accuracy

Accuracy measures the closeness of the ratio to the true ratio of an element. A ±5% accuracy (our normal accuracy goal) indicates that the ratio is within 5% of the true element time. If a true ratio of one task to the total job is 25%, the ±5% accuracy would allow the observer to record any ratio between 23.75% to 26.25% (±1.25) and be within the accuracy tolerance. An elemental ratio of 10% with an accuracy goal of ±5% equals a range of from 9.5% to 10.5%. Even though the normal accuracy goal is 5%, 10% is a cheaper goal, and there are many times when this is good enough. Some small nonproductive times may cost more to set than we could save, so a 10% accuracy is good enough. We will discuss how much we save by reducing our accuracy a little later.

Confidence Level

The confidence level refers to how positive (confident) the work sampler wants to be about the resulting ratios. At the beginning of a study, estimates are required. They are based on very little information, and there is little confidence in these ratios. But as data is collected, our confidence builds. With more observations being collected every day, the ratios are more consistent and our confidence grows. The question at the beginning of the study is, How many observations do we need to take to achieve a specific level of confidence? A 95% confidence level indicates that our ratios are accurate (within the ±5% range) 95% of the time. The remaining 5% of the time we will be off one way (over) or the other (under), but not by much.

In summary, the industrial engineering profession feels comfortable with ±5% of the objective 95% of the time. We can produce time standards that are ±1% accurate 99% of the time, but they would be too costly. Table 11-1 shows accuracy, confidence levels, and the required number of observations. This table will help us understand the cost of accuracy and confidence when we discuss this topic later.

Sample

A sample is one observation of one operator once. The observation of 100 operators once each is exactly the same as observing one operator 100 times. Both produce 100 samples. Work sampling is observing enough employees enough times to collect the number of samples required to achieve the accuracy and confidence designed into the study. An observation of the operator must be made at the first moment of sight, and this observation must be made at random times, without any preplanning. The science of sampling is based on the theory that a sample taken at random tends to have the same characteristics as the whole population. The work sampling of an operator, then, will be an indicator of how the operator spends his or her day. One sample won't do it, but many samples will. Table 11-1 is a list of sample sizes. We need a few more definitions before we work with this important table.

Randomness

Randomness is a requirement of sampling. The exact time of an observation must be completely random (based on chance only), or the study's accuracy and confidence will be

Table 11-1 Sample Size Table.

HOW MANY SAMPLES ARE REQUIRED?[a]

90% CONFIDENCE Z = 1.645 LEVEL OF ACCURACY				95% CONFIDENCE Z = 1.960 LEVEL OF ACCURACY				99% CONFIDENCE Z = 2.575 LEVEL OF ACCURACY		
P^b	1%	5%	10%	P^b	1%	5%	10%	P^b	1%	5%	10%
1	2,678,965	107,159	26,790	1	3,803,184	152,127	38,032	1	6,564,319	262,573	65,643
2	1,325,952	53,038	13,260	2	1,882,384	75,295	18,824	2	3,249,006	129,960	32,490
3	874,948	34,998	8,749	3	1,242,117	49,685	12,421	3	2,143,902	85,756	21,439
4	649,446	25,978	6,494	4	921,984	36,897	9,220	4	1,591,350	63,654	15,914
5	514,145	20,566	5,141	5	729,904	29,196	7,299	5	1,259,819	50,393	12,598
6	423,944	16,958	4,239	6	601,851	24,074	6,019	6	1,038,798	41,552	10,388
7	359,515	14,381	3,595	7	510,384	20,415	5,104	7	880,926	35,237	8,809
8	311,193	12,448	3,112	8	441,784	17,671	4,418	8	762,522	30,501	7,625
9	273,609	10,944	2,736	9	388,428	15,537	3,884	9	670,430	26,817	6,704
10	243,542	9,742	2,435	10	345,744	13,830	3,457	10	596,756	23,870	5,968
15	153,341	6,134	1,533	15	217,691	8,708	2,177	15	375,735	15,029	3,757
20	108,241	4,330	1,082	20	153,664	6,147	1,537	20	265,225	10,609	2,652
25	81,181	3,247	812	25	115,248	4,610	1,152	25	198,919	7,957	1,989
30	63,141	2,526	631	30	89,637	3,585	896	30	154,715	6,189	1,547
35	50,255	2,010	503	35	71,344	2,854	713	35	123,140	4,926	1,231
40	40,590	1,624	406	40	57,624	2,305	576	40	99,459	3,978	995
45	33,074	1,323	331	45	46,953	1,878	470	45	81,041	3,242	810
50	27,060	1,082	271	50	38,416	1,537	384	50	66,306	2,652	663
55	22,140	886	221	55	31,431	1,257	314	55	54,251	2,170	543
60	18,040	722	180	60	25,611	1,024	256	60	44,204	1,768	442
65	14,571	583	146	65	20,686	827	207	65	35,703	1,428	357
70	11,597	464	116	70	16,464	659	165	70	28,417	1,137	284
75	9,020	361	90	75	12,805	512	128	75	22,102	884	221
80	6,765	271	68	80	9,604	384	96	80	16,577	663	166
85	4,775	191	48	85	6,779	271	68	85	11,701	468	117
90	3,007	120	30	90	4,268	171	43	90	7,367	295	74
95	1,424	57	14	95	2,022	81	20	95	3,490	140	35
99	273	11	3	99	388	16	4	99	670	27	7

[a]Most work sampling studies use a 95% confidence level and a 5% accuracy. A sample is one observation.

[b]P = the percent of the total that this element represents.

destroyed. The opposite of randomness is routine, which means predictability, and both destroy the study. Randomness can be developed many ways, for example:

1. Random number tables
2. Random number calculator button
3. Drawing numbers from a hat
4. Using the last four digits of telephone numbers in a telephone book.

If we need to take 10 trips through the department today, we use the random numbers to select the starting time of each trip. If we are sampling all day long, we vary our path to give randomness. It is important not to build predictability into your study.

Sample Size: The Number of Observations

Sample size is the number of observations required to attain the accuracy and confidence we want, and it is determined with a combination of accuracy, confidence, and element percentage from Table 11-1. This table was developed from the following formula:

$$N = \frac{Z^2(1 - P)}{(P)(A^2)}$$

N = the number of observations needed.

Z = the number of standard deviations required for a specific confidence level (can be found in any statistics book). A 95% confidence is used for most time studies, and Z for 95% is 1.96.

CONFIDENCE LEVEL	Z
99.5	3.25
99.0	2.575
95.0	**1.960**
90.0	1.645
80.0	1.245
75.0	1.151

P = the percent of the total time that employees perform one element of work. Also, the elemental percentage. A job may have several elements, but only the smallest element of an operation is used. Earlier in this chapter, we listed seven elements, one of which was material handling. Material handling was 4% of the machinist's day, and machine time was 25% of the day. All the elements must add up to 100%, a full accounting of the day's activities. For this reason we often have a miscellaneous element.

A = the accuracy desired. Most of the time, the study uses ±5% accuracy.

EXAMPLE:

How many observations would we have to make for our example of ±5% accuracy and 95% confidence on a job that accounts for only 4% of the workday?

$$N = ? \qquad Z = 1.96 \qquad P = .04 \qquad A = .05$$
$$N = \frac{(1.96)(1 - .04)}{.04(.05)} = 36,897 \text{ observations}$$

Check Table 11-1: 95% ± 5% for a 4% job. Did you find the answer was 36,897 observations? Check the table for a 5% job (inspection): 29,196 observations are needed. Only the smallest element percentage is used because all higher numbers will require fewer observations. Making 36,897 observations sounds like a lot of work, but if you have a plant of 100 employees, it would take only 369 trips through the plant at 10 minutes per trip, or 6 trips per hour, and 48 trips per day. Each day you could collect 4,800 observations, so we are looking at less than 8 days of work. If we give ourselves a month (22 workdays) to perform this job, we would need to commit about 36% of our time to this project. This is about 3 hours per day, or 18 trips through the plant per day. We will use this later to design a random schedule.

The new technologist often asks, Why ±5%, 95% of the time? The answer is the cost of quality. An uninformed manager may ask for 99% ± 1% standards. From the sample size table (Table 11-1), an elemental ratio (P) of 1% would call for 6,564,319 observations. At a rate of 800 observations per hour (a good rate for a plantwide study), 8,205 hours of the work sampler's time will be required. If the pay rate for a work sampler is $20.00/hour, $164,100.00 would be the cost of the study. A 95% ± 5% study on a 2% job would require only 75,295 samples and would take 94 hours to collect the data, and the cost would be a modest $1,880.00. The cost of quality in this case was nearly $162,220.00. Do we need to be this accurate? Let's be more realistic and cost conscious.

Probability and Normal Distribution

Work sampling is based on the laws of probability. Review the normal distribution curves in Chapter 9 (Figures 9-15 and 9-16). The normal distribution curve is used to describe the laws of probability. For example, a welder working on a job has the cycle times (the time to make one part) shown in Figure 11-1. We grouped the times into cycle time groups of .03 minute. The number of times the welder produced a part within each of these time categories is listed, and a frequency graph is produced. The frequency percent is the probability that the next part produced will take that amount of time or will be produced within a group of times. For example, there is an 85% chance that the next part produced will be within the time range of 0.93 to 1.07 minutes. The chances of one observation being below 0.93 or above 1.07 are 7.3% each. The more observations we take, the closer our estimated average will be to the real (true) average.

A tried and true explanation of probability theory, throwing a pair of dice, comes from the gaming industry. The question here is, What are the odds, chances, or *probability* that we can throw a particular number? From Figure 11-2, a 7 can be thrown 6 different ways, and a 2 can be thrown only 1 way. Only 36 total combinations can be thrown, so the odds of throwing a 2 are 1 time out of 36 tries, or 2.78%. The odds for throwing a 7 are 6 times out of 36 tries, or 16.67%. Which has the largest probability of success? Seven, of course.

Step-by-Step Procedure for an Elemental Ratio Delay Study

Step 1:   Identify the subject.

Step 2:   Establish the purpose and goal of the study.

Step 3:   Identify the elements.

Step 4:   Estimate the ratio percent of the elements.

Step 5:   Determine the level of accuracy and confidence.

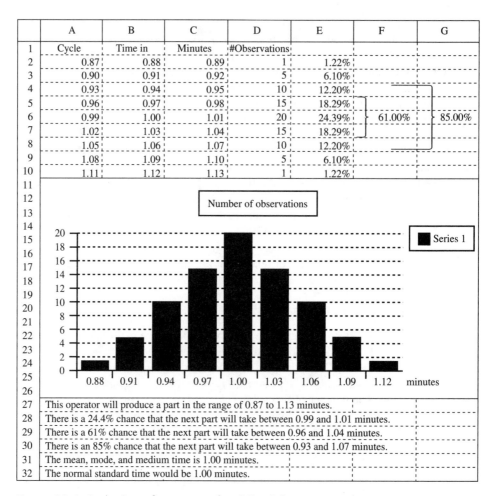

	A	B	C	D	E	F	G
1	Cycle	Time in	Minutes	#Observations			
2	0.87	0.88	0.89	1	1.22%		
3	0.90	0.91	0.92	5	6.10%		
4	0.93	0.94	0.95	10	12.20%		
5	0.96	0.97	0.98	15	18.29%		
6	0.99	1.00	1.01	20	24.39%	61.00%	85.00%
7	1.02	1.03	1.04	15	18.29%		
8	1.05	1.06	1.07	10	12.20%		
9	1.08	1.09	1.10	5	6.10%		
10	1.11	1.12	1.13	1	1.22%		

27	This operator will produce a part in the range of 0.87 to 1.13 minutes.
28	There is a 24.4% chance that the next part will take between 0.99 and 1.01 minutes.
29	There is a 61% chance that the next part will take between 0.96 and 1.04 minutes.
30	There is an 85% chance that the next part will take between 0.93 and 1.07 minutes.
31	The mean, mode, and medium time is 1.00 minutes.
32	The normal standard time would be 1.00 minutes.

FIGURE 11-1 Cycle times frequency of welding job.

Step 6: Determine the number of observations needed to achieve the quality goals.

Step 7: Schedule the observations.

Step 8: Talk with everyone involved.

Step 9: Collect the data.

Step 10: Summarize and state conclusions.

We will look at five cases to illustrate the procedures in this list. One conclusion drawn from ratio delay studies is that some cases will require continued study to establish time standards resulting from work sampling. We will discuss this in the last section of this chapter.

Case 1: The subject of this case is (1) an oil company with a quart canning machine that produces at the rate of 1,000 cases per hour, yet the machine puts out only 3,500 cases per

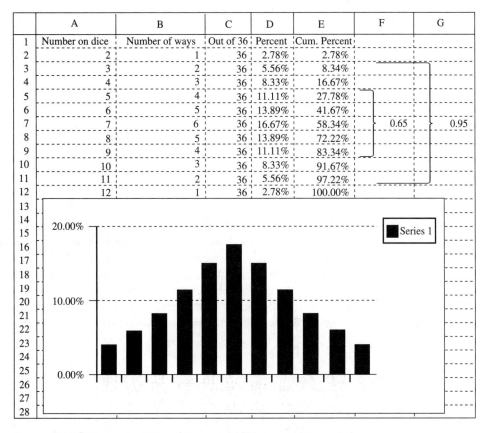

	A	B	C	D	E	F	G
1	Number on dice	Number of ways	Out of 36	Percent	Cum. Percent		
2	2	1	36	2.78%	2.78%		
3	3	2	36	5.56%	8.34%		
4	4	3	36	8.33%	16.67%		
5	5	4	36	11.11%	27.78%		
6	6	5	36	13.89%	41.67%		
7	7	6	36	16.67%	58.34%	0.65	0.95
8	8	5	36	13.89%	72.22%		
9	9	4	36	11.11%	83.34%		
10	10	3	36	8.33%	91.67%		
11	11	2	36	5.56%	97.22%		
12	12	1	36	2.78%	100.00%		

FIGURE 11-2 The probability of throwing numbers on two dice.

shift. The goal (2) is to identify the problems and their extent (size). The elements (3) are working, idle (operator caused), setup and changeover, material handling, quality control, maintenance, and bad material. After an initial look at the operation it was *estimated that the ratio* (4) percents were 30%, 20%, 15%, 5%, 5%, 5%, 12%, and 8%, respectively. The level of *accuracy* will be ±5%, and the *confidence level* (5) will be 95%. The *number of observations* (6) will be based on the 5% element (the smallest). We will look in Table 11-1 under the 95% confidence ± 5% column, then look down the *P* column to the 5% element. We find 29,196 observations are required. We see that ±10% requires only 7,299 observations, and since our goal is only a preliminary study, we will start with this more modest size job. *Scheduling the observations* (7) will be done by use of a random sampling beeper, which will give us 250 samples per hour. (This is only one way of scheduling; we will investigate other ways in later examples.) We need to *talk with everyone* (8) involved in our study:

1. The supervisor and the employees must be involved with the purpose and goal of the study. If they know the design of the study and the problems we are trying to solve,

they will help. The study may require the people to assist in the recording of production counts, absences, reassignments, and other data input.

2. The supervisor and employees should be informed that being idle or doing nothing is expected 10% of the time because of personal, fatigue, and delay time allowances.

3. The payroll department should be consulted to record the number of hours involved in the study. The study period should be the same as the payroll period.

We will be located in the area of the machine. When the beeper goes off, we will *collect and record* (9) a tally mark behind the appropriate element name for this one observation. A total of 7,299 observations divided by 250 observations per hour will require 30 hours of our time. The results of the study will look like Table 11-2.

Case 2: The subject (1) is a fork truck that seems to be overloaded. The supervisor has asked management to purchase another truck. Good management requires hard data before making such an expensive decision. They have asked for your help. The goal (2) is to see how busy the fork truck and driver are. We need to see how the truck and driver spend the day. The elements (3) of this study will be driving loaded, driving empty, and idle. We estimate the *ratios* to be 40%, 40%, and 20%, respectively. With an *accuracy* (4) of ±5% and a *confidence level* (5) of 95%, we look up the *number of observations* (6) in Table 11-1. Under

Table 11-2 Oil Canning Line Work Sampling Study Results.

	A	B	C	D	E
	A	B	C	D	E
1	Elements	#Observations	Ratio	Hours	
2	Work	2500	33.78%	67.57	
3	Idle	1500	20.27	40.54	
4	Setup	800	10.81	21.62	
5	Material handling	450	6.08	12.16	
6	Quality control	300	4.05	8.11	
7	Maintenance	550	7.43	14.86	
8	Bad material	1300	17.57	35.14	
9	Total	7400	100.00	200.00	
10			Total hours worked 200 hours		
11	Questions:				
12	1. How are we going to increase the machine utilization?				
13	2. 20% idle time is excessive. How can it be reduced?				
14	3. Why do we need to stop the machine for quality control,				
15	material handling, maintenance?				
16	4. How can we prevent bad material from getting to the machine?				

a P of 20%, 6,147 observations are required. We may need some help with this job to do it in a reasonable amount of time. With four work samplers (who are doing other engineering or management work at the same time), we could each do 50 observations per day. An observation is seeing the truck and driver once. With four observers collecting 50 samples per day each, we will collect 200 per day, and our study will take 31 days. We will *schedule the random observations* (7) using a random number generator. Many calculators have a random number function, and when depressed, it will show a 3- or 4-digit random number. We need to decide which 2 digits we want to use and stick to it. It should not take us longer than 5 minutes to find the truck and another moment to record our observation, so we need to divide the day into 5-minute increments (96 in an 8-hour day). If 7 A.M. is 1, then 12 would be 8 A.M. and 14 would be 8:10, etc.

We now collect the first 50 random numbers from our random number generator, convert them to the time of day, and put them in time order. This is one observer's *schedule*. Each observer must independently determine his or her schedule and make observations. When we finish one day of data collection (9), it will look something like this:

WORK SAMPLE FORK TRUCK	TUESDAY, JAN. 28, XX	TOTAL	%
40% transport loaded	ᚺᚢ ᚺᚺ ᚺᚺ ᚺᚺ ᚈᚺ ᚈᚺ ᚺᚢᚺᚺ ᚺᚢ ᚺᚺᚺᚺ ᚈᚺ ᚺᚺ ᚈᚺᚢ ᚷᚷ	72	37.5
40% transport empty	ᚺᚢ ᚺᚺ ᚺᚺ ᚺᚺ ᚈᚺᚢ ᚈᚺ ᚈᚺᚢ ᚺᚺ ᚷᚺᚢ ᚺᚢᚢ ᚺᚢ ᚈᚺᚢ ᚷᚷᚷ	63	32.8
20% idle	ᚈᚺᚢ ᚷᚺᚢ ᚷᚺᚢ ᚺᚺᚈᚺ ᚷᚺᚢ ᚷᚺᚢ ᚺᚺᚢ ᚺᚢ ᚈᚺᚢ ᚈᚺ ᚷᚷ	57	29.7
	Total	192	100.0

The pencil marks are called tallies, and there is one per observation. On this date, we collected only 192 observations. Each day we add today's totals to the previous totals and keep a running ratio accounting. As we progress, the volatility of changing ratios will lessen until the fluctuations are less than 5%. We will have started toward our goal on the first day, and each subsequent day makes us more comfortable and confident. In *conclusion* (10), if the final ratios look anything like our first day's data, the supervisor will not hire a new driver. Also, we should look at how we can reduce the transporting empty element by scheduling the driver with back hauling of return material, trash, or scrap.

Case 3: The *subject* (1) is the tools crib and supply room. The *goal* (2) is to see if we need to add attendants to service the windows. Employees have been complaining that they must wait too long for service. In this case, we are looking only at how many people are in line waiting for the attendants to serve them. Our *elements* (3) will be the number of people in line (0, 1, 2, 3, 4, 5, 6). We would not expect more than 4 people to be waiting, but it doesn't cost much to be prepared. The *percent (ratios)* (4) we expect to find are 25%, 35%, 20%, 9%, and 1%, respectively. We decide that $\pm 10\%$ *accuracy* with a 90% *confidence* (5) will be our quality goal, and this will be for a P of 15% or larger. Therefore, 1,533 *observations* (6) will be needed. This tool room is just outside our office, so we plan to observe it 100 times per day for 16 days. We will use the telephone book and the last three numbers as our random number generator for our *schedule* (7) of observations. Because we can observe this operation instantaneously, we will use every minute of the 8-hour (480 minutes) day as a potential observation time. We will *collect* (9) the first 100 phone numbers between 0 and 480 as we go down the list, then put them in time order to set up our schedule. We make our 1,533 observations and summarize them as follows:

NUMBER WAITING AT TOOL CRIB WINDOW	NUMBER OF OBSERVATIONS	% OF TOTAL	WEIGHTED AVERAGE
	WORK SAMPLE TOOL CRIB, **11/18/xx** TO **12/10/xx**		
0	350	22.8	0
1	625	40.8	625
2	250	16.3	500
3	225	14.7	675
4	75	4.9	300
5	8	0.5	40
Total	1,533	100.0	2140 ÷ 1533 = 1.4

In *conclusion* (10), we have 1.4 people waiting for tools or supplies. We could add a tool crib attendant to eliminate this one person waiting in line, but we will never get the number down to zero because people will always need tools and supplies unless you change the method of supplying the employees. Adding a new tool crib attendant could pay for itself.

Case 4: The *subject* (1) is the fact that the maintenance manager wants to establish a satellite maintenance station in a remote area that will require a lot of maintenance. The maintenance manager thinks his employees spend too much time walking that could be better spent working on machines. He has already determined that $1\frac{1}{2}$ maintenance people are needed in this area. The *goal* (2) is to determine if there is enough saving of time to justify the expense of moving part of maintenance to this area. The *elements* (3) that we are interested in are maintenance, walking, and idle or away. We *estimate the ratios* (4) to be 60%, 20%, and 20%, respectively. The 20% *P* with ±10% *accuracy* and 95% *confidence level* (5) will require 1,537 *observations* (6). We have 15 maintenance people, and it will take us 15 minutes to travel the area to find all 15 people. We can collect 60 observations per hour, so the schedule (7) for this job will take 26 days. We will use the random path technique to ensure random sampling. When sampling full-time, we can vary our path through the plant. When finished with the first trip, we go anywhere and start sampling again. As long as we vary our path with no predetermined plan in mind, we will maintain randomness. This is the hardest type of work sampling, full-time.

RESULTS OF THE SATELLITE MAINTENANCE SAMPLING (8) AND (9)			
ELEMENTS	NUMBER OF OBSERVATIONS	% OF TOTAL	HOURS
Maintenance	875	57	1,601
Walking	355	23	646
Idle	307	20	561
Total	1,537	100	2,808

In one month, 15 maintenance people spent 646 hours walking to and from jobs, tools, materials, and their work areas. Almost anything we do will reduce this. A satellite area is only one option, but it is a start. After making the changes, we conclude (10) that we need to work sample again and see if we increased the hours working on maintenance.

Case 5: This is the most frequently used type of ratio delay study. The *subject* (1) is a plant of 200 people, and our *goal* (2) is to see how efficient they are. We may want to implement a motion and time study program. How much potential do we have? The *elements* (3) are working or not working, and the *estimated ratio* (4) percent of the two elements are 60% and 40%, respectively. This is what we expect to find in a plant without time standards. This initial study will have an *accuracy* of ±5% and a *confidence level* (5) of 95%. With the smallest P of 40%, an accuracy of 5%, and a confidence level of 95%, from Table 11-1 we find that we need to take 2,305 observations (6). To *schedule the observations* (7), we must go through the plant 24 times collecting 200 observations per trip (200 people). Each trip will take us $\frac{1}{2}$ hour, so we are talking about a day and a half of work. We have other work, so we will do this study over a week and we need to make 5 trips per day. To ensure the randomness of our observations, we will divide the day into half-hour increments, write these times on a piece of paper (one time per piece), fold the time sheets, place them in a hat, pull out 5 pieces of paper, unfold, and record the 5 times on a sampling form. *Talk with everyone* (8) involved so they know what you are doing.

To collect the data (9), we will take our 5 trips through the plant recording working or not working by placing a tally mark behind the element on the data collection sheet as follows:

WORK SAMPLING STUDY "OUR PLANT" WEEK ENDING 11/19/xx				
ELEMENT	NUMBER OF OBSERVATIONS	TOTAL	%	HOURS
Working		1,375	58	4,611
Not working	(On the daily sheets, this area would be full of tallies.)	1,000	42	3,339
Totals		2,375	100	7,950

When we finish the first day of observations, we *summarize and state our conclusions* (10). Each day we raise the confidence level. It is very common that the first day's results are very close to the final report, but we need to have all the observations to ensure quality. Our study shows that the plant employees work only 58% of the time. With a 10% allowance, that would be 63.8% labor performance. With a performance control program, this plant could be working at 85% performance, an increase of 33.2% (($85\% - 63.8\%) \div 63.8\%$). An 85% performance has 10% allowances included, so if we were to sample this plant under a performance control system (discussed in Chapter 13), we would show a 77.3% working element ($77.3\% \times 110\% = 85\%$). Our study showed only 4,611 hours of work per week. If we increased this number to 77.3%, we would increase the hours worked to 6,145 hours per week. This translates into a savings of 1,534 hours per week $\times$ 52 weeks = 79,768 hours $\times$ \$14.00 per hour average pay rate = \$1,116,752.00 per year savings. We can save over \$1 million a year by instituting a performance control system, which requires time standards. We will talk about this program in Chapter 13.

PERFORMANCE SAMPLING STUDIES

Performance sampling requires rating the operator when observing him or her. Rating was a major subject of stopwatch time study, and that is exactly what must be done in performance sampling. The observance of an operator happens in a moment, and in that moment the observer must judge the speed and tempo of the operator.

Operator speed and tempo vary from person to person, and for an individual operator the speed and tempo can vary from minute to minute. For work sampling, performance sampling fine tunes the ratios, making them more accurate. For example, in a bag-packing operation, five independent work stations are available. From an elemental ratio study, the operation is divided into four elements:

OPERATION DESCRIPTION	OBSERVATION (%)								
	60	70	80	90	100	110	120	130	140
1. Bag packing	׀׀׀׀	׀׀׀׀ ׀׀׀׀	׀׀׀׀ ׀׀׀׀ ׀׀׀׀	׀׀׀׀ ׀׀׀׀ ׀׀׀׀ ׀׀׀׀	׀׀׀׀ ׀׀׀׀ ׀׀׀׀	׀׀׀׀ ׀׀׀	׀׀׀׀ ׀	׀׀	׀
2. Load hoppers			׀׀	׀׀׀׀	׀׀׀	׀׀	׀		
3. Away	׀׀׀׀ ׀׀								
4. Idle	׀׀׀׀ ׀׀׀׀	׀׀׀׀	׀׀׀׀						

Elements 1 and 2 are productive and can be rated by placing a tally mark under the proper % heading. Ten percent increments are normal, and most rating is between 70% and 130%. Elements 3 and 4 are nonproductive and therefore cannot be rated. The step-by-step procedure is exactly the same as for elemental ratio studies until Step 10, the last step. The data for this bag-packing example is summarized in Table 11-3.

As each day's work is summarized and added to the previous totals, the ratios come closer to the accuracy and confidence level being sought. During the one-month study of these five bag-packing stations, 825 hours of labor were used. How many hours were spent on each element of the job?

Table 11-3

	WORK SAMPLING				BAG PACK			FRI., FEB. 18, XX		
	60%	70%	80%	90%	100%	110%	120%	130%	140%	TOTAL
A. Bag packing										
No. observations	5	10	15	20	15	8	6	2	1	82
Leveled observ.[a]	3	7	12	18	15	8.8	7.2	2.6	1.4	75
B. Load hoppers										
No. observations	—	—	2	5	3	2	1			13
Leveled observ.			1.6	4.5	3	2.2	1.2			12.5
C. Away—nonproductive	7									7
D. Idle—nonproductive										20

Total observations: 82 + 13 + 7 + 20 = 122

A. Bag-packing ratio $\quad$ 75/122 = 61.5

B. Load hopper $\quad$ 12.5/122 = 10.25%

C. Away $\quad$ 7/122 = 5.75%

D. Idle $\quad$ 20/122 = 16.4%

% productive $\quad$ (75 + 12.5)/122 = 71.7

Average rating of bag packing $\quad$ 75/82 = 91.5

Average rating of loading hoppers $\quad$ 12.5/13 = 96.2%

[a]Leveled observations are comparable to normal time (% × number of observations).

	%	HOURS	EXPLANATION
Bag packing	61.50	507.38	(.615 × 825 hours)
Load hoppers	10.25	84.56	(.1025 × 825 hours)
Away	5.75	47.44	(.0575 × 825 hours)
Idle	16.40	135.30	(.1640 × 825 hours)
	93.9	774.68	
Efficiency loss	6.1	50.32	
	100	825	

What if during this 825 hours of bag packing, the operators packed 35,392 bags? A time study has been performed, and a time standard can be set.

$$\frac{\text{Hours used}}{\text{Bags packed}} = \frac{507.38 + 84.56 \text{ hours}}{35,392 \text{ bags}} = .01673 \text{ hours/bag}$$

$$+10\% \text{ allowance} \quad .00167$$
$$\text{Standard time} \quad .01840$$
$$\text{Bags/hour} \quad 54$$

TIME STANDARD DEVELOPMENT STUDIES

Work sampling can be used to develop time standards accurately and quickly. Time standard development studies pull together all the techniques of work sampling, and it is the ultimate use of work sampling.

The step-by-step procedure is exactly the same as for the elemental ratio study and the performance sampling study. The additional data needed are units produced and allowances. The time standard development system starts after the other two techniques are completed.

EXAMPLE: A maintenance department was work sampled with the goal of setting time standards to a level of 95% ± 5% on 3% elemental ratio (P) jobs. The resulting data was collected and is summarized in Table 11-4.

If 40% of the maintenance time is millwright, then the maintenance department spent 1,536 hours working on millwright jobs (40% × 3,840 hours = 1,536 hours). This is pure 100% work—no idle time.

This is where elemental ratio studies and performance sampling studies would leave us. Now the production counts and allowances would be added. Table 11-5 shows how we extend the data from Table 11-4 to create time standards.

① These element numbers refer to Table 11-4 element numbers. Only the productive elements have time standards.

② Hours were calculated in Table 11-4.

③ Work order counts were collected and given to the industrial technologist by the maintenance supervisor.

Table 11-4

ELEMENTAL NUMBER	JOB DESCRIPTION	NUMBER OF OBSERVATIONS[a]	RATIO %[b]	HOURS[c]
1	Millwright	20,000	40	1,536
2	Welder	5,000	10	384
3	Electrician	2,500	5	192
4	Machine repair	8,500	17	653
5	Carpenter	2,000	4	154
6	Walking	4,000	8	307
7	Away	3,000	6	230
8	Idle	5,000	10	384
	Total	50,000	100	3,840

[a]Number of observations resulted from the observations over a one-month period.
[b]Ratio % is the number of observations of one element divided by the total observations (20,000/50,000 = 40%).
[c]Hours: Total hours for the one-month study were made available from payroll (3,840 hours).

④ Hours per work order are calculated by dividing hours ② by work orders ③.

⑤ Plus 15% allowance. A management decision placed 5% of the 8% walking ratio into the allowance with 10% personal and fatigue time, creating a 15% personal, fatigue, and delay allowance. One hundred fifteen percent times the hours per work order ④ equals standard time ⑤.

⑥ Work orders per hour: Work orders per hour is the $1/x$ of standard hours ⑤; or divide standard hours ⑤ into the whole number 1. An entire plant can be work sampled and time standards set in one month. The size of the plant will determine the number of observers, but no other system can develop a total plantwide time standard system faster than work sampling. Table 11-6 shows the daily data collection sheet, and Table 11-7 shows the summary of the study after one month. Notice that time standards were developed for all jobs, and the totals have the labor hour total for labor costing. This is the ultimate use of work sampling.

Table 11-5

① ELEMENT NUMBER	② HOURS	③ WORK ORDER	④ HOURS PER WORK ORDER	⑤ PLUS 15%	⑥ WORK ORDERS PER HOUR
1	1,536	825	1.86	2.14	0.47
2	384	475	0.81	0.93	1.08
3	192	150	1.28	1.47	0.68
4	653	55	11.87	13.65	0.07
5	154	30	5.13	5.90	0.17

Table 11-6 Example: Work Sampling Data Collection Sheet: Work Sampling Observations.

	TIME OF SAMPLING					
	8:15	9:10	9:50	10:10	11:30	TOTAL
Die cut	/	/		/	//	
Roll inner end	/	//	/	//	//	
Tack fiberglass	/	(	//	/	/	
Roll	/	/	/	//	///	
Nylon	/	)	/	'	/	
Material handling		//	///	//	//	
Pull mandril	////	++++	////	//	'	
Denylon	/	/	//	//	//	
Drill			/	'		
Saw			/	'		
Centerless grind	///	//	/	//	(	
Silk screen	++++	//((	++++	///	//	
Spray and clean	/	/		//	///	
Assemble	++++ /)	++++	++++ (	++++ //	///	
Packout	//	////	//	(		
Total productive						
No activity	++++ (	++++ ///	++++	++++ ///	++++ ++++ ++++	
Walking	//	///	////	(/	///	
Talking	//	(	)/	(/)		
Away	///	/	(/)/	//	///	
Sheet spread						
total nonproductive	13	13	15	15	21	77
Total observations	41	43	45	45	44	218
% performance	68%	70	67%	67	52	65%
						5
					+5% Allow.	70%

PROCESS EFFECTIVENESS STUDIES

The process effectiveness study permits the analyst to evaluate several parameters about the operation, process, or plant. It combines the various ratio delay studies into one standard format and permits comparison of studies taken on the same processes at various times. In most cases, a large number of subjects will be observed during each observational period.

Because such studies often require precision on units of work that occur less than 5%, large numbers of observations are required. Often, several analysts are required on various shifts. As in any extended observational technique, care must be taken to ensure that the categories of data collected are the same in each observation sequence.

Table 11-7 Summary of Work Sampling Data After One Month and Time Standards.

	A	B	C	D	E	F	G	H
1	Operations	Number of observations	% of total	Hours	Number of units produced	Hours/unit	+ 10% allowances	Units/hour
2	Die cut	500	3.55	70.92	7,500	0.00946	0.01040	96.1
3	Roll inner	800	5.67	113.48	7,500	0.01513	0.01664	60.1
4	Tack	600	4.26	85.11	7,500	0.01135	0.01248	80.1
5	Roll	800	5.67	113.48	7,500	0.01513	0.01664	60.1
6	Nylon	500	3.55	70.92	7,500	0.00946	0.01040	96.1
7	Material handling	900	6.38	127.66	7,500	0.01702	0.01872	53.4
8	Pull mandril	1,600	11.35	226.95	7,500	0.03026	0.03329	30.0
9	Denylon	800	5.67	113.48	7,500	0.01513	0.01664	60.1
10	Drill	200	1.42	28.37	7,500	0.00378	0.00416	240.3
11	Saw	200	1.42	28.37	7,500	0.00378	0.00416	240.3
12	Centerless grind	900	6.38	127.66	7,500	0.01702	0.01872	53.4
13	Silk screen	1,900	13.48	269.50	7,500	0.03593	0.03953	25.3
14	Spray and clean	700	4.96	99.29	7,500	0.01324	0.01456	68.7
15	Assembly	2,800	19.86	397.16	1,500	0.26478	0.29125	3.4
16	Packout	900	6.38	127.66	1,500	0.08511	0.09362	10.7
17	Total	14,100	100.00	2000	1,500	1.33333	1.46667	

The collection of data on various types of nonproductive time permits separate analysis of the causes of unproductive activities. The collection of the more significant work elements permits the analysis of how working time is spent on value-added or support activities.

During the study, information will be collected about the type of work being performed, the amount of contribution to core processes, and the operator pace speed. It will be collected by individual random observations over a representational time period. Sample size and confidence intervals will be calculated for the activity requiring the smallest portion of the work period; as stated earlier, the data on all other activities will have at least this level of confidence. A format for recording details is below.

Steps in the Process Effectiveness Study

Step 1: Identify the subject. Apply the rules of who, what, where, when, how, and why to the problem and define the processes, operations, and people that need to be studied to gather the information.

Step 2: Establish the purposes or goals of the study. Know what is to be accomplished by the study. The design must be able to accomplish the goals. A lot of time and effort can be lost due to faulty design.

Step 3: Identify the elements to be studied. If the study is to quantify muda, categories that show different types of muda must be included. I usually place several subcategories under the main elements of Unobserved, Avoidable Delay, Unavoidable Delay, Inspection, Transport, Setup, Internal Work, and Operation. Examples are given so that observed elements may be categorized into one of these classes.

a. Unobserved. Subject is not in any appropriate work area.

b. Avoidable Delay. Subject is idle or conducting personal activity.

c. Unavoidable Delay. Subject is waiting for material, parts, or instruction.

d. Inspection. Subject is engaged in measuring, observing, or counting nonconformities.

FRED MEYERS & ASSOCIATES   **PROCESS EFFECTIVENESS STUDY**

Preliminary   Detailed

Organization _____ Subjects _____

SAMPLE TIMES _____, _____, _____, _____, _____, _____, _____, _____,

_____, _____, _____, _____, _____, _____, _____, _____, _____, _____,

_____, _____, _____, _____.

Estimated Smallest Observed Activity _____

Estimated Percent Occ. P _____   Calculated Accuracy: A _____

Calculated Confidence: Z _____   Confidence Interval _____ % $N = \dfrac{Z^2(1 - P)}{(P)(A^2)}$

Observations needed _____

  e. Transport. Subject is moving object or self.

  f. Setup. Subject is preparing machine, tooling, or parts for production.

  g. Internal Work. Subject is accomplishing other tasks while machine tending.

  h. Operation. Subject is performing a manual task without a pace restriction.

Step 4: Estimate the ratio percents of the elements. This usually requires taking a guess, calculating a sample size, and designing a one-day to one-week study of the process. Very rarely will prior experience be sufficient to develop a final study.

Step 5: Select the smallest occurring value and use it to calculate the accuracy and confidence intervals for a random study.

Step 6: Use the calculated values to determine the number of samples required to meet the study confidence and accuracy criteria.

Step 7: Design the study.

  a. When will the study be taken? The number of samples is just the beginning. Each sample must be equally representative of the universe the study is to define. If a day and night shift are to be compared, samples on each shift must be taken. If conditions such as weather are involved, the sample must contain appropriate proportions of samples.

  b. Where will the study be taken? The subjects must be available for systematic observation in accordance with the plan.

  c. What will the study involve? Familiarity with the work and its elements will allow that occasional element to be recorded identically by all observers.

  d. How will the study be taken? What are the logistics to ensure that the analysts and the subjects will be able to complete the study?

  e. Who will take the study? How many observers are needed? What training will they need? And when will they be available? Who will calculate and edit the data?

  f. Why is the study taken? The goals of the study greatly affect the design. Before conducting a long study, a review of the plan and, particularly, of the ability of the collected data to support the plan is appropriate.

Step 8: Talk with everyone involved. Once the design is complete, everyone should be told about the plan and support for the outcome developed. Any questions about the purpose or the project must be resolved. Management must be supportive and understand the cost of collecting irrelevant data. Employees must understand the type and frequency of data collection.

Step 9: Collect the data. Data will include observations of each subject at the random observation time. This may include from 25 to 100 times per day, and each time may include from 10 to 100 individual subjects. Each observation requires marking the appropriate element and rating. It is not uncommon for an analyst to collect over 1,500 observations per day.

Step 10: Summarize and state conclusions. Once the data are collected, a report is developed that explains the evaluation of the goals. The report will list each goal and describe in a short paragraph what was found and what needs to be modified to improve. In addition, the report will include tables showing the summarized data.

Of interest on a process effectiveness study is the percent of time each element is performed and the following categories:

a. Unobserved. Is this large enough to invalidate the study? (about 3–10%)

b. Avoidable Delay. Supervision is not effectively managing work group.

c. Unavoidable Delay. Should compare with Delay allowance in Standard.

d. Inspection. Is there appropriate documentation, certification, and training?

e. Transport. Is there justification for process changes?

f. Setup. Is SMED system functioning?

g. Internal Work. What percent of cycle? Is it equitably distributed?

h. Operation. What is the performance level? What is the process effectiveness (rating X percent observed)?

QUESTIONS

1. What is work sampling?
2. What are the four techniques of work sampling, and how do they differ?
3. What is an elemental ratio, and how is it estimated?
4. Define these terms as they relate to work sampling:
   a. Confidence level
   b. Accuracy
   c. Sample
   d. Randomness
   e. Sample size
   f. Probability
   g. Normal distribution
5. What are the 10 steps of a work sampling program?
6. How does performance sampling improve elemental ratios?
7. What is the time standard for the following:

JOB	NUMBER OF OBSERVATIONS	%	HOURS	UNITS PRODUCED
1	5,000			2,900
2	10,000			8,800
3	20,000			25,000
Idle	15,000			
Total	50,000	100%	4,250	

Add 10% allowances. What is the efficiency?

CHAPTER 12

Indirect Labor and Motion and Time Study

Direct labor and touch labor have the same meaning. Touch labor is production labor that can be reasonably and consistently related directly to a unit of work being manufactured, processed, or tested. It involves work affecting the composition, condition, or production of a product. It may also be referred to as hands-on labor, or factory labor. Machine, weld, assembly, and packout are just some of the typical direct operations. Indirect labor is all other labor. See Table 12-1 for a list of indirect labor classifications.

The number of people and the cost of indirect labor are determined by using ratios of indirect labor categories to direct labor based on the previous year's actual head count. If past practices are good enough, then this is a plan. But should we be satisfied with present cost, methods, procedures, systems, or labor power? We have already discussed the fact that people not covered by time standards typically work at about 60% performance (provable by work sampling). Is this acceptable? No. Indirect labor is controllable, and the use of motion and time study techniques can make a big difference in the quality and quantity of indirect labor.

The techniques of motion and time study used for direct labor are also used for indirect labor. In this chapter, we will discuss indirect labor in detail and provide examples of how categories of indirect labor can be affected by motion and time study.

The example of indirect ratios in Table 12-1 shows what type of work is considered indirect, and the ratios indicate national industrial averages. Twenty-one percent of direct labor (210/1,000) or 17.4% of the total people (210/1,210) are indirect. Many of these indirect laborers earn more than direct laborers, so cost percentages can be higher.

When a new budget is prepared for the following year, these ratios will be used. If business is good during the year and more employees are hired for direct labor, the indirect labor should

Table 12-1 Factory Labor Analysis: Previous Year.

CATEGORY	NUMBER OF PEOPLE	PERCENT (%) OF TOTAL LABOR
1. Material handling/control	60	6.0
2. Quality control	30	3.0
3. Manufacturing, plant, and industrial engineering	14	1.4
4. Supervision/management	45	4.5
5. Maintenance/tooling	17	1.7
6. Warehouse/shipping	18	1.8
7. Receiving/stores	7	0.7
8. Factory clerical	10	1.0
9. Miscellaneous	9	0.9
10. Total indirect	210	21.0
11. Direct labor	1,000	79.0
12. Total labor	1,210	

also increase. The ratios must be maintained—meaning that an increase or reduction in each category is maintained as direct labor increases or decreases. Ratios are very important to labor estimating and control, and I would not eliminate them. However, I would try to reduce them and their costs. Several indirect categories in Table 12-1 are quite large, and the potential savings through the application of motion and time study techniques can be spectacular.

Incentive systems create another problem for indirect labor categories. When the direct laborers are working 42% faster (using 42% more material per unit of time, demanding 42% more services) with the same indirect crew size, do we add more people, or do we put indirect laborers on incentive and increase their work output by 42% too? Are we confident that material-handling equipment can handle a 42% increase in workload? These questions are very difficult to answer if we do not know what indirect laborers can do. Let's look at some of the indirect areas.

DEVELOPING MEASURES FOR INDIRECT LABOR

The elements done in indirect labor are often done by a very few individuals or are done only occasionally. Procedures for indirect labor rarely describe the task to be done. The efforts put into developing either methods or standards for indirect labor provide the same returns as those for direct labor. The primary purpose for measures of indirect labor is the planning and scheduling of work. In contrast, the primary purpose of measures for direct labor is the improvement of work force efficiency. These goals require quite different measurement costs and measurement accuracy. Infrequent or nonrepetitive activities require measures that are easy to apply and may be less accurate than measures for mass production.

Steps for Measuring Indirect Labor

Step 1: Conduct a process study. All indirect labor evaluations begin with a traditional methods analysis. Each process should be defined and documented with the process chart and process flow chart. Planned improvements are implemented and tested. This is sufficient for defining sequences, providing routing documents, and training employees. Any such operations that affect quality require similar documentation, training, and record availability. The process study may be sufficient for very infrequent and not very complex processes.

Step 2: Conduct an operations study. Using estimates by employees or managers, or data from similar activities, estimate the time for the steps of the more complex operations. Develop multiactivity charts for these operations and use the results to develop and document them. Use the time cycle estimates to plan indirect labor analysis and to calculate monthly and annual labor budgets. This may be sufficient for all infrequent processes.

Step 3: Develop time standards using work sampling. Many activities have frequent and non-repetitive tasks. Although work elements are similar, the sequence of performance changes. The results of a work sample can be used to determine the overall workload and provide a more precise measure than the methods and procedures analyses. If work samples are taken under various conditions, a regression analysis comparing the workload of the indirect employee to a common direct labor measure can be done. The number of standard hours generated by direct labor assemblers provides a guide to the number of indirect labor employees needed to supply materials.

Step 4: Develop benchmark standards by time study or standard data. Typical processes, such as changing a tire or changing oil can be time-studied and, even though there are many variations of size and mounting, a general standard can be developed. For example, depending on the model, it takes 4 to 10 minutes to change the filter and oil in a vehicle. Hence, the 10-minute service advertisement. With a benchmark, crew loading, facility size, and cost can be estimated. Although this type of benchmark reflects the actual local conditions, it requires that all of the typical jobs be actually observed by the technician before the data are used.

Step 5: Develop benchmark standards by PTS systems. Application of these systems to indirect labor is similar to using time study for measuring direct labor, except that one of the benefits is that the PTS system permits visualization of the method rather than the observation of actual operators. For activities that have operations charts, the process of developing benchmarks is usually one of coding the motion patterns and calculating the standard times. There could be an application time saving, especially if one of the simpler PTS systems is used. The technologist must, of course, be skilled at applying the PTS system.

Step 6: Develop slotting estimates from benchmark standards. One can calculate a standard time by summing the elemental times; such times will have the same variability. However, to estimate to what extent a simple job compared to a complicated job might vary would provide quite a different answer. Slots are groupings of standards that are of similar difficulty and require similar times to complete. The groupings are made on a nonlinear scale, similar to the way one estimates initially. As the degree of difficulty increases, both the mean time value and the range increase. Each cell is called a *slot*. Benchmark standards are placed into the appropriate estimating slot. If enough estimates are provided, someone familiar with the work can find a slot that is similar, more difficult, and less difficult for each new task. The scheduler can then estimate the time for any job within the work unit, not just for a particular calculated standard. Schedulers sequence work for various crafts and trades. The slotting system is useful for large maintenance crews, municipal workers, and other nonrepetitive jobs. A scheduler can handle about 50 workers and can double their effectiveness.

Step 7: Develop a work control system to track performance. Performance reporting is necessary to ensure that the work is being accomplished in accordance with the standard. The goal is to be sure that predictions of completion times are being met rather than to log individual performance.

Step 8: Audit the entire system. Work sampling, methods review, standards auditing, control system auditing, and a complete review of all calculations is conducted cyclically to ensure that the system is operating correctly and to suggest improvements to the system.

MATERIAL HANDLING

Methods work (motion study) can improve material handling more than anything else. Several points to consider are listed below:

1. Can we eliminate any moves? The process chart or flow process chart will help identify moves. We can make improvements, like rearranging the equipment to eliminate or reduce moves.

2. Can we automate moves? Once we pay for the automatic equipment, the move is free of cost and manual labor. Modern material-handling equipment has taken the drudgery out of work.

3. Can we combine operations to eliminate the move between operations? By putting all the equipment to produce a part into one cell, we eliminate most movement between machines. This is a big part of the savings to justify cells.

4. Can we move more parts at a time? Piggyback trains are a good example of moving hundreds of semitrucks with one railroad train engine. Taking two at a time instead of one is also an example of this. An industrial jeep can pull 10 trailers through the plant at the same time.

5. Can we put the transporting empty element of industrial trucks to good use? This is called backhaul, and we should always try to load backhauls with some useful work, like removing empty containers, sending material back, or removing trash.

6. Can we manufacture parts in line using gravity slides?

7. Can we move machines closer together?

8. Are we using the best material-handling device? We do not want to use larger equipment or raise cost higher than we need to. Using a big fork truck when a hand cart would suffice is an example.

9. Can we manufacture the part next to the assembly line to reduce the amount of transport?

10. Can we use moving storage, like overhead conveyors from paint to assembly, or gravity chutes between machines, or skate wheel slides to move boxes from point to point, or powered conveyors to feed parts to where they are needed?

After we have reduced material handling to the minimum that is also economically feasible, time standards can be set. Each move must be listed (each part and the container count) and the distance estimated (averages are sufficient). A foot-per-minute travel time (remember that a person walks 264 feet per minute) can be determined for each piece of equipment, and a time for maneuvering (set down, pick up) can be determined. Any miscellaneous job can be time-studied or an allowance can be built into the estimated standard. An example of standard data can be applied to fork trucks. Fork trucks travel at 5 miles per hour, which multiplied by 5,280 feet per mile equals 26,400 feet per hour, divided by 60 minutes per hour equals 440 feet per minute. Dividing 440 feet per minute into 1 minute equals .00227 minute per foot. A 1,000-foot trip would take 2.27 minutes (1,000 × .00227).

The objective of material-handling standards is not to create performance reports but only to see if the person is loaded properly with work. Standard data is the most economical time standard, and once the data is built, application is quick. The job standard needs to be reviewed only when the workload changes. Material handling doesn't add one bit of value to our finished product, and we should minimize its cost.

QUALITY CONTROL

Quality control varies greatly from one company to another, and the need for employees varies considerably. The highly labor-intense operations of quality control are the main subject of this section.

1. Line inspectors: Many assembly lines have inspectors assigned permanently to the end of the line. They have work stations, tools, and gauges just like any other operator. Most major appliance manufacturing companies have large work areas at the end of their assembly lines that actually run the appliances to ensure quality operation. If there is a problem, the unit is pushed off the line for rework and then looped back in front of inspection for reinspection. The motion and time study technique of PTS allows the technologist to develop the most efficient method and to set a time standard for that method, just like any other job. On assembly lines, the inspector is balanced like any other operator. In many companies, this is considered direct labor because inspectors are working on an identifiable product.

2. Department inspectors: A department inspector will roam the area randomly to check material and operations. The purpose is to develop proper operation. Department inspectors may be required to approve setups before the operator starts production, and they may be required to inspect a part every hour and make entries on the control charts. The quality requirements must be determined, and elemental inspection times must be applied to each job element. The total time requirement of the quality control job is the result of adding all the task times together.

3. Inspection department: Material is moved to the inspection department for 100% inspection. This is the least productive method of inspection, but expensive tooling and special gauges may dictate it. Motion and time study of this type of inspection is exactly like that of any other mass-production job.

MANUFACTURING, PLANT, AND INDUSTRIAL ENGINEERING

The highly professional people in manufacturing, plant, and industrial engineering need goals too. The industrial engineer can use expert opinion time standards to set standards for themselves, and the backlog control system can control their workload. Methods improvements are also valuable in these areas. Standard data is the most efficient method of setting standards, and industrial engineers should have the goal of setting standard data for ever-increasing coverage of an area. The number of standards set per labor hour is a good indication of efficiency. In consulting, I'll bid a job using one-half hour per time standard. Personal computer applications of motion and time study techniques are big time savers.

Plant engineering can be made efficient with the use of computer-aided design (CAD) systems showing overlays on the plant layout for air, water, power, and computer cables; heat and air conditioning; lighting; etc. Preventative maintenance computer programs help us maintain our equipment in good working order. These programs can collect repair and maintenance costs for each machine, which helps us justify replacement.

Manufacturing engineering can use computer-aided design (CAD), computer-aided manufacturing (CAM) for tool and fixture design. New software is available for sourcing tools, machines, repair parts, and components needed in the design of new facilities.

SUPERVISION

Supervision can be assisted by good methods work, locating supervisors close to their laborers and the services needed, and designing good systems and procedures for controlling their needs. A supervisor ratio of 20 to 25 laborers per supervisor in manufacturing is good. This ratio drops to 10:1 in offices, engineering, and upper management. Supervision is probably one area where time standards would be inappropriate. Ratios are useful.

MAINTENANCE AND TOOLING

Maintenance and tooling cannot be talked about in general. Maintenance and tooling is composed of routine maintenance, ordinary maintenance, and emergency maintenance. Each category must be handled separately.

1. Routine maintenance is often called scheduled maintenance, because the maintenance is required on specific intervals and needs to be scheduled. Oiling machinery; changing bearings, bulbs, oil, and coolant; and inspecting equipment is a small list of routine maintenance

chores that can be time-studied and scheduled. Using the computer, a list of jobs can be printed out each morning for each routine maintenance person, giving them 8 hours of work. If the routine maintenance person finds something else wrong, he or she writes up a work order for ordinary maintenance or calls for emergency help at once. However, the routine maintenance person must continue to the next job.

2. Ordinary maintenance (planned maintenance) is the kind that can be planned for, scheduled, and completed in a modest amount of time. Building new items, modifying existing facilities, or making major repairs fall into this category. The expert-opinion time standard system and backlog control system are used on this category of maintenance.

3. Emergency maintenance needs to be done immediately. Any qualified craftsperson working on ordinary maintenance jobs can be called away from that job to get the plant running again. Emergency maintenance is the most expensive maintenance, because people are waiting and there is no time to plan. For this reason, emergency maintenance must be discouraged and replaced. Good routine maintenance and inspection can predict breakdowns that can be corrected as ordinary maintenance projects. A performance report for maintenance would surely include how much emergency maintenance was performed, with the goal of elimination. The newsprint business has nearly eliminated breakdowns by predicting and repairing problems before the breakdown. The following are a few methods for improving maintenance:

1. Vibration analysis
2. Heat recordings
3. Machine records
   a. Type of repair
   b. Oil consumption
4. Automatic lube systems
5. Instrumentation for flow, temperature, amp, etc.
6. Replacement part inventory with reorder points and reorder quantity
7. Plug-out/plug-in replacement units
8. Backup equipment.

WAREHOUSING AND SHIPPING

Warehousing is the safekeeping and issuing of the company's finished product. Great work has been accomplished in warehousing methods. A few examples of warehousing philosophies are as follows:

1. Locate a small amount of everything in a small area to reduce the travel time to pick orders. Locate backup stock close to picking area.

2. Have a fixed location for every finished product to aid pickers in learning the locations of parts.

3. Place the highest volume (most popular) products in the most convenient locations.

4. Design the location of all products to be available immediately without having to move other products.

5. Place the highest-volume products closest to the shipping door.

6. Pick orders and pack at the same time.

7. Design all material-handling equipment to work together—for example, picking carts the same height as packout conveyors for rolling material off the cart and onto a roller conveyor.

A major-brand tool company has six product lines (different name brands) and over 5,000 different stock-keeping units (SKUs). The tools were stored on seven high shelf racks, 3 feet wide, in part-number order. Thirty-six back-to-back rows of shelves, 30 feet deep, were needed. A person picking an order of tools needed to pass in front of every shelf, a trip of over 2,100 feet. The new layout placed the tools in order of popularity and classified each tool as an A, B, C, or D item. A items accounted for 80% of the sales dollars but only 20% of the part numbers. These were very popular tools that moved fast. These items were placed close to shipping. On the other end of the scale, D items were slow-moving tools and accounted for 5% of sales but 40% of part numbers. These tools were located at the storage area farthest away from shipping, because the picker needed to go that far for only 5% of the orders. B and C items were in between. The travel distance was reduced to 350 feet, a savings of 1,740 feet/order. This savings is astronomical, but to put a dollar figure on it, time standards are needed. In most of warehousing, three things cause time to vary:

1. The number of orders picked

2. The number of lines per order

3. The number of individual items per line.

In the tool warehouse, the warehouse person received an order at the supervisor's desk and gave it to an inspector. This work took 2 minutes. Between each line, the operator had to record the number of items picked and walked to the next location. Average distances were used to develop a standard of .3 minute per line. Once at the location, an operator could pick a unit every .2 minute (a unit may have been one socket, but they were packed in boxes of six). A time card would look like the example in Table 12-2.

Consider another example. An oil company warehouse was set up according to the A, B, C inventory classification system. The pickers drove a fork truck. Cases of oil were the units of sale. (See Table 12-3.)

Shipping, on the other hand, usually includes closing containers, weighing containers, addressing containers, writing the case number weight and carrier's name on the shipment, making out the bill of lading, and loading trucks. Individual time standards could be calculated for each of these functions, but a much easier way is to keep a count of the weight

Table 12-2 Daily Performance Report: Picker #12.

	ORDER NUMBER	NUMBER OF LINES	NUMBER OF PIECES
1.	123	25	100
2.	149	55	300
3.	175	28	250
4.	201	35	150
5.	222	15	500
6.	251	45	400
7.	300	10	600
Total	7 orders	213 lines	2,300 pieces
	×	×	×
Standard	2.0 minutes	.3 minute	.2 minute per piece

$$\text{Earned minutes} = 14 + 63.9 + 460 = 538$$

$$\frac{538 \text{ earned minutes}}{480 \text{ actual minutes}} = 112\%$$

being shipped every day and divide this by the total labor hours used by everyone in the shipping department.

$$\text{Performance} = \frac{480,000 \text{ lb}}{48 \text{ labor hr}} = 10,000 \text{ lb/labor hr}$$

If we calculate this daily, a meaningful trend line should soon develop. This is a standard on which one could base a 50/50 incentive plan. A year's worth of history would provide a great base standard.

Table 12-3 Daily Performance Report: Oil Warehouse.

	ORDER NUMBER	NUMBER OF LINES	NUMBER OF CASES	EARNED MINUTES
1.	167	30	250	74
2.	250	1	900	95.5
3.	950	20	400	74
4.	1295	15	300	56.5
Total	4 orders	66 lines	1,850 cases	
	times	times	times	
Standard	4 minutes/order	1.5 minutes/line	1 minute/case	
	equals 16	equals 99	equals 185	

Earned minutes: 16 minutes + 99 minutes + 185 minutes = total earned minutes of 300

800 cases makes one truckload (sixteen pallets/trailer)

$$\text{Efficiency} = \frac{300 \text{ earned minutes}}{480 \text{ actual minutes}} = 62\tfrac{1}{2}\% \text{ performance}$$

Drivers

Some companies use their own drivers. An oil company warehouse had 20 drivers. A driver performance report was based on 50 mph on the highway, 25 mph in the city, 20 minutes per stop, plus 2 minutes per pallet. The drivers were loaded with a known amount of work and were expected back at a specific time for their next load or inside work.

A lumber company in Portland, Oregon, had 33 trucks and drivers delivering lumber. All 33 were loaded by yard crews overnight, and at 7:00 A.M., all trucks left the yard to make their first run. At 10:00 A.M., trucks were lined up waiting for their next runs to be loaded. The drivers were idle during this loading time. A methods improvement was to eliminate 6 drivers and use their trucks as backups to the others. When a driver returned from a delivery, he or she left that truck and picked up the next loaded truck in line—no waiting. Time standards were developed for each run by the dispatcher, and an estimated time of arrival (ETA) was placed next to that driver's name on the dispatch board. Drivers had no problem if they were late once in a while, but consistent lateness earned them some coaching. This may sound a little harsh, but this company saved over $300,000 per year with this system.

RECEIVING AND STORES DEPARTMENT

The receiving department's responsibilities are to unload carriers, create a receiving report, and check in all material as to quantity, quality (visual only), and correctness of part numbers. Receiving is a smaller department than shipping, but it can develop a pounds-per-labor-hour standard or a trucks-received-per-labor-hour standard. Sometimes companies combine receiving and shipping.

The trucking industry delivers products coming into the area in the morning and picks up finished products from shipping in the afternoon. This is called less-than-truckload (LTL) shipping or common carriers shipping. All the less-than-truckload freight for your company is delivered to a central trucking company station in your town and is collected until the next morning. The local common carrier will load your material and several other companies' materials onto a trailer and make its deliveries. This high morning workload for receiving is combined with a high afternoon workload with shipping, making one easily manned, combined department. The combined department will use pounds shipped per labor hour as a standard because everything shipped must have been received, and in the long run, shipping and receiving must be equal.

Common equipment is used in both shipping and receiving; therefore, fewer pieces of expensive material-handling equipment may be needed if shipping and receiving are combined. Fork trucks, scales, and dock doors are a few examples of this expensive equipment.

Employee qualifications are similar for both receiving and shipping: attention to detail, equipment operations, and responsibility for company assets.

Stores' responsibilities include locating, safekeeping, and delivering production materials to the first production operation. A stores clerk must put stock away so that it can be readily available when needed and must retrieve that stock when ordered by production. Storekeepers must keep accurate records of locations and inventory on hand.

All movement of material must be authorized to ensure that inventory records are correct. Unauthorized movement of material could cause a plant shutdown because the material needed is not there, even though the inventory list said it was.

Methods improvement ideas for storerooms include the following:

1. Random locations: Put the pallet or box into the first open space the driver comes to and record that location on a locater ticket. This reduces driving time and reduces the storage space requirements of the storeroom by one-half.

2. Create addresses for every location in the warehouse. Each row could be a number, each pallet location back from the main aisle could have a letter (A, B, C), and each shelf would have a number (floor 1, 2, 3, 4). An 11-F-4 location would be the eleventh row of racks, the sixth pallet location down that row, and on the top tier. F would always be the sixth location back, and 4 would always be the top.

3. Locate all slow-moving and obsolete inventory in the back of the storeroom.

4. Use narrow-aisle vehicles to save space.

Time standards for storekeepers are based on the number of pallets put away and the number of pallets retrieved. With a little planning, one can be put away and one can be retrieved on every round trip, making for great efficiency. A locater ticket must be made out for every item put away, and a locater ticket must be pulled for every item withdrawn from the storeroom. These locater tickets can be collected and multiplied by a standard hour per move to develop an earned hours figure:

$$\text{Efficiency} = \frac{\text{earned hours}}{\text{actual hours}}$$

FACTORY/CLERICAL

Many factory clerical employees are staff assistants to busy factory managers and thus are like extensions of the manager. These positions are justified by saving the manager time for more important duties. No attempt will be made to measure their performance, only their supervisor's performance.

Other factory clerical employees may be required to keep records of activities, inventory, equipment, cost, etc. These activities and transaction times can be determined. Otherwise, how do we know if the person has enough work—or too much?

Computer input clerks are often monitored by their computers, and keystrokes per hour are printed out each day for each person. This is very useful data, but the different jobs must have different expectations (standards).

OTHER INDIRECT AREAS

1. Utilities companies

2. Mining companies

3. Hospitals

4. Retail stores

5. Insurance companies

6. Airlines

7. Hotels

8. Restaurants

All of these areas are very different, but they have much in common with direct labor in manufacturing plants. If they have too much or too little work, cost efficiency is lowered. They need time standards to ensure correct manning and to balance the work among employees. They need to control costs, and they must continue to improve work methods.

QUESTIONS

1. What is direct and indirect labor?

2. How are numbers of indirect employees calculated in budgets?

3. List eight areas of indirect labor.

4. List two methods improvement ideas for each category of indirect labor.

5. When would you use the stopwatch, PTS, or expert-opinion time standards?

Performance Control Systems

The motion and time study application that affects more people than any other is the performance control system. A performance control system is very personal because it judges people, and for this reason it receives more attention than any other use of time standards. Our discussion of performance control systems is divided into four sections:

1. The functions of any control system
2. Expert opinion standards system
3. Backlog
4. Time card system.

THE FUNCTIONS OF ANY CONTROL SYSTEM

Quality control, inventory control, production control, cost control, attendance control, and performance control all have the same required functions:

1. Planning or goal setting
2. Comparison of actual to goal
3. Tracking results
4. Variance reporting
5. Corrective action.

Unless each of these functions is performed properly, there is no control system. The following discussion is intended to develop an understanding of what is required of each step of any control system.

Planning or Goal Setting

What is to be achieved? In every control system, the goals must be set; most important, these goals must be measurable and achievable. The planning techniques used in performance control systems are the time-standard–setting techniques studied in this book:

1. PTS
2. Stopwatch
3. Standard data
4. Work sampling
5. Expert opinion (discussed later in this chapter).

In addition to these planning and goal-setting techniques, the managers of the operation to be controlled must have an acceptable performance goal in mind:

1. With time standards and a performance control system in place, the average performance is 85%, not 100%. An expectation of much higher than 85% is not practical and the chance of success is small; therefore, planning and goal setting must include standards and reasonable expectations.

2. Operators on incentive work at 120% performance. When planning labor requirements, 120% must be used, not 100%, or there will be too many operators with not enough work to keep them busy.

Without goals (or standards), we are without direction, and achieving our potential is impossible.

Comparison of Actual to Goal

The second function of any control system is to compare the actual results to the planned results. A quality control system would ask us to measure a product (actual results) and compare it to a blueprint (goal or standard). If the results are in tolerance, the product is classified as a good one. Performance control systems do exactly the same.

An operator will produce a number of units in a period of time. This is compared to the number of units the time standard asked for. The result is a percent performance. Every job the operator does during a day is collected and compared to create a daily performance report. This operator is combined with all the other operators to produce supervisor, department, shift, and plant performances for each day, week, month, and year.

Tracking Results

Tracking could be called graphing. It is plotting results against a horizontal time line. In a cost control system, the goal line is the planned expenses per period. The actual expenses are plotted against this goal for each period, and any casual observer can see how the plan is going. No single period is important, but the trend or direction is important.

In performance control systems, percent performance and percent indirect are plotted weekly. The trend on percent performance (productivity) should be going up, while at the same time the indirect percent must be held flat or be diminishing. The trends are what is important. Percent indirect is indirect hours divided by total hours.

Variance Reporting

When actual performance does not live up to goals (expectations), a variance from standard exists. For example, in an attendance control system, employees are allowed to be absent twice a month without being criticized, but if they exceed this number, a variance exists and a report would be generated by the payroll department. This is a variance report. In a production control system, if a part is behind schedule by 4 hours, it is placed on an expedite list—a variance report. Nearly every actual result will be different from the goal. The magnitude of these variances is what is important. The larger the variance, the bigger the problem. Managers should attack the largest problems (variances) first and fix these problems.

In performance control systems, those individual jobs that reported performances lower than 70% and above 130% are collected and printed out on the variance report. Each job of the variance report becomes a project, and the reasons for the variances must be determined. Each variance can be assigned to a person for investigation and corrective action.

Corrective Action

Solving problems and implementing these solutions is the reason control systems exist. Corrective action makes it all happen. Talk to the person who is having an attendance problem and find a solution acceptable to both employee and supervisor, adjust the machine to bring quality back to standard, expedite a purchase order if material level drops below target, work overtime if behind schedule, and fix the problems causing poor performances. Corrective action in performance control systems is wide-ranging:

1. Machine maintenance problems
2. Material problems
3. Management problems such as lack of assignments, lack of instruction
4. Poor time standards
5. Poor operator effort.

Performance control systems hold problems up to public scrutiny, and problems get solved. Without performance control systems, operators know that management does not care about productivity and supervisors don't want to be bothered with problems.

Performance control systems raise the productivity of departments and plants by 42%. Taking corrective action is how that happens. A 42% decrease in direct labor cost is a significant savings, but there is no easy way. Performance control systems are hard work, but hard work is what successful people do.

EXPERT OPINION STANDARDS SYSTEM

Another technique for setting time standards is the expert opinion technique. An expert is someone who has a great experience base, and because of this experience can estimate requested work in his or her area of expertise with acceptable accuracy. I can ask you how long it takes to drive to your parents' home, and you can make an accurate estimate. The next time you make that trip, it will be a little longer or shorter than the estimate, but the average will probably be accurate. I can make the estimate for you using 50 miles per hour average, but I won't know the small towns, the construction zones, the coffee stops, etc. required by the trip, so my standard (engineered standards) would not be as good as yours.

Many engineers, maintenance managers, office managers, and other people not on machines say, "You can't set time standards on my job." They are correct. We can't. But they can! The maintenance department is a good example. As an industrial technologist, you have designed a new work station for a cost reduction. Before the cost reduction return on investment can be calculated, the cost must be estimated. You take the drawings to the maintenance supervisor and ask how much it will cost. The material cost, as well as labor cost, must be estimated, but we are talking only about labor right now. The maintenance supervisor looks at the job, mentally compares it with other jobs completed in the past, and zeros in on the estimate. The maintenance supervisor breaks down big jobs into small jobs and estimates their cost. The estimates are written on work orders and given back to you for approval. The approval process will require upper-level management's approval as well, but when the work orders are approved, they are returned to the maintenance supervisor for scheduling and completion. The maintenance supervisor enters the job on the backlog, orders material, and schedules the workers.

BACKLOG

Each service department maintains a list of jobs that have not yet been completed. This is the backlog, and the estimated hour total of all jobs is the backlog hours (see Table 13-1). The backlog control system, incorporating expert opinion standards, can be used for any service work such as engineering, maintenance, and office functions. A backlog of work is needed to give time for planning and efficient scheduling. Management and the department manager determine what an acceptable backlog is (number of days worked), and the department manager maintains that much work.

Every job in the maintenance backlog is listed and estimated. As jobs are completed, the completion date and actual hour columns are filled in. As new jobs come in, they are estimated and added to the backlog. At the end of every week, the backlog hours are totaled and

Table 13-1 Maintenance Backlog List—Steps ① and ② of the Step-by-Step Procedure in Section 2 of This Chapter.

JOB NO.	DATE RECEIVED	JOB DESCRIPTION	HOURS REQUIRED	DATE COMPLETED	ACTUAL HOURS
101	11-15-XX	Overhaul fork truck	16	11-25-XX	15
102	11-18-XX	Build safety cage	12	11-24-XX	13
103	12-1-XX	Build supervisor office	40		
104	12-3-XX	Build conveyor system	132		
105	12-10-XX	Repair machine 12y #1576-05	24		
106	12-12-XX	Build work station #1576-10	10		
107	12-12-XX	Build work station # 1576-15	7		
108	12-12-XX	Build work station #1576-20	15		
109	12-12-XX	Build work station #1576-25	4		
110	12-13-XX	Move machine 64	2		
Etc.	50 more jobs				
		Total hours	1,324		

recorded on the control graph. This maintenance department has 10 maintenance workers. Ten times 40 hours per week means that maintenance has 400 hours per week available. Not all work is covered by backlog:

1. Emergency maintenance has accounted for 10% of the maintenance time in the past, so 360 hours are still available.

2. Routine maintenance or scheduled maintenance is scheduled just like production work, and one of the crew spends full time oiling, changing bulbs, etc. on a routine schedule.

3. Ordinary maintenance has 320 hours per week available. How many weeks of backlog does management think appropriate? Let's say, three weeks. Three weeks times 320 hours per week equals 960 hours of backlog. What should the maintenance manager do?

A backlog control graph (Figure 13-1) results from totaling the backlog list (Table 13-1) each week. The trend line is important.

Step-by-Step Procedure for Developing a Backlog Control System

Step ① List all jobs waiting to be completed in the department. List oldest jobs first.

Step ② Estimate the time required for each job, and total the time. This is the beginning backlog.

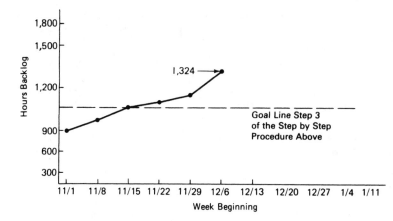

FIGURE 13-1 Backlog control graph.

Step ③ Establish a backlog hour goal. A sufficient amount of backlog is needed to allow for efficient planning and scheduling. (See Figure 13-1, Step ③.)

Step ④ Establish a control graph and a control chart. Plot the beginning backlog. (See Figure 13-1, Step ④.)

Step ⑤ During each week, jobs are added to the backlog, and times required are estimated. At the end of the week, the total hours added is totaled and entered onto the backlog control chart. (See Table 13-2.)

Step ⑥ During each week, jobs completed are totaled and entered onto the backlog control chart. (See Table 13-2.)

Step ⑦ At the end of each week, the new job hours are added to the beginning backlog hours, and the hours completed are subtracted, resulting in the ending backlog

Table 13-2 Backlog Control Chart—Variance Report.

WEEK BEGINNING	BEGINNING BACKLOG HOURS	HOURS ADDED	HOURS COMPLETED	ENDING BACKLOG HOURS	CHANGE VARIANCE
11/1	890	400	310	980	90
11/8	980	360	340	1,000	20
11/15	1,000	460	335	1,125	125
11/22	1,125	375	350	1,150	25
11/29	1,150	450	400	1,200	50
12/6	1,200	444	320	1,324	124
12/13	1,324	380	340	?	?
12/20	?				

Note: ? = fill in and plot on graph (Figure 13-1).

hours. This number is recorded on the backlog control graph and brought forward to the beginning backlog hours for the next week.

Step ⑧ The beginning backlog hours are subtracted from the ending backlog hours, resulting in the change or variance. Table 13-2 is an example of a variance report.

    a. A positive variance means the backlog is growing and some action may be required.

    b. A negative variance means the backlog is shrinking and other actions may be required.

Step ⑨ Corrective action:

    a. If the backlog trend is growing, such as in Table 13-2, it may be necessary to add people.

    b. If the backlog trend is coming down by at least 40 hours per week, a reduction of people may be required.

    c. If the backlog is flat but too high, overtime, temporary help, or farming out work would help.

    d. If the backlog is flat but too low, loaning people out, taking vacation time, or sending people to school would correct the situation.

TIME CARD SYSTEM

The performance control system for most of industry is based on the individual time card. The math and the layout of forms for department, shift, and plant performance reports (daily, weekly, monthly, and yearly) are designed after the daily employee time card. The time card takes a different form than the weekly payroll card and is punched upon arrival and leaving the plant. This time card is used only for performance control. Table 13-3 is an example of a completed time card.

Step-by-Step Procedure for Time Card Calculation

Steps ①–⑦ are completed by the operator. Steps ⑧–⑬ are completed by data processing.

Step ① Name: The operator's name goes here. In large systems or computer systems, an operator number may be used, but for good employee relations, the name should always be included.

Step ② Date: The date the work is performed goes here. As always, the complete date should be used.

Step ③ Department: Each employee is assigned to a department, and all employees of a department are combined for a department performance report. If an operator works in two departments on a given day, two time cards are used. Department numbers are often used instead of department names.

Table 13-3 Time Card Example.

Shift 7 A.M. to 3:30 P.M.
Lunch 11 A.M. to 11:30 A.M.

OPERATOR ① NAME: MARY				② DATE: 12/13/xx			③ DEPARTMENT: PRESS			
TIME IN	TIME OUT	JOB NO.	OPER. NO.	TIME STD.	PIECES PRODUCED	ACTUAL HOURS	EARNED HOURS	%	INDIRECT TIME
7:00	8:45	1660	10	100	200	1.75	2.00	114	
8:45	10:15	1700	15	167	250	1.50	1.50	100	
10:15	12:15	1660	15	850	1,500	1.50*	1.76	117	
12:15	1:00	Meeting				—	—	—	.75
1:00	2:06	1750	15	45	55	1.10	1.22	111	
2:06	3:30	1800	10	750	1,000	1.40	1.33	95	
Total						7.25	7.81	108	.75
④	④	⑤	⑤	⑥	⑦	⑧ ⑫	⑨ ⑫	⑩ ⑫	⑪ ⑬

*Lunch deducted.

Step ④ Time in, time out: A time clock could be used, but having the operators write the time in is the preferred practice. Decimal hours are preferred, and six minutes is .1 hours. Well-conceived systems allow only one column for time in, and the time in on one job is the time out on the previous job. This eliminates gaps in time that are impossible to figure out after the operator goes home.

Step ⑤ Job no. and oper. no.: Having the operator fill in the job number and operation number ensures that the operator knows the correct time standard. The information can be retrieved from a traveling route sheet or operating instructions at the supervisor's desk. The data processing systems must also look up these numbers to check the time standard. The operator must know what is expected, or we have no performance control system.

Step ⑥ Time std.: The time standard is the engineering time standard in pieces per hour. The hours per unit or 1,000 units is sometimes used, but pieces per hour is easier for the operator to use.

Step ⑦ Pieces produced: The number of units made during the run of the job goes here. The count can come from a counter mounted on the machine or work station, or it can be predetermined by standard container counts. Pieces produced always means good pieces. Scrap cannot be allowed in production figures or efficiency.

The operator is required to fill in all the information up to this point. The performance control system must audit this information just like any good management system.

Step ⑧ Actual hours: The actual hours are the number of clock hours the operator actually works on the job. Mathematically, actual hours are the ending time less the beginning time. A normal day is 8 hours, and the total of actual hours should be 8 (more on this in Step ⑫).

Step ⑨ Earned hours: Earned hours are the amount of work the employee did compared to the time standard. Earned hours are calculated by dividing the time standard (Step ⑥) into the pieces produced (Step ⑦).

Step ⑩ %: Percent performance is the relationship between actual hours and earned hours. If the operator earned more hours than used, the percent performance is over 100%. If the operator earned fewer hours than actually used, the percent performance is under 100%. Percent performance is calculated by dividing earned hours (Step ⑨) by actual hours (Step ⑧).

Step ⑪ Indirect hours: Indirect hours are hours worked that are not covered by the time standard. Indirect hours are kept separate from direct labor hours. Indirect hours are neither in the actual hours nor the earned hours, but are recorded in this column. Indirect labor hours must be controlled by a reason code. Each indirect labor charge has a reason code number, and time is totaled and summarized by code number. Example reason code numbers follow:

01 Maintenance

02 Quality

03 Safety

04 Material handling

05 Clean-up

06 Setup

07 Rework

08 Miscellaneous

The number of indirect codes can be large. Each cause of indirect labor must be controlled so inflation does not occur. Achieving 100% performance would be easy if production were run while charged out on an indirect code. In well-managed plants, indirect labor is kept under 20% of total labor.

Step ⑫ Totals: Total actual hours, the total of column ⑧, is divided into the total earned hours (Step ⑨) to calculate total percent performance (Step ⑩). The only way to calculate percent performance (whether for a job, person, department, plant, day, or year) is to divide total actual hours into total earned hours.

Step ⑬ Total indirect hours: Total indirect hours are added to total actual hours to figure how many total hours were paid for. (Total indirect hours are discussed later in this chapter.)

Step ⑭ (Not shown in the table.) The time card hours are compared to the payroll hours. These two numbers must come out the same. This process is called *justifying* time card and payroll hours. Step ⑭ is in addition to Table 13-3. The total actual hours in ⑫ must agree with payroll hours. Remember that payroll is another system.

Figure 13-2 shows examples of daily performance reports. The step-by-step procedure that corresponds to Figure 13-2 shows how to calculate department performances.

Step-by-Step Procedure for a Department Performance Report (Daily)

Step ① Daily performance report—Dept.: One performance report per department and one summary for all departments is needed. If the plant has 10 departments and 200 people, there will be 11 daily performance reports—one for each department supervisor and one for the plant production manager. The plant performance report will look just like the one in Figure 13-2, except with supervisors' names where the employees' names are.

Step ② Date: The date the work was accomplished.

Step ③ Supervisor's name: The department performance report measures the supervisor's performance. Productivity performance is one of the top two measures of production management, although there are many more.

Step ④ Operator name: The time keeper maintains preprinted daily performance reports with operator names already included. The time keeper will put the time cards in name order and then transfer the information for columns ⑧, ⑨, ⑩, and ⑪ onto the daily departmental form.

Daily Performance Report   Assembly

Daily Performance Report   Dept. Paint

Daily Performance Report   Dept. Press

Date 12/13/XX   Supervisor's Name   Dale ③

Operator Name	Actual Hours	Earned Hours	%	01	02	03	04	05	06	07	08	Σ	%
Mary	7.75	7.81	108								.75	.75	9
Fred	8.0	9.25	116								—	—	
Jim	1.25	1.50	120		4				2		.75	6.75	84
Bob	—	—					6	2				8	100
Lynn	8.0	10.0	125										
Tommi	7.25	8.00	110								.75	.75	9
Pat	8.0	12.0	150										
Bill	—	—		3				2	2.25	.75		8	100
Dee	8	6	75										
① Total	47.75	54.56	114	3	4	—	6	4	4.25	.75	2.25	24.25	34
④	⑤	⑥	⑦				⑧				⑨		⑩

Indirect Codes

FIGURE 13-2 Performance control system: daily department reports, one for each department plus one for the total plant.

Step ⑤ Actual hours ⎫ These three columns are transferred
Step ⑥ Earned hours ⎬ from the time card totals to the
Step ⑦ %   ⎭ daily performance report as they are.

Step ⑧ Indirect codes: Indirect time charged on individual time cards is coded and listed to the right of the operator's name.

Step ⑨ Σ: The sum of each individual's indirect hours is placed here.

Step ⑩ %: The percent indirect hours is calculated by dividing the indirect hours by the total hours worked (direct actual hours plus total indirect hours).

Step ⑪ Total: Each column is totaled, except the two % columns.

    a. % Direct: Step ⑦ is calculated by dividing earned hours by actual hours.

    b. % Indirect: Step ⑩ is calculated by dividing hours by actual hours plus indirect hours.

The daily performance reports for operators and supervisors are good indications of how their days go, and a good work record is built on many good days.

Weekly Performance Reports

Weekly performance reports (see Table 13-4) are exactly the same as daily reports, except the day of the week replaces the operator's name and daily totals are placed in each block. Notice in Table 13-4 that the information for Mon. 12/13/XX has been transposed from the bottom of the daily performance report in Figure 13-2. Not one change has been made. At the end of the week, each column will be totaled except the two percent columns (just like the daily report), and the percent direct labor and the percent indirect are calculated just like the daily performance report. Notice how the columns in Figure 13-2 and Table 13-4 line up exactly.

Table 13-4 Performance Control System: Weekly Performance Reports, One for Each Department Plus One for the Total Plant.

| | | WEEKLY PERFORMANCE REPORT | | | | | DEPT. PRESS | | | | | | |
| | WEEK ENDING: 12/18/xx | | | | | | SUPERVISOR'S NAME: DALE | | | | | | |

| | | | | INDIRECT CODES | | | | | | | | | |
DAY	ACTUAL HOURS	EARNED HOURS	%	01	02	03	04	05	06	07	08	Σ	%
Mon.	47.75	54.56	114	3	4	—	6	4	4.25	.75	2.25	24.25	34
Tue.													
Wed.													
Thurs.													
Fri.													
Sat./Sun.													
Total													

Table 13-5 Example: Actual Performance Report.

Crain Enterprises, Inc.
Plant Performance Report

EMPLOYEE	WEEK ENDING/PERCENT EFFICIENCY								
	02 MAR	09 MAR	16 MAR	23 MAR	30 MAR	07 APR	14 APR	21 APR	28 APR
Bailey	N/A	N/A	N/A	N/A	N/A	N/A	N/A	N/A	N/A
Baird	77	90	99	78	99	84	104	97	110
Cook	52	63	95	83	94	47	56	53	90
Cross	73	95	95	107	VAC	85	91	92	95
Curley	N/A	N/A	N/A	N/A	N/A	N/A	46	63	66
Dover	N/A	128	116	196	87	160	112	N/A	140
Goins	107	105	118	128	115	120	128	111	99
Goins	85	107	107	74	83	100	112	108	102
Hannan	140	137	120	115	108	112	140	115	118
Isom	111	106	98	94	87	107	118	90	100
Jones	66	68	65	65	84	103	93	101	91
Kennedy	66	75	93	95	101	120	116	120	127
Marrs	74	100	111	82	84	90	103	82	91
Monan	87	95	100	N/A	N/A	N/A	N/A	86	87
Polley	VAC	128	124	116	115	113	119	119	123
Stigall	N/A	141	111	131	115	121	129	101	101
Taylor	112	135	130	123	129	126	150	131	109
Taylor	85	104	95	69	74	91	127	104	99
Wilson	107	131	130	130	97	131	132	130	128
Plant total:	87	104	105	97	98	108	110	100	103
Percent direct hours	75	75	69	57	59	62	66	67	66

There will be the same number of weekly, monthly, and yearly reports as there are daily reports. The time it takes to calculate any of these reports by hand is only a couple of minutes per time period because only transferring information is required.

The monthly report will have week-ending dates instead of days of the week. The yearly report will have months instead of days of the week. Everything else is the same. On a yearly report, 12 months of information is available on one $8\frac{1}{2} \times 11$ sheet of paper per department. This information becomes valuable in planning budgets for the following year.

Tracking

A graph of weekly performances will be useful to management. If anything is going right or wrong with productivity, the tracking will highlight it. Three pieces of information should be tracked as in Figure 13-3 (the data for tracking came from Table 13-5):

1. Percent performance

2. Percent indirect

3. Output per laborhour.

					PERCENT EFFICIENCY						
05 MAY	12 MAY	19 MAY	26 MAY	02 JUN	09 JUN	16 JUN	23 JUN	30 JUN	07 JUL	14 JUL	21 JUL
N/A	N/A	N/A	N/A	N/A	N/A	69	85	95	Closed	97	113
103	100	104	127	150	73	N/A	N/A	N/A	Closed	N/A	N/A
90	89	81	91	119	118	114	N/A	N/A	Closed	N/A	N/A
87	82	121	106	112	103	104	102	104	Closed	107	122
104	118	115	118	123	LOA	LOA	LOA	108	Closed	123	109
N/A	115	127	123	VAC	131	112	117	96	Closed	114	132
110	116	128	124	119	96	122	VAC	121	Closed	119	113
99	132	125	138	134	96	104	110	117	Closed	110	116
128	140	142	110	127	130	130	135	147	Closed	123	136
99	103	104	108	107	117	110	111	98	Closed	N/A	N/A
101	107	124	117	109	108	108	80	107	Closed	114	108
VAC	97	119	109	144	123	137	104	130	Closed	136	137
82	112	114	69	83	81	86	98	88	Closed	IND	93
104	123	121	106	107	97	85	VAC	91	Closed	94	119
114	113	120	114	126	110	130	118	118	Closed	102	107
N/A	118	120	128	124	121	113	LOA	106	Closed	122	104
131	130	118	122	114	121	135	93	112	Closed	125	123
100	122	127	119	107	VAC	118	117	122	Closed	118	111
121	N/A	139	VAC	138	138	133	112	133	Closed	142	145
104	112	121	115	121	112	113	106	114		116	120
57	58	70	62	60	65	73	74	65		76	69

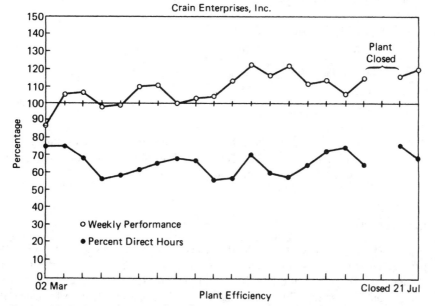

FIGURE 13-3 Example: actual performance graph (same data as in Table 13-5).

Percents can be measured on the left side of a graph, while output per laborhour can be measured on the right side. The output per laborhour is a good check of the percent performance accuracy. Both should go up and down at the same time if time standards are meaningful.

Variance Reporting

The time cards can be returned to the operator once the daily performance report is calculated. If an operator's performance is not normal, the supervisor can talk to the operator to find out what happened. The goal is to correct problems. The operator is the best source of information, but the operator should notify the supervisor when he or she is having trouble. Often, the operator does tell the supervisor of a problem, but the supervisor forgets to assign someone to fix it. Supervisors are often bombarded with problems from many directions at the same time, and forgetting one problem is easy.

Computer programs are also used to highlight variances that are beyond acceptable limits. Limits must be set, and variance beyond the limits will be listed. For example, any work less than 70% or over 130% performance needs to be reviewed. If the computer lists variances in order of magnitude, the worst problems will be on top of the list.

Corrective Action

The reason for variance reporting is to identify the problems so they can be fixed and eliminated. Each variance becomes a project, and a project engineer can be assigned. The project engineer is the person who can best fix the problem, someone like the

1. Industrial technologist,
2. Supervisor,
3. Maintenance person,
4. Operator,
5. Tooling manager, etc.

A project control system is needed, and the most important aspect of this system is to motivate the project engineer to get the job done. The project list should be distributed to all engineers and managers and the top manager every week.

Conclusion

A properly designed and operated performance control system can save thousands of dollars per week. The productivity of the plant increases an average of 42%, and cost decreases proportionately. Manufacturing and distribution companies cannot afford *not* to have performance control systems. Table 13-6 and Figure 13-3 show actual performance reports and control charts.

Table 13-6 Time Card Problem.

Shift 7 A.M. to 3:30 P.M.
Lunch 11 A.M. to 11:30 A.M.

TIME IN/OUT	JOB NO.	OPER. NO.	TIME STD.	PIECES PRODUCED	ACTUAL HOURS	EARNED HOURS	% PERFORMANCE
7:00	1600	20	125	300			
9:00	1610	25	1,200	2,000			
10:45	1500	15	45	120			
2:00	1700	25	300	600			

QUESTIONS

1. Why is a performance control system given so much attention?

2. What are the five functions of any control system?

3. What is proper expectation for productivity?

4. What percent performance will you achieve if you tell the operator that 85% performance is acceptable?

5. What is an expert opinion standard?

6. Who can use expert opinion standards?

7. What is a backlog? Why do we need a backlog? How big should a backlog be?

8. What are the nine steps of a backlog control system?

9. What is the percent performance for the time card problem shown in Table 13-6?

10. What is the relationship between individual time cards and daily, weekly, and annual performance reports?

CHAPTER 14

Wage Payment Systems

Wage payment is a measure of value; therefore, it is more important than mere economics would indicate. Maslow's hierarchy of needs (discussed at greater length later in this chapter; see Figure 14-1) shows us that, at the fourth level, people are motivated by reputation, self-respect, achievement, recognition, challenge, responsibility, competence, and challenging work. Many of these motivators are reinforced by the pay scale. Incentive pay plans build on these motivators and form a major part of the discussion in this chapter. People are classified as hourly or salaried and exempt or nonexempt. People are paid by the hour, week, month, or the piece, and they can earn extra money through bonus plans, commissions, and profit sharing.

This chapter builds on the performance control system of the previous chapter and the time standards developed throughout this book. So far, we have developed the goals and reported the percent performance, but how much do we pay people? There are three ways of paying employees:

1. Salaries: paid by the hour, week, or month
2. Incentive and commission (including time off)
3. Bonus and profit sharing.

SALARIES: HOURLY, WEEKLY, OR MONTHLY

How much is each position (job) worth per hour, week, or month? Each job has different requirements for education, work experience, responsibility, and several other factors.

To pay equitably for these differences, we use a job evaluation program. To evaluate what each job is worth, a job evaluation must be made. Each position earns a point value: the higher the point value, the more complicated the position. Groups of positions with similar point values are pulled together into a job classification. Once we know the relative value of these job classifications, a local wage and salary survey is used. Management must decide in what wage percentile its company will compete. It may choose to pay the average rate, above the average rate, or even below the average. But once the wage percentile has been determined, the rates are set using that percentile from the wage and salary survey. A salary range may be specified to allow for longevity and experience.

Exempt Versus Nonexempt

An exempt category of employee means that these employees are exempt from the wage and hour laws. Supervisors, engineers, and managers are in this classification and have no protection under the law. A nonexempt employee will be paid overtime for work over 8 hours per day and 40 hours per week, must be paid at least the minimum wage, and cannot be discriminated against because of race, religion, sex, national origin, or age. Exempt employees are salaried and are often paid a fixed sum per week or month, no matter how many hours are involved. Nonexempt employees can be either hourly or salaried but are paid for overtime on an hourly basis.

A rate of pay has been calculated for each position, and a time card is legally required for every nonexempt person. The time clock hours multiplied by the rate per hour is the weekly pay. This is true of hourly or salaried nonexempt employees. Exempt employees are paid automatically until terminated.

Measured Day Work

A measured day work system is a performance control system based on time cards and time standards. The time card has nothing to do with payroll unless it is used in combination with an incentive system. A measured day work system is for productivity measurement and control and will generate a large savings in labor. About 75% to 85% of all factory employees use job time cards for measured day work. Measured day work is one step required before starting an incentive system and was illustrated in Chapter 13 as the performance control system.

Day Work

The term *day work* refers to being paid by the hour, week, or month. Day work has no time standards or performance calculation. Day work pay systems are, in fact, no system at all.

U.S. Government Bodies That Regulate and Influence Wages and Salary Practices

1. Wages and Hours Division, Department of Labor: administrates laws governing minimum wage, overtime, etc.

2. National Labor Relations Board: regulates labor practices.

3. Equal Employment Opportunity Commission: investigates charges of discrimination.

Job Evaluation Technique

Job evaluation for wage and salary determination is the process of determining the value of each specific job or position in a company and comparing this value to every other job or position in the company. The purpose is to set wage and salary levels for all employees.

During World War II, the following organizations worked together to design a fair and equitable method of determining wages and salary:

1. National Electrical Manufacturers Association

2. National Metal Trades Association

3. General Electric Company

4. Westinghouse Electric Company

5. U.S. Steel.

The results of their work were similar to the factors and point values shown in Table 14-1. A description of each degree is helpful in making proper assignments of point values for each factor, but this is a subject for an advanced course.

A technologist making the evaluations must have the following information:

1. A good understanding of all jobs

2. A good knowledge of the company

Table 14-1 Wage and Salary Point System.

	DEGREES AND POINT VALUES					
FACTORS	1	2	3	4	5	6
1. Work experience needed	22	44	66	88	110	120
2. Essential knowledge and training needed	14	28	42	56	70	90
3. Initiative and ingenuity	14	28	42	56	70	
4. Analytical ability	30	40	50	60	70	
5. Personality requirements	20	30	40	50	60	
6. Supervisory responsibility	5	10	15	20	25	
7. Responsibility for loss	10	20	30	40	50	
8. Physical application	10	20	30	40	50	
9. Mental and visual application	5	10	15	20	25	
10. Working condition	10	20	30	40	50	
11. Dexterity	60	120	180	240		
12. Character of supervision required	10	20	30	40	50	
13. Responsibility for confidential matters	5	10	15	20	25	

3. Direction from top management

    a. Number of wage classifications

    b. Company's wage percentile goal

4. An area wage and salary survey

Once armed with this information, the six-step guide to making a job evaluation can be used.

Steps in Making a Job Evaluation

1. List all the jobs to be evaluated on a single sheet of paper. Make 14 copies of this list—one for each *factor* and one total sheet.

2. Assign the number value for each job, one factor at a time. This will allow you to think about only the work experience required for each job, thereby doing some ranking and grouping of jobs. Go through all 13 factors, but do every job for one factor at a time.

3. After all factors for all jobs have been evaluated, use the fourteenth copy of the job list to total all the numerical values for each job—the higher the point value, the more complicated the job.

4. List all the jobs, starting with the highest point value, down to the lowest point value.

5. Group the jobs into pay grades—depending on how many ranges the management has recommended. For example, some firms have only four different wage rates; others have as many rates as they have people. It is better to have as few as possible to eliminate excessive job bidding and transfer.

6. Compare your wage rate with the community's and recommend a wage and salary plan.

Some words of caution for making job evaluations:

1. Do not think about the people in the jobs now. Think only of job content.

2. Make your pay grade groupings logical. After totaling all the points for each job, we list those jobs and points in decreasing order. Look for large gaps in the totals and pick the midpoints of these number gaps as your pay grade breaking points. One point should not be the difference between two pay grades because we are not that accurate. For example, if the total points for several jobs were 300, 299, 297, 280, 279, 277, 275, we would pick 288 as the breakpoint between pay grade 1 and pay grade 2.

3. Do not make the evaluations public knowledge, or you will spend the rest of your career justifying your position.

4. If management recommends 10 wage scales and 9 or 11 would be better, tell them.

INCENTIVE SYSTEMS AND COMMISSIONS

Incentive systems are as old as humanity, and they work because they satisfy Maslow's hierarchy needs of survival, recognition, and ego gratification. They are not used often enough because incentive systems are a lot of work, but the savings can be spectacular. A 41%

increase in productivity is typical when going from a performance control system to an incentive system. In a 100-person plant, such an increase is equivalent to having 41 more people and can give a company a very competitive advantage. It usually means more business and more employees, not less. We believe in incentive plans.

A properly designed incentive system will

1. Reduce unit cost,
2. Increase equipment use,
3. Promote team spirit,
4. Increase employee pay for increased effort,
5. Improve job satisfaction,
6. Recognize outstanding employees, and
7. Create a cost-conscious, motion-conscious work force.

Maslow's hierarchy of needs explains much about what people need and want in life. It states that people strive to satisfy lower-level needs until those needs are satisfied, then they strive to satisfy higher-level needs. This is important to understand in industry so that we can get high-quality participation from our work force. The pyramid in Figure 14-1 illustrates Maslow's thoughts. Maslow says that we need to satisfy the physiological needs before we look after the safety needs, and safety needs need to be satisfied before we strive for the social needs. These three lower-level categories of needs have been described as maintenance factors and not motivators for industrial workers. They must be satisfied before we can concentrate on the top two categories of needs. Incentives promote the ego needs of category D, the fourth level. Management wishes all employees were in the self-actualization stage—the employees would supervise themselves!

Every manager has seen what can happen when one employee really cares about his or her job. That employee is eventually promoted and/or given other recognition and rewards. Just think what can happen if a whole plant cared. An incentive wage plan will do this. The National Science Foundation found that when workers' pay is linked to their performance, the motivation to work is raised, productivity is higher, and the employees are likely to be more satisfied with their work. A Mitchell Fein Study of 400 plants in the United States found that incentive systems increased performance by 42.9% over measured day work systems, and 63.8% over plants with no standards.

Managers in the United States believe in incentives for managers. Eighty-nine percent of manufacturing companies have executive bonus plans, and managers earn an average of 48% over their base pay. Even more important is that their companies earn an average of 43.6% more pretax profit. Yet only 26% of U.S. labor has an incentive system.

Sales management uses incentives more than any other area of business. Incentives in sales are known as commissions, and salespeople usually receive a minimum salary as a base plus a percent of their sales. Sales commission incentives have been very successful, and no one would consider eliminating them, so why do managers reject the fact that incentives will work in operations? A Bureau of Labor statistics study in 1976 of 1,711 union con-

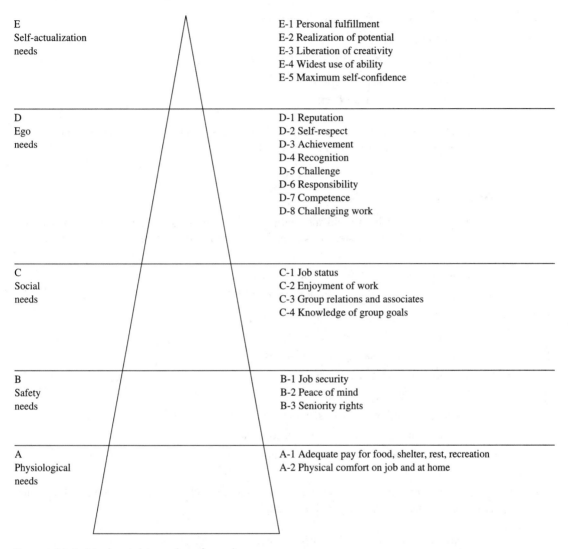

E
Self-actualization
needs

E-1 Personal fulfillment
E-2 Realization of potential
E-3 Liberation of creativity
E-4 Widest use of ability
E-5 Maximum self-confidence

D
Ego
needs

D-1 Reputation
D-2 Self-respect
D-3 Achievement
D-4 Recognition
D-5 Challenge
D-6 Responsibility
D-7 Competence
D-8 Challenging work

C
Social
needs

C-1 Job status
C-2 Enjoyment of work
C-3 Group relations and associates
C-4 Knowledge of group goals

B
Safety
needs

B-1 Job security
B-2 Peace of mind
B-3 Seniority rights

A
Physiological
needs

A-1 Adequate pay for food, shelter, rest, recreation
A-2 Physical comfort on job and at home

FIGURE 14-1 Maslow's hierarchy of needs.

tracts (7.6 million workers) showed that 1,504 contracts allowed time standards (5.5 million workers) and 476 contracts allowed incentives.

Types of Incentive Systems

Incentives can take many forms:

1. Percent of sales (commission)
2. Piecework and differential piece rate (Taylor piecework)

3. Earned hour plans, standard hour plan

4. Time off with pay

5. Productivity sharing, Improshare, and Scanlon

6. Suggestion system

7. Profit sharing, bonus systems.

Each plan is discussed in detail later in this chapter, including the advantages and disadvantages. There is an incentive plan for every situation.

Goals and Philosophies

Incentive systems design should start with a good list of goals and philosophies so everyone will understand why the system was implemented. Goals of an incentive system are to

1. Reduce cost (must be first and foremost),

2. Increase productivity—produce more with less,

3. Increase employee earnings,

4. Improve employee morale,

5. Improve labor/management relations,

6. Reduce delays and waiting time,

7. Improve customer service,

8. Develop motion consciousness and cost consciousness,

9. Reduce need and type of supervision, and

10. Increase plant and machine utilization.

Philosophies spell out the way we want the system to work and are used to direct future discussions and problem resolution. For example:

1. No one will be laid off because of the incentive system.

2. Extra pay for extra effort—120% is normal.

3. No credit is earned for producing rejects or scrap.

4. Increased wages can be earned only for work done on standard. No average pay for nonstandard work.

5. No time standard will be changed unless the job changes, and time standards will be kept current.

6. Complaints will be answered as soon as possible.

7. High earnings by employees are good because of savings in overhead and fringe benefits.

Potential Problems

There are some problems to be aware of when using any incentive system. These problems should be understood to prevent errors in the design of a new system:

1. Motion and time study technicians are needed to set and maintain time standards and to investigate problems. This cost can be from 1 to 3% of the cost of the production crew being covered.
2. Time standards that are not kept current will lead to increased cost.
3. Union grievances will increase. It will happen, so be prepared.
4. Management must put more effort, labor, and cost into
   a. Quality control.
   b. Production control.
5. If time standards are too tight, the employees will lose motivation. Average earnings are 120%, and each person in the plant must have the opportunity to earn a 20% bonus.

These problems must be recognized and communicated to top management so realistic expectations are set. These problems will not cost anywhere near the savings potential of an incentive system, but if top management is surprised by any of the increased problems, the technologist will have trouble.

THE TWO TYPES OF INCENTIVE SYSTEMS: INDIVIDUAL AND GROUP

Individual incentive systems have a longer industrial history than do group or team incentive systems. Group and team incentive systems agree with the new lean manufacturing philosophy that promotes teamwork and cooperation and pays for reducing costs. There is a place for both, and the industrial engineer needs to know when and where to use each. Management may have different ideas about which incentive system to use, and you must be able to communicate the advantages and disadvantages of each system.

Individual incentives are used when there is no connection between operators. When two or more people are tied together in work cells (the pull system of production control, assembly lines, multiple operator machines, or other team work where individual effort depends on others), a team or group incentive system should be used. To promote cooperation and teamwork, entire plants have been put on team incentive plans. The philosophy here is to give the employees a position of equal responsibility and reward with that of management.

Modern manufacturing is changing its organization from process layout (grouping all similar equipment together) to product layout (where a product may be completely fabricated and assembled in one area). This is changing the individual incentive needs to group incentive. The reduction in travel (move) costs and inventory carrying costs, and the simplification of the manufacturing process, greatly favors this new process (product-oriented) plant layout and operation philosophy. Teams need to be grouped carefully so that they have control of their destiny. A product or product line should be independent of other products or product lines to give the group some control. We will now discuss each type of incentive.

Individual Incentive Systems

Individual incentives are great motivators and are easier than team incentives to sell to the employee. Employees feel they have more control over their earnings when they are on individual incentive programs. The following list and discussion of individual incentive systems is limited to only the most popular and easy-to-use systems. The subject continues to grow, and the technologist must keep up with developments.

Commissions Commissions are used extensively in sales. The salesperson will get a percent of the sales dollar as his or her salary. Salespeople will have a minimum salary, called a *base*, but this salary (if indeed there is one) would not sustain the person at any desired level. If the salesperson wants a better lifestyle, he or she must go out and work for it. Most salespeople would not want it any other way, and most of us recognize that salespeople in any organization are the top earners. Most of management recognizes the pressure put on salespeople to produce (some of them envy the high earnings of salespeople), but no one would suggest removing the commissions because they earn more money than they cost.

A type of commission called *royalties* is used for other industries such as publishing, film making, and inventing. With royalties, individual effort can produce increased income. The employee receives a commission depending on the success of the mission. Royalties are effective because motivation to succeed is great. Socialist countries do not have the creative output of capitalist societies because there is no reward for effort. Examples of commissions or royalties include:

a. Accountants and lawyers earn a percent of the billable hours.

b. A barber earns a percent of his or her gross income in someone else's shop.

c. A sharecropper keeps a percent of the crops.

d. Any service person is awarded a share of the income he or she produces for the company to motivate him or her to work harder and earn more money.

Piecework A piecework incentive system is one in which an employee earns a specific dollar per unit produced. In farming, a crop picker earns so many dollars per pound, box, or bag. A fishing crew earns so many dollars per pound of fish. If these workers don't produce anything, they earn nothing. In manufacturing, employees are guaranteed an hourly rate, but they are also given a per unit price for each part they produce. If their incentive earnings are greater than their guarantee, they are given this bonus.

With *straight piecework,* after the standard pieces per hour is reached, the earnings are increased at the unit rate:

$$\text{Unit rate} = \frac{\text{(wage) \$12.00/hour}}{\text{(standard) 100 pieces/hour}} = \$.12 \text{ per unit}$$

For example, if in 8 hours an employee produced 1,050 units, the employee would earn $126.00 (.12 × 1,050) for the day. The employee would never earn less than $96.00 (8 × $12.00), the base rate, so a bonus of $30.00 was earned on this day.

Each part has a unit rate. This has been a problem in the past, before computers—when the base rate changes, every unit rate changes. Therefore, management developed the concept of cost of living adjustment (COLA). The incentive pay continued on the original base rates, and the COLA was added to the pay at the end of the pay period. After many years, the COLA grows and the base rate becomes such a small part of the hourly rate that it no longer motivates employees to work harder. The incentive (motivation) to work becomes less and less as the COLA grows. An example is a base rate of $5.50 per hour plus $6.00 per hour for the COLA. The incentive is based only on the $5.50 per hour rate, instead of $11.50, and is only approximately one-half of what it should be.

Differential Piece Rate This piece rate system uses two rates:

1. For production up to 100%, a smaller unit rate, for example, $.50 per unit

2. For production over 100%, a larger unit rate, such as $.75 per unit.

For example, one production worker produces 190 units and a second produces 240 units. The first standard is 24 per hour, so 100% is 192 per day. The first person would earn $95.00 at the bonus rate of $.50 each, so he or she would be given the guaranteed hourly rate of $12.00 per hour, or $96.00. The second person is paid $96.00 for the first 192 units and $.75 each for the 48 units ($36.00 bonus) over 100%, for a daily total of $132.00. This incentive plan tends to be more effective as a motivator than straight piecework, but it never gained the popularity of the straight piecework plan because of the complexity of calculating the pay rate.

Taylor Multiple Piecework Plan Taylor's plan was to discourage poor performers while attracting and keeping high performers. No minimum salary was provided, and only 50% of the unit rate was paid until 100% was achieved. Workers operating over 100% performance were paid at the rate of 125% of the per-unit rate. This large disparity created a strong incentive to produce over 100%. Weak and marginal employees soon left the company.

This plan would not be acceptable today, but it is still very interesting. For example, two employees perform the same job. One produces 1,000 units and the second produces 1,500 units. The standard is 100 units per hour. The incentive rates would be $.06 per unit for production under 100% and $.15 per unit for production over 100%. These rates were based on $12.00 per hour base rate divided by 100 units per hour equals $.12 per unit. The first worker would be paid $60.00 ($.06 per unit times 1,000 units) and the second worker would be paid $225.00 ($.15 times 1,500 units). The 50% more work earned the second operator almost four times more money. This is a real incentive to produce over 100%.

Earned Hour Plan or Standard Hour Plan These two plans are the same and both are still the most popular incentive plans used in manufacturing today. The earned hour plan is built on the performance control system discussed in Chapter 13. The earned hours (hours per unit times number of units produced) plus indirect hours (hours not on jobs covered by time standards) multiplied by the employee's current hourly pay rate equals the day's pay. For example, an employee produced 800 units on a job that had a time standard of 100 units per hour and spent the last two hours cleaning the shop. This person earned 8 hours on the production job (800 units divided by 100 units per hour equals 8 earned hours), plus 2 hours time for the indirect job, for a total of 10 hours pay. No matter what the pay rate, this person will earn 10 hours pay for this day's work.

These plans guarantee that the employee will never earn less than the base rate and, for every percentage point over 100%, the employee will earn a 1% bonus: 125% will earn a 25% bonus, even though the percentage number never enters the calculation. No incentive is earned on time spent off standard (clean-up, material handling, waiting, etc.), but no penalty is charged either. Employees soon learn that the only way to earn an incentive is to stay on jobs that have time standards.

EXAMPLE: An employee works 7 hours on a job with a time standard of 250 per hour. The operator produces 2,000 units and spends 1 hour on preventive maintenance. At the rate of $12.00 per hour, how much does this person earn?

$$\frac{2,000 \text{ produced}}{250 \text{ standard}} = 8 \text{ earned hours}$$

$$\text{Plus 1 hour maintenance} = 9 \text{ total hours pay due}$$

$$\text{Employee's present salary} = \$12.00/\text{hour}$$

$$\$108.00/\text{day}$$

$$\% \text{ performance} = \frac{8 \text{ hours earned}}{7 \text{ hours actual}} = 114\%$$

Look back at the time card problem in Chapter 13 (Table 13-6). Does this employee deserve a bonus? (The answer is in the back of the book under Chapter 13, #9.) The performance control system described in Chapter 13 is the technique most commonly used to pay incentive wages.

Time Off with Pay Additional production may not be needed, and instead of laying off part of the work force and paying the remaining employees a bonus, the company lets the employees go home once the goal has been achieved. This technique was used at an oil company that had produced 2 million gallons of oil packaged in quart cans using the wrong formula. The oil was recalled and sent to a newly rented warehouse, where it was to be removed from the cartons and cans. The oil was sent back to the refinery in tank cars, the cardboard was scrapped, and the cans were sold as scrap metal. A production line was designed, built, and manned. The production standard was set at 2,000 cases per shift. The average after a month was 1,200 cases per shift, and 1,500 cases was the best day. The technician was absolutely convinced that 2,000 was a good standard and sold management on the first incentive system ever used at this company. The technician worked with the crew on the first day of the program, and 2,000 units were produced in $7\frac{1}{2}$ hours. The crew went home early, and the supervisor and the technician cleaned up. The crew knew they could achieve the standard. From that day on, the crew finished early and cleaned up the plant. Before the end of the project, the employees were going home two hours early and receiving pay for 8 hours. The company got a 42% increase in production, and the employees got 2 hours of additional pay without putting in the time. Everyone can be a winner in a good system.

Group Incentive Systems

Productivity sharing plans are partnerships entered into by labor and management to share in any savings produced by working smarter and harder. Productivity sharing plans are group plans and promote teamwork. Everyone can participate in the incentive bonus—hourly, salaried, exempt, nonexempt, clerical, craftspeople, etc.—and they can share equally. There are many plans, but we will discuss only two.

Improshare by Mitchell Fein and Associates Improshare measures productivity as it has been over the past year using common output per laborhour figures. This is called the base period, and any improvement over the base period is shared 50/50. Fifty percent of the savings are split by the employees, and 50% of the savings stay in the company to reduce cost, reduce price, or increase profits.

An example of this technique is a fiberglass bathtub/shower stall plant that produced 100 units per day with 50 people. The only person not in the 50-person count was the plant manager. Everyone in this plant knew that over 100 units was a good day and under 100 units was a bad day. The measure of productivity was hours per unit:

$$\frac{50 \text{ employees} \times 8 \text{ hours/day}}{100 \text{ units}} = 4 \text{ hours/unit}$$

The evening before the start of the incentive program, the employees were told that anything they could do to reduce the hours per unit figure would be shared with them 50/50. They were told that the standard would not be changed unless an expense of over $10,000 was required to make productivity improvements, and then the standard would change only by 80% of the normal change. The 20% left in the standard would allow the employees to share the improvement, but the company needed the reduced cost to remain competitive.

The advantages of Improshare are as follows:

1. The standards are set to the present production output—not engineering standards.
2. Savings can start immediately.
3. Everyone participates; everyone is on the same team.
4. Everyone gets the same bonus, and it is based only on the number of hours worked.
5. Employees become equal partners with the company on productivity gains.
6. The system is based only on quality output.
7. The program is easy to set up and to maintain.
8. A production employee can make the calculations.

The calculations are easy, as shown in Table 14-2. Complete Friday, Saturday, and total. How much does every person take home per hour? The math can be explained as follows:

1. Units produced are counted as finished parts and are moved into the warehouse. Only good units are moved into the warehouse, and if a customer sends one back, it is subtracted until it is reworked and sent back to the warehouse as a first-quality unit. A big cost improvement was realized by the company when seconds (not quite first quality) were eliminated.

2. Standard hours per unit are always the same.

3. Earned hours equals units produced times the standard hours per unit (line 1 × line 2).

4. Clock hours are the actual time card hours. If 50 employees worked all 8 hours, then 400 hours would be paid. A time keeper would provide this figure, and it is one of the most reliably kept statistics in all of industry.

5. Bonus hours: If the clock hours are less than the earned hours, a bonus is earned. Negative numbers do not need to be extended because no bonus is earned and no pay will be subtracted (line 5 = line 3 − line 4).

6. Fifty percent of the bonus hours are the employees' bonus (50% of line 5 = bonus hours).

7. Average labor rate is calculated monthly by adding all the base hourly wage rates and dividing by the number of employees. It is simply the arithmetical average hourly wage rate. It is constant for at least one month, and it doesn't change much.

8. Employee bonus equals the bonus hours (line 5) times the average wage rate (line 7). This is the money to be divided among the employees. To make everyone equal partners in the incentive plan, hours worked is the only reason for a bonus to vary.

9. Bonus per hour equals the employee's bonus (line 8) divided by clock hours (line 3). The decimal to two places means cents (.35 = 35 cents).

The potential of Improshare is outstanding. Team development and idea sharing are key to high earnings. Management of a plant using the Improshare system changes from a

Table 14-2 Weekly Rockville Plant Incentive Plan.

	MON.	TUE.	WED.	THUR.	FRI.	SAT./SUN.	TOTAL
1. Units produced	100	110	100	110	115	—	—
2. × stand. hours/unit	4	4	4	4	4	4	4
3. = earned hours	400	440	400	440	—	None	—
4. − clock hours	408	400	384	384	376	16	—
5. = bonus hours	—	40	16	56	—	—	—
6. × 50%	—	20	8	28	—	—	—
7. × avg. labor rate	7.00	7.00	7.00	7.00	7.00	7.00	7.00
8. = employee bonus $	—	140	56	196	—	—	—
9. Divided by clock hours = $ bonus/hour	—	.35	.15	.51	—	—	—

pusher/controller to a problem solver/implementer of ideas. Improshare is a system where everyone can win. Improshare leads to some interesting questions:

1. Who pays for overtime for maintenance?
2. Who decides if we replace the person who quit?
3. What happens if management doesn't implement our suggestion?
4. What do we do with the person who is not pulling his or her weight?

Scanlon Plan An older group productivity sharing plan that involves the work force in the cost reduction efforts is the Scanlon plan, developed by Joe Scanlon in 1929. Three factors must be in place for a Scanlon plan to work:

1. Bonus payment
2. Identity with the company's problems
3. Employee involvement.

Like Improshare, the Scanlon plan starts with a measure of productivity. In Scanlon, it is the base ratio:

$$\text{Base ratio} = \frac{\text{payroll cost of all involved}}{\text{value of production produced}}$$

As an example, let's say a 110-person company has a weekly payroll of $44,000 and sales of $550,000 per week:

$$\text{Base ratio} = \frac{44,000}{550,000} = .08$$

The ratio has been 8% during the previous several months. If this week the payroll was $42,500 and we produced $600,000 worth of product, the labor budget would have been $48,000 (8% of $600,000). We used only $42,500, a savings of $5,500. This savings would be shared with the employees according to some plan like Improshare's 50/50; therefore, each of the employees would receive a portion of $2,750.00.[*] Any method of sharing must be fair to all, and the number of hours worked seems to be the best. If payroll said we used 6,071 hours last week, then the employees' share of the bonus pool would be divided by this number:

$$\frac{\text{Bonus dollars}}{\text{Hours worked}} = \frac{\$2,750.00}{6,071} = \$.45/\text{hour}$$

Each employee will receive $.45 for each hour worked last week.

---

[*]$2,750.00 is 50% of the total savings of $5,500.00.

Quality circles or employee/management meetings must be frequent. The meetings will generate ideas to improve production efficiency. The savings from these ideas are shared with all employees. Continued company training programs are needed to communicate the company's needs and problems.

The Scanlon plan can be expanded into the cost-of-goods-sold concept; and any material, supplies, or other overhead expenses would be fair game for cost reduction. This would make managers out of every employee.

For example, one very successful team incentive plan was implemented by a trailer manufacturing company in Kentucky. The company standardized one of their product lines to enable a cell-type/just-in-time plant layout. The plant was designed to produce eight flatbed over-the-road trailers, but it had a difficult time getting up to speed. The owner instituted a simple incentive plan: Pay a bonus of $.50 per hour for every trailer produced over five trailers per day. Six trailers per day earned the employees $4.00 a day extra, seven trailers per day earned them $8.00 per day, and eight trailers per day earned them $12.00 per day. There was still a problem with continually producing eight trailers per day, so they hired an industrial engineering consultant to develop a plan to achieve their goal. After time studying every cell, the industrial engineer found the plant was not balanced well. Some cells had over an hour's work, while other cells had about a half hour of work. After balancing the plant and giving everyone between 40 and 45 minutes of work, a new target and schedule for nine trailers per day was established.

The first day of the new schedule produced only seven trailers, but the second day produced nine trailers one half hour before the end of the shift. The next three days produced nine trailers as well, for a total of 43 trailers the first week of the new program. A bonus of $72.00 per week was paid to every employee, and a $144.00 bonus was paid to the three supervisors. It took about two months to do this consistently because of normal problems. Material supply, machine maintenance, and absences still created poor days, but these problems were addressed and greatly reduced overtime. The employee's average hourly rate increased over $1.65 per hour, and the company is doing well too. The company can split the fixed overhead between eight-and-a-half trailers instead of five trailers, a 70% reduction in cost.

Suggestion Systems

Suggestions are a part of the previous two incentive systems, but they can be used with any pay plan. Suggestion systems can be the entire incentive plan or one part of an individual incentive plan. A typical suggestion system will say, "The company will pay the employee whose suggestion has been accepted one-half of the first year's savings." This can be a considerable amount of money. There are many variations, but 50% of the first year's savings is normal. Large savings may be paid in three phases: first payment due upon acceptance by management, second payment made upon successful implementation, and final payment after six months, if the suggestion truly saved the amount claimed. This payment system can help the employee spread the tax liability.

Suggestion systems provide rewards for creative ideas. Without such rewards, there is much to benefit the company but little for the employee. If the company has no wage incentives, the employee gains nothing. On wage incentives, the employee may earn more by using the idea and decreasing workload. The fact that the company reduces cost and maybe saves jobs has little direct effect on the employee. The rewards for suggestions must, therefore, be important enough as well as large enough to have the individual employee feel rewarded. Some successful suggestion systems have been based on standard awards of $10, $25, and $100. This system is based more on the reward of recognition than the reward of money.

Suggestion systems are many, but not too many are considered successful. Success is measured by the number of suggestions submitted and implemented. One reason for the low number of suggestions is the reluctance of production employees, including supervisors, to put anything in writing. A company should encourage employees to bring their ideas to a person who has some ability to write.

If an idea is rejected, management must take great care to explain to the submitter why the idea will not work. Every effort should be made to implement employee suggestions. Any ties or close calls should be implemented; otherwise, employees will become discouraged.

Winners of cost reduction awards should be asked if they want to make their good fortune public. Many employees worry about the social pressure resulting from helping management. They may wish to tell only a few close friends.

Bonuses and Profit Sharing

Bonuses and profit sharing are paid on an annual or semiannual basis and are tied to company performance. Bonuses may be used when a department surpasses its goal but cannot affect profit. Manufacturing is a good example. Manufacturing may perform at superefficient rates, surpassing all previous performance, but business conditions called for reducing product price, resulting in lower profits. Should manufacturing be penalized for doing so well? Most think not, and the bonus system results. Most bonus systems are tied to specific objectives.

Profit sharing sets a part of the company's profits aside to be divided among eligible employees according to base salary or job grade. Distribution of profits can be made quarterly, annually, or even deferred until retirement. However, the primary disadvantage of bonus and profit sharing is the lag between effort and reward.

Bonuses and profit sharing increase employee morale, reduce turnover, reduce grievances, and promote the feeling of being part of the company, but neither system is a motivator of increased productivity. The employee sees his or her efforts too far removed from the result. Small companies may find more success with either system than larger companies.

Organized labor dislikes individual incentive systems that promote competition among employees and reward individual productivity. They prefer plans that are universally applied throughout the plant and apply equally to all workers, such as bonuses and profit sharing. (Personally, I dislike rewarding nonproductive people in any way.)

At a meeting of the Profit Sharing Council of America, Mr. Arthur Wood, then chairman of Sears Roebuck, stated the following:

> Profit sharing is not the first step in building a program of sound employee relations, but the last step. If a company has a good employee selection program . . . if a company provides training for employees so that they become productive on the job quickly, and have an opportunity to prepare themselves for additional responsibility; if a company pays its employees fairly; if a company has developed personnel policies that provide the individual with security against arbitrary acts of supervision; if a company has demonstrated a concern about the morale of its organization . . . if a company provides its employees with an opportunity to participate in benefit programs that offer protection against the hazards of life, . . . then that company should consider profit sharing.

Profit sharing can be rewarding for employees, and these rewards will bring returns of loyalty and higher morale. If profits dwindle, so will loyalty and morale.

Factors Other Than Labor

Employees can and do affect cost based on much more than their labor. Material utilization can affect cost greatly. Machine utilization can cost 100 times more than the operator's cost. Quality or reject rate can affect not only cost but customer satisfaction and repeat business. If an employee can be motivated properly, he or she can save money on each of the aforementioned items. The present cost of material can be calculated, and a savings program can be established to share the savings with the employees. Every dollar given to employees will return another dollar to the company if a 50/50 split is used.

Machines can have a $10,000 per hour operating cost. If we know the history of downtime, we can build a savings program and share with employees. The experience with an oil quart canning line showed that the machine had a capacity of 24,000 cans per hour, yet only 84,000 cans were packed on an average day ($3\frac{1}{2}$ hours). A performance improvement program was developed to pay the employees a 10% bonus for every hour of additional runtime squeezed out of the machine. Forty-five percent was the maximum bonus potential, and if employees accomplished even one hour of additional production, an overhead of $1,000 per hour was saved.

An incentive system can be designed for any cost. What are we waiting for?

QUESTIONS

1. What are the three ways of paying people?
2. What is the difference between:
   a. Salaried and hourly?
   b. Exempt and nonexempt?
   c. Day work and measured day work?
3. How is a job evaluation system used?
4. What are the six steps of a job evaluation study?
5. What will an incentive system do?
6. What are the two basic types of incentives?
7. What did the Mitchell Fein study of incentive systems show?
8. What percent of the U.S. labor force is on incentive pay systems?
9. Review the seven philosophies of an incentive system.
10. What problems can be anticipated when starting an incentive system?
11. Why are these problems important to the industrial technologist?
12. List four individual incentive systems and describe each.
13. List three group incentive systems and describe each.
14. When are bonuses and profit sharing useful?
15. What factors other than labor can be included in incentive systems?

CHAPTER 15

Time Management Techniques

Most successful adults have had to become good time managers. To do everything expected of us, we have had to prioritize and delegate and be very selective on how we spend our most valuable commodity, time. As students, we have experienced the pleasure of accomplishing superhuman amounts of work in limited amounts of time when faced with exams, student work, family obligations, and so on. What if we could produce at this high level 52 weeks per year? You may consider this funny now, but successful people do exactly that. The time management techniques outlined in this chapter will help you produce many times your normal results. Industrial engineers are efficiency experts, and the best way to live up to this title is to become efficient yourself. Two truisms are important:

1. If you want something done on time, give it to a busy person. Problem solvers who get things done are the kind of people every industrial manager wants.
2. There is always time to do what you truly want to do. This truism tells everyone your attitude about those things you didn't do, so be careful. What do we really say when we say that we don't have time?

The industrial engineer's time is valuable, and you should know what your time is worth. This will keep you from spending your time on work not worth what you cost the company. Knowing what your time is worth and doing only what is worthwhile is an attitude that will make you more valuable as well as give you more time. Time management techniques are ways of getting more out of life. It may be true that we all have only 24 hours in a day, but we have all seen people who can do more than everyone else. These time management techniques are tools to help you determine what's important and to keep you from wasting irreplaceable time. There are eight basic time management techniques:

1. Be creative.
2. Be selective.
3. Delegate.
4. Do it now.
5. Set time standards for yourself.
6. Eliminate the unnecessary.
7. Respect other people's time.
8. Do not rationalize.

BE CREATIVE

Creativity is finding new ways to do the work. The industrial engineering profession is great at finding better methods for others, but we should also think about our own work. Some ideas include the following:

1. Become a personal computer expert. Do all reporting on the computer to reduce the amount of routine reporting projects.
2. Use spreadsheet programs for standards, cost estimating, and cost reduction. Most math work becomes automatic and more accurate.
3. Use computer-aided design programs for layout and work station design.
4. Promote the use of standard data for time standards. It is 20 to 30 times faster.
5. Develop computer forms and formulas for costing to speed up the process and to improve the quality of your work.
6. Look into computer time study for establishing standard data. Again, the quality of your work will improve.
7. Investigate on-line performance reporting to automate the labor performance reporting system.
8. Investigate automated downtime monitoring to monitor equipment for problems.

Whatever the job, we can eliminate, combine, change the sequence, or simplify to save time and money. The more routine the job, the better the chances that we can automate.

BE SELECTIVE

Selectivity is important in every area of human endeavor. Every engineer and manager must make decisions on what is the most important at this moment. Several problems (job problems, family problems, and social problems) may occur at the same time. Which one should you work on first? There is no single absolute rule; each situation is different, each person is different, and the choice is yours alone to make. You can ask for advice, but the real choice is yours. Problems compound when choices are made without all the information needed to set priorities.

The industrial engineer must know what is important and give the most important tasks top priority. Knowing what is important requires an understanding of what a supervisor wants and the other work demands. The supervisor's requests take top priority. At times you may not agree with this, so talk to the management. They may not know what you are doing and that it is important, so discuss priorities. A good relationship with management starts with good communications. Tell your supervisor what you think will make life easier for both of you. I would suggest you tell your supervisor something like, "Okay, I'll do it right away, but do you know I've been asked to do—? Should I drop that for now and do what you are asking?"

The routine and special jobs that make up your backlog of work should be listed and prioritized. When additional projects are added to your backlog, be sure you understand the priority. Once all jobs have been listed, then do the most important job first.

Take notes at meetings and add requested or promised work to your backlog list. Do what you promise and do it on time.

Learn to say no. People will ask you to complete tasks that you should not do for various reasons: Maybe someone else could do it better, maybe it is social (coffee break) and you don't have time, maybe it is some activity they want you to join, or maybe it is some project. The more successful you become, the more demands will be made on your time. There will always be only 24 hours per day, so be selective and learn to say no diplomatically.

Selecting what is important and concentrating on that problem will make you a valued employee. Keeping a proper balance among work, family, and social demands will make you a successful person. Good time management techniques application will allow you to get more out of life.

DELEGATE

Delegation is assigning other people some of your work. As your workload grows, your employees can lend assistance. As your ability to solve problems increases, more problems and responsibility will be added. Management rewards results with more resources: a secretary, an assistant, a department, etc. With the addition of the first part-time help, you must decide what you will give away in the form of work.

Routine tasks are the first jobs to delegate. Writing daily, weekly, and monthly reports; collecting routine information; or providing routine services can be delegated.

Priorities are set for all projects. ABC priority designations could mean that the A priorities are the ones you do, and the B and C priorities you will delegate because they are not as important.

Screen calls and mail. Much time is wasted on meaningless phone calls and junk mail, so delegate answering the phone and opening the mail to someone else. Develop in this person an ability to prioritize.

Enjoy what you do. Do what you do best and recruit people to complement and augment your skills so you do not duplicate yourself. This is called team building.

Whenever you delegate anything, you give up the work but not the responsibility. Keep informed on what has been delegated through briefings.

DO IT NOW

W. Clement Stone, a Chicago self-made billionaire, is known for saying, "Do it now!" He said it at the top of his voice. An application of this statement is, "Don't put that project down until it is finished." An example is opening a report and reading it now, instead of putting it in a basket to read it later. Read it now, or throw it away, or file it permanently. Putting things off (procrastination) is a disease. Like the IRS recommends—do your taxes early and stop the worry; it's not so bad. Many unpleasant jobs cause us more headaches and heartaches when we put them off. Do it now, and stop worrying.

Many of us have put something off so long that it is embarrassing when we finally do it. Writing, returning something borrowed, returning a phone call, and turning in a report are familiar examples. Doing it now is a habit that will save time because we handle it, think about it, and worry about it only for a short time.

SET TIME STANDARDS FOR YOURSELF

Time standards are goals, and each new project should be estimated and the time recorded. These time standards are expert opinion standards and do not need to be as accurate as direct labor standards. Time standards will limit the amount of time given to any project. Sometimes we study a problem too long, and a time standard tells us to move on.

The backlog control system discussed in Chapter 13 can be very useful to the industrial engineer. The backlog hours tell management how much work needs to be accomplished. If this time frame is too long, management will provide help. The backlog control graph is the best way to communicate to management just how much work we are doing. The backlog control system will save much more time than it takes to complete the forms. Systematic approaches to management are always better than seat-of-the-pants approaches.

Time standard estimates are another way of ensuring communications between you and your supervisor. Your supervisor may have been thinking about a 15-minute project, and your estimate of three days indicates a misunderstanding. Cost and relationships can be saved by developing and communicating time standards.

The most important employee you will ever need to control is you. Set time standards for your work.

ELIMINATE THE UNNECESSARY

Knowing what is unnecessary is as important as knowing what is most important. We want to eliminate insignificant tasks and nonproductive activities. Insignificant tasks are jobs that can be done more economically by other employees at lower pay rates. Preoccupation, remembering errors, and planning too much are examples of nonproductive activities, and they waste your time.

Preoccupation can take many forms: personal problems, marital problems, financial problems, people problems at work, and past errors are all examples of preoccupation that even successful employees can lose themselves in. Preoccupation must be controlled for your own peace of mind. The best rule is, "Do what you can now and move on to the next project."

Remembering past unpleasant problems is also a waste of time and energy, and it harms your attitude in the present. Put mistakes behind you as soon as possible. Learn from the mistakes and talk about them with your supervisor, but do not waste one moment worrying about how you didn't do it right. No one is perfect, everyone has made mistakes, and we will make mistakes again. We must try not to make the same mistake twice. Talking with your supervisor is important. Tell him or her what you did and how bad you feel, and then ask if you can go on from here. Taking responsibility for your actions is a sign of maturity, and every supervisor appreciates this.

Thinking ahead too much can also be nonproductive. Planning is good and necessary, but planning for every consequence of every action will only slow you down. People that plan every detail make very few mistakes, but they also do very little work. The challenge for every engineer is to know how much planning is enough. At some point, planning must stop and implementation must start.

RESPECT OTHER PEOPLE'S TIME

If you were deep in concentration and someone stopped to talk about last night's baseball game, your concentration has been disrupted. Much time has been lost getting back to where you were before the interruption. Socialization is important to a well-balanced life, but keep it to social times (lunch, breaks, happy hour). Before and after regular hours are good times to be social with your supervisor.

When you visit other people's offices, check with them or their secretary for a meeting time. Set up some signals like open door, office hours, or schedules with secretaries for meeting times.

Don't expect your supervisor to do your work. Don't bounce everything off him or her. Get good instructions up front and turn in completed reports. If your supervisor doesn't like something, he or she will let you know. You can let your supervisor know what you are doing by providing well-planned project status reports. Keep your supervisor informed, but respect his or her time.

DO NOT RATIONALIZE

Rationalizing is coming up with a good excuse for doing something wrong. Excuses are nonproductive and a waste of everyone's time. Doing something unnecessary because you do it so well is rationalizing. Putting something off because someone did not do what they were supposed to is rationalizing. Not doing something your supervisor thought was important because you did not have time is rationalizing. There is no good excuse other than your inability.

Time management techniques will allow you to do more in the same amount of time. You can get $1\frac{1}{2}$ years of experience every year if you manage your time. You will go farther and faster than your fellow employees because you have more time.

QUESTIONS

1. What are the eight basic techniques of time management?
2. Give two personal examples of each of these basic techniques.
3. Is there *always* time to do what you truly want to do? Why?

CHAPTER 16

Attitudes and Goals for Industrial Engineers

An industrial engineer and technologist solves problems and reduces costs. The industrial analyst works with everyone in the organization and must possess human relations skills. No graduate of a production management course has all the information needed to deal successfully with the broad range of problems in industry today. To help new industrial engineers cope with these problems, and to keep them from reinventing the solutions to common problems, we discuss attitudes and goals in this chapter. Try some of these attitudes on prospective employers. You will be grateful to know that this is what they are looking for. To achieve anything, we need goals. And to achieve goals, we need the right attitudes. Control your attitudes, and you will achieve all your goals.

ATTITUDES OF AN INDUSTRIAL ENGINEER

I Can Reduce the Cost of Any Job

1. I can eliminate the unnecessary. I will eliminate muda (any cost that does not add value to our product).
2. I can combine jobs and eliminate movement, storage, delays, and some handling time.
3. I can change the sequence of work elements to create a more efficient job, operation, or plant.
4. If I can't take the first three steps, I can always simplify the job by moving items closer and downgrading work elements to produce more work in less time with less effort; there is always a better way to do any job.

Cost Is Number One—The Most Important Consideration

Cost consciousness is one of the most important attributes of a manager, and industrial engineering is the perfect training ground for production management. It is said that an engineer who doesn't know the economic consequences of decisions is worthless to industry. To ensure that this does not pertain to us, we develop the attitude that cost is the deciding factor. If there are any other factors to consider, upper management should make these decisions, not new engineers. An idea is developed that will reduce total cost. The deciding factors are a combination of savings (return) and expense (investment). The return on investment must be sufficient to motivate management to take a chance on your idea. A 100% return on investment will pay for itself in the first year, and 100% is the most common objective for methods work.

The savings (return) is from labor time standards created before and after motion and time studies. The investment (cost) is the total cost of implementing the methods improvement (machine, installation, tools, utility requirement changes, etc.). The best methods improvements are those that reduce cost with no investment.

Growth Gives Meaning and Purpose to Life

Industry is getting more complicated every day. An industrial engineer must keep up with the growth of his or her profession and industry. Become a member of a professional society, attend its annual conferences, and read its monthly publications. Never stop learning. When starting a new job, you will learn faster and more than ever before. You must learn the company's product, people, systems, and procedures. Learn as much as you can about your new company, and never stop learning.

Problems Are Opportunities—Without Them, We Would Not Have a Job

Problem solving is the job, and a proper attitude toward problems helps to sustain the industrial engineer and manager. Problem solving is also how the industrial engineer and manager grow professionally. The more problems a person can solve, the more top management depends on that person to solve more and bigger problems. Big problems are the biggest opportunities to show top management what you can do. There are people who hate and avoid problems, and if they can't avoid them, they are incapacitated by them. There are routine tasks for these employees, but not top engineering and management positions. Your attitude about problems will make the difference. Learn to love problems because they will give you the opportunities to do your best work. Eventually, you will find that no problem will be too big.

I Walk at 125%; I Work at 110%

Productivity improvement and a proper work ethic are two attitudes the industrial engineer must project at all times. "Do as I do and as I say" is an easy sell. Walking at 125% demonstrates what you expect of yourself. Your work and walking pace are habits that will not go unnoticed. Working 10% faster than your competition isn't asking much of yourself and will win any race. Walking pace is the most observable characteristic of a work ethic, so walking fast (125%) is a good way to project your sense of urgency and your attitude toward work ethic.

Anything Done the Same Way for More Than a Year Is Obsolete

Selling change is a tough job and must be done continually. Every time you have the chance to promote change, do so. Companies do not live in a vacuum. A company is in competition with other companies around the world, and if it allows itself to become satisfied with present methods, it will lose the competitive fight for its life. The company must become better and better every day or it starts to die. One of our jobs as industrial engineers and managers is to sell change as the only way to stay competitive and to retain jobs. Modern management includes all employees in the cost reduction—the cost improvement efforts of the company. This can only make the company better, and our job of selling change easier.

The Person Doing the Job Knows More About It Than I Do

The opposite view is not only old-fashioned, it is wrong. We cannot know as much about a job as someone working on it even for only a short time. The person assigned to the job has probably worked on it for a long time (8 hours per day, week after week) and has developed a wealth of information. The industrial engineer who includes the operator's cost reduction ideas will be many times more effective than the engineer working on his or her own. Talking to and asking the operator's opinion shows respect, and we all want to be respected. The lean manufacturing concept understands this and promotes training production employees in the areas of cost reduction attitudes and techniques so they can share their ideas.

The Best Ideas Are Developed in Groups

One person cannot know as much as two. If people talk to each other, there is no limit to what can be done. This is called synergism. The back-and-forth sharing of ideas leads to better solutions than any one of us could have produced. U.S. management has grown up with the Lone Ranger complex (working alone to do the job), but today's world is too complicated for the loner. Teamwork is the order of the day.

Education Is Part of My Job

Motion consciousness and cost consciousness are two areas of expertise we can share with others. If we teach people what we are doing, they can use these techniques and give us ideas for review. We can respond with feedback to make their ideas even better. There are many techniques in industrial engineering that we can share with employees. The more we share, the better for our company.

I Will Be Open and Honest with Everyone

Having hidden ulterior motives and playing games are dangerous to good relationships. You can disagree with people without being disagreeable. Be an honest person. Tell people you would do nothing in the world to hurt them, and mean it. Help them whenever you can.

All My Standards and Goals Are Attainable

Be confident in your standards. If you don't believe in them, no one else will. Later in this chapter, we discuss the only way to set time standards, which is fair both to the company and to the employee. You must guard against being closed-minded; you could have made a mistake. When challenged about a time standard, I review it with the person and ask if I might have left something out (which is the most common error). If I haven't left something out, I may time the job with the person's help. I start the watch, lay it down in front of him or her and have him or her turn it off when finished. We then review the results. Most of the time, the other person has finished in less than 75% of the time that I have allowed. I ask that the other person try for a while. If he or she continues to have trouble, I'll come back.

I Will Be Patient and Understanding Toward People

The motion and time study job has a history of antagonism from production employees. Understand that their negative opinion is not personal, and you will do better working with the people than your predecessor.

If They Don't Like Me, They Don't Know Me

We cannot get a group of people to know and like us; we can work on only one at a time. If you have the proper attitude toward people, and if you do your job professionally (honestly), people will like you—one at a time. People must know you as a friendly, happy person.

If People Don't Understand My Ideas, I'm Not Communicating

I take responsibility for communicating my ideas by giving and listening to feedback. Many techniques of motion and time study perform the same function and communicate the same information. By using two or three techniques, we may be able to communicate better with more people. If I take responsibility for the communications, I'll try harder. Eighty percent of engineering and management is selling ideas, and this is the most overlooked part of industrial engineering education. Successful people are those who can sell their ideas.

No One Wants to Know How Good You Are, But They Want to Hear How Good They Are

A conversation with either employees or supervisors should be directed toward them and not you. You will be considered wise and all-knowing when you listen to and applaud their accomplishments. This is especially true with supervisors.

Successful People Do What Other People Dislike to Do

1. They work hard. They know that there is no easy way. They collect all the data before making decisions, ask for help, and take advice.

2. They work long hours. You can compete successfully with people who work fewer hours than you do, and over the years of working longer you build up a large lead on the competition for a big new job. The boss has learned this and is usually around after normal hours. The boss is more accessible after hours than during the day, and the employee who is also around may pick up valuable contact time.

3. They get involved. Volunteer for more work, be a part of the solution, have opinions, and take stands. This commitment to getting involved in problem solving puts you on the firing line where shots can be taken at you.

4. They take criticism. Use criticism as feedback to make you better and stronger. Without criticism, we go on thinking that everything is okay. Do not be afraid of criticism, and do not be overly embarrassed by it. A good attitude toward criticism will make life easier, and you will grow. To avoid criticism, do nothing, say nothing, and be nothing.

5. They constructively criticize others. This is more difficult than being criticized, but if you think someone is saying or doing something wrong, it is your responsibility to say so. The proper attitude toward criticizing others is that if they persist, harm will come to them, and you want the best for them. Older, experienced people are able to criticize much too easily. One must never lose sight of the basic attitude that criticism is to help. Once a boss stops criticizing you, he or she has made a decision to remove you from the organization. Don't seek perfection; be reasonable and helpful.

6. They can be aggressive. There is no room for timidity in an aggressive organization. You must love the involvement and aggressively take an active part. Have opinions, share opinions, think outwardly, and have fun being creative. As Franklin D. Roosevelt said, the only thing we have to fear is fear itself. Fear is not knowing and not being certain. So, if we learn all we can, we become certain and fear disappears.

I Will Not Wait on Others to Deliver the Goods

You cannot just ask for or order something and then sit back and wait. You must follow up continually until you get what you need. The squeaky wheel does get the grease. You can never blame others for not performing; only children do this.

I Will Confirm My Instructions by Asking Questions

Be sure you understand what is being asked of you. Ask, "Is this what you want?" and put it in your own words. The results will be apparent.

I Will Complete Everything I Start

Don't leave tasks half done. Some type of job list check-off system is needed to ensure completion. There is always an answer, so don't give up too soon.

I Will Say Nothing as Fact Unless I Can Back It Up

Opinion must be stated as opinion.

I Will Never Forget Who I'm Working For

The boss may not always be right, but he or she is always the boss and deserves your respect. You will never win by fighting with the boss. Who would want you working for them if you have a reputation of fighting? Keep the boss informed and let him or her know that you know who is boss. Sell your boss to everyone you meet. What the boss wants takes priority over what you want to do. The boss sets priorities. Never go over or around the boss. No one will respect or admire your initiative. Getting the boss promoted is your best way of getting yourself promoted, and following the boss up the ladder can be very rewarding. Bosses sometimes go to other companies. If you've been a good supporter, they will ask you to join them.

Never Take Myself Too Seriously

A good sense of humor and a little self-deprecation will help your cause. Greet people with a smile, be friendly, and have some fun. A person is as happy as he or she wants to be, so learn to be happy.

I Will Look and Act Professional

Clean, neat business attire, polished shoes, and clean language are important to you and your career.

GOALS

To achieve great things, one must set great goals. To achieve your goals, you must first write them down and commit to them. If your goals are not written down, you have no goals. Each person must strive for something in his or her personal life and professional life, and personal goals must be in line with professional goals. Professional goals must be aligned with company goals.

Goals must be reviewed periodically to check progress. Measurable goals are the best because progress can be measured. The review may also indicate a revision of goals. Everything changes, and goals are no exception. Time standards are goals, just like a return on investment is a goal for an engineer and profitability is a goal for a company president.

Some Goals for the Motion and Time Study Department

1. To set fair and equitable time standards for the company and the employees. The number of grievances or complaints about time standards is a useful measurement of satisfaction, and a comparison of your plant's performance to industrial averages is a good measure. The image of your motion and time study department is at stake.

2. To have full-time standards coverage. How many hours of work were performed on standards divided by total hours worked by a group is a good measure of coverage. The performance control system gave us a summary of all hours and classification of indirect work, and this total was divided by total hours worked to produce a percentage of indirect labor.

3. To promote productivity.

$$\% \text{ performance} = \frac{\text{earned hours}}{\text{actual hours}}$$

From our performance control system discussed in Chapter 13, the percent performance can be calculated for every person, department, shift, and plant total. Another useful technique of productivity measurement is units per laborhour, dollars per laborhour, pounds or tons per laborhour, or board feet per laborhour. These are usually old measures accepted industry-wide, and if they improve at the same rate as our percent performance, the time standards credibility improves.

4. To develop motion economy and cost consciousness in all employees, both labor and management. Each plant should have a cost reduction program, and each motion and time study person should save five times their annual salary. Developing the best possible methods is part of this.

5. To develop time standards for indirect labor. Setup, material handling, quality control, rework, distribution, data processing, filing, etc. are all being studied and included in time standards systems. Expand the influences of motion and time study.

6. To develop standard data. Standard data is faster, more accurate, and easier to explain than any other method of setting time standards. The percent of standards set using standard data is a good method of measuring standard data development and makes the department count the number of time standards set per time period.

7. To teach employees and supervisors techniques of motion and time study. The number of people trained per year should be reported against a previously set goal.

8. To keep up with the developments in the field by attending conferences and subscribing to journals.

9. To prepare technologists for supervisory positions. What better place for training a supervisor?

10. To expand continually our knowledge of our company's technology and to become the expert in manufacturing methods, machines, techniques, and time standards.

11. To be ready to serve whenever asked and to be ready with the answer in our area of expertise.

QUESTIONS

1. Review the attitudes of an industrial technologist and discuss those attitudes until you are comfortable with them. (If you understand them, you will make them a part of you.)

2. List the goals of an industrial engineer.

3. Why must goals be written and measurable?

Appendix A:
Forms

1. Multiactivity chart—2 sides
2. Process chart—2 sides
3. PTSS form—2 sides
4. SIMO form—2 sides
5. Time study form—2 sides
6. Rater trainer form—2 sides
7. Long cycle time study form—2 sides
8. Assembly line balance form—2 sides
9. Vertical time study form—2 sides
10. Work cell load chart form—1 side

These forms are for your use and you have the author's permission to duplicate them as needed.

FRED MEYERS & ASSOCIATES

MULTI ACTIVITY CHART
☐ OPERATOR/MACHINE   ☐ GANG   ☐ MULTI MACHINE
☐ LEFT HAND/RIGHT HAND   ☐ OPERATIONS

OPERATION NO.	PART NO.	OPERATION DESCRIPTION:
DATE:	TIME:	
BY I.E.:		

ACTIVITY	TIME IN MINUTES	ACTIVITY
	.05	
	.10	
	.15	
	.20	
	.25	
	.30	
	.35	
	.40	
	.45	
	.50	
	.55	
	.60	
	.65	
	.70	
	.75	
	.80	
	.85	
	.90	
	.95	
	1.00	
	1.05	
	1.10	

TOTAL UTILIZATION _____
% UTILIZATION

TOTAL UTILIZATION _____
% UTILIZATION

TIME STUDY CYCLE			COST:		
			HOURS PER UNIT _____	TOTAL NORMAL TIME IN MINUTES PER UNIT	
			DOLLARS PER HOUR _____	+ ___ % ALLOWANCE	
TOTAL			DOLLARS PER UNIT _____	STANDARD TIME	
OCC				HOURS PER UNIT	. _ _ _ _ _
AVG. OCC			LAYOUT & MOTION PATTERN ON NEXT PAGE	PIECES PER HOUR	
LEV FACT					
NORM. TIME					

WORK STATION DESIGN

LAYOUT   SCALE = _____

PRODUCT OR PART SKETCH   SCALE = _____

FRED MEYERS & ASSOCIATES PROCESS CHART

☐ PRESENT METHOD ☐ PROPOSED METHOD DATE:_____ PAGE___OF___.

PART DESCRIPTION:

OPERATION DESCRIPTION:

SUMMARY	PRESENT NO.	PRESENT TIME	PROPOSED NO.	PROPOSED TIME	DIFF. NO.	DIFF. TIME	ANALYSIS:		FLOW
○ OPERATIONS							WHY	WHEN	DIAGRAM
⇨ TRANSPORT.							WHAT	WHO	ATTACHED
☐ INSPECTIONS							WHERE	HOW	(IMPORTANT)
D DELAYS									
▽ STORAGES							STUDIED BY:		
DIST. TRAVELED		FT.		FT.		FT.			

STEP	DETAILS OF PROCESS	METHOD	OPERATION	TRANSPORT	INSPECTION	DELAY	STORAGE	DISTANCE IN FEET	QUANTITY	TIME HRS/UNIT .00000	COST PER UNIT	TIME/COST CALCULATIONS
1			○	⇨	☐	D	▽					
2			○	⇨	☐	D	▽					
3			○	⇨	☐	D	▽					
4			○	⇨	☐	D	▽					
5			○	⇨	☐	D	▽					
6			○	⇨	☐	D	▽					
7			○	⇨	☐	D	▽					
8			○	⇨	☐	D	▽					
9			○	⇨	☐	D	▽					
10			○	⇨	☐	D	▽					
11			○	⇨	☐	D	▽					
12			○	⇨	☐	D	▽					
13			○	⇨	☐	D	▽					
14			○	⇨	☐	D	▽					
15			○	⇨	☐	D	▽					
16			○	⇨	☐	D	▽					
17			○	⇨	☐	D	▽					

STEP	DETAILS OF (PRESENT/PROPOSED) METHOD	METHOD	OPERATION	TRANSPORT	INSPECTION	DELAY	STORAGE	DISTANCE IN FEET	QUANTITY	TIME (000.00)	COST PER UNIT	COST PER UNIT	TIME/COST CALCULATIONS
18			○	⇨	□	D	▽						
19			○	⇨	□	D	▽						
20			○	⇨	□	D	▽						
21			○	⇨	□	D	▽						
22			○	⇨	□	D	▽						
23			○	⇨	□	D	▽						
24			○	⇨	□	D	▽						
25			○	⇨	□	D	▽						
26			○	⇨	□	D	▽						
27			○	⇨	□	D	▽						
28			○	⇨	□	D	▽						
29			○	⇨	□	D	▽						
30			○	⇨	□	D	▽						
31			○	⇨	□	D	▽						
32			○	⇨	□	D	▽						
33			○	⇨	□	D	▽						
34			○	⇨	□	D	▽						
35			○	⇨	□	D	▽						
36			○	⇨	□	D	▽						
37			○	⇨	□	D	▽						
38			○	⇨	□	D	▽						
39			○	⇨	□	D	▽						
40			○	⇨	□	D	▽						
41			○	⇨	□	D	▽						
42			○	⇨	□	D	▽						

FRED MEYERS & ASSOCIATES PREDETERMINED TIME STANDARDS ANALYSIS

OPERATION NO.	PART NO.	OPERATION DESCRIPTION:
DATE:	TIME:	
BY I.E.:		

DESCRIPTION-LEFT HAND	FREQ.	LH	TIME	RH	FREQ.	DESCRIPTION-RIGHT HAND	ELEMENT TIME

TIME STUDY CYCLE			COST:		
			HOURS PER UNIT _____	TOTAL NORMAL TIME IN MINUTES PER UNIT	
				+ ___ % ALLOWANCE	
TOTAL			DOLLARS PER HOUR _____	STANDARD TIME	
OCC					
AVG. OCC			DOLLARS PER UNIT _____	HOURS PER UNIT	. _ _ _ _ _
LEV FACT				PIECES PER HOUR	
NORM. TIME					

338

LAYOUT   SCALE =

MOTION PATTERN

SEARCHING FOR A BETTER METHOD   ELIMINATE-COMBINE-CHANGE SEQUENCE-SIMPLIFY

FRED MEYERS & ASSOCIATES

SIMO ANALYSIS

OPERATION NO.				PART NO.							OPERATION DESCRIPTION:	
DATE:				TIME:								
BY I.E.												

DESCRIPTION-LEFT HAND	5	4	3	2	1	TIME	1	2	3	4	5	DESCRIPTION-RIGHT HAND	ELEMENT TIME

TIME	STUDY	CYCLE		COST:		TOTAL NORMAL TIME IN MINUTES PER UNIT	
				HOURS PER UNIT _____			
				DOLLARS PER HOUR _____		+ _____ % ALLOWANCE	
TOTAL				DOLLARS PER UNIT _____		STANDARD TIME	
OCC						HOURS PER UNIT	
AVG. OCC							
LEV. FACT.						PIECES PER HOUR	
NORM. TIME							

LAYOUT
SCALE =

MOTION PATTERN

SEARCHING FOR A BETTER METHOD   ELIMINATE-COMBINE-CHANGE SEQUENCE-SIMPLIFY

FRED MEYERS & ASSOCIATES — TIME STUDY WORKSHEET

OPERATION DESCRIPTION

PART NUMBER	OPERATION NO.	DRAWING NO.	MACHINE NAME	MACHINE NUMBER

OPERATOR NAME	MONTHS ON JOB	DEPARTMENT	TOOL NUMBER	FEEDS & SPEEDS.

PART DESCRIPTION:

MATERIAL SPECIFICATIONS:

MACHINE CYCLE TIME

☐ SNAP BACK
☐ CONTINUOUS

☐ QUALITY OK ?
☐ SAFETY CHECKED ?
☐ SETUP PROPER ?

NOTES:

ELEMENT #	ELEMENT DESCRIPTION		READINGS										TOTAL CYCLES TIME	AVERAGE TIME	% R	NORMAL TIME	FREQUENCY	UNIT NORMAL TIME	RANGE	R/X	HIGHEST
			1	2	3	4	5	6	7	8	9	10									
		R																			
		E																			
		R																			
		E																			
		R																			
		E																			
		R																			
		E																			
		R																			
		E																			
		R																			
		E																			
		R																			
		E																			

FOREIGN ELEMENTS:

NOTES:

R/X	# CYCLES
.1	2
.2	7
.3	15
.4	27
.5	42
.6	61
.7	83
.8	108
.9	138
1.0	169

TOTAL NORMAL MIN.
ALLOWANCE + ___ = ___
STANDARD MINUTES
HOURS PER UNIT
UNITS PER HOUR

ON BACK
WORK STATION LAYOUT
PRODUCT SKETCH

ENGINEER: _____ DATE: _ / _ / _

APPROVED BY: _____ DATE: _ / _ / _

WORK STATION DESIGN

LAYOUT   SCALE = _____

PRODUCT OR PART SKETCH   SCALE = _____

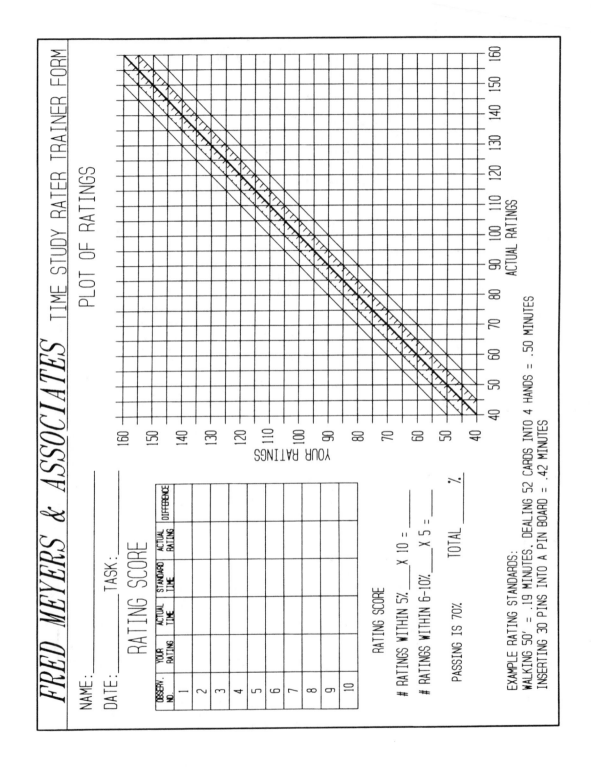

TEN FUNDAMENTALS OF PACE RATING

1. Healthy people in the right frame of mind easily turn in 100% performance on correctly standardized jobs. For incentive pay, good performers usually work at paces from 115% to 135%, depending on jobs and individuals.

2. For most individuals *it is uncomfortable to work at a tempo much below 100% and extremely tiring to operate for sustained periods at paces lower than 75%;* our reflexes are naturally geared to move faster.

3. Poor efficiency on a correctly standardized job usually results from stopping work frequently—"goofing off" for a variety of reasons. Specifically, *substandard production seldom results from inability to work at a normal pace.*

4. Some standards of 100%:

    a. Walking 3 mph or 264 ft/min.

    b. Dealing cards into four stacks in .5 min.

    c. Filling the pinboard in .435 min.

5. Very seldom can true performance of over 140% be found in industry.

6. When an operator comes up consistently with extremely high efficiencies, it is usually a sign that the method has been changed or the standard was wrong in the first place.

7. An operator's work pace during a time study does not affect the final standard. His or her actual time is multiplied by the performance rule to give a job standard that is fair for all employees.

8. Inasmuch as healthy employees can easily vary work pace from approximately 80% to around 130%—through a range of 50%—reasonable inaccuracies in the setting of standards should be accepted.

9. Ineffective foremen usually fight job standards. Good supervisors, however, sincerely help in the standard-setting effort, realizing that such information is their best planning and control tool.

10. Methods usually influence production more than work pace. Don't ever get so absorbed in how quickly or slowly an operator "seems" to be moving that you fail to consider whether or not he or she is using the right method.

Courtesy of Tampa Manufacturing Institute.

FRED MEYERS & ASSOCIATES LONG CYCLE TIME STUDY WORK SHEET

PART NO. _____	OPERATION DESCRIPTION:
OPERATION NO. _____	
DATE/TIME _____	MACHINE; TOOLS, JIGS:
BY I.E. _____	MATERIAL:

ELEMENT #	ELEMENT DESCRIPTION	ENDING WATCH READING	ELEMENT TIME	% R	NORM. TIME

FRED MEYERS & ASSOCIATES MACHINERY & EQUIPMENT LAYOUT DATA SHEET

DESCRIPTION OF MACHINERY & EQUIPMENT: DATE:

DESCRIPTION OF OPERATION:

COMPANY NAME: LOCATION:

DESIGNED BY:

PHOTO

LAYOUT DRAWING

SCALE =

Reference Notations/Changes

MACHINE SPECIFICATIONS ON BACK.

FRED MEYERS & ASSOCIATES

ASSEMBLY LINE BALANCING

PRODUCT NO.: _____
DATE: _____
BY I.E.: _____

PRODUCT DESCRIPTION: _____

NUMBER UNITS REQUIRED PER SHIFT _____

"R" VALUE
CALCULATIONS

EXISTING PRODUCT = $\dfrac{365 \text{ MINUTES}}{\text{UNITS REQ'D/SHIFT}}$ = "R"

NEW PRODUCT = $\dfrac{300 \text{ MINUTES}}{\text{UNITS REQ'D/SHIFT}}$ = "R"

NO.	OPERATION/DESCRIPTION	"R" VALUE	CYCLE TIME	# STATIONS	AVG. CYCLE TIME	% LOAD	HRS./1000 LINE BALANCE	PCS./HR. LINE BALANCE

ASSEMBLY LINE LAYOUT

SCALE 1/4" = 1'

SEARCHING FOR A BETTER METHOD   ELIMINATE-COMBINE-CHANGE SEQUENCE-SIMPLIFY

	Vertical Time Study Form						
Element \ Cycles							
1							
2							
3							
4							
5							
Total time \ Number of cycles							
Average time							
Rating %							
Normal time							
Allowances							
Standard time							
Notes							

Motion and time study number

Sheet _____ of _____

Part number _____ Operator number _____

Part name _____

Date _____ Department _____

Length of experience: this operation _____

Description of operation _____

Machine number _____

Machine name _____

Machine Element	Surface feet/ minute (S.F.M.)	R.P.M.	Feed
1.			
2.			

Group _____ Operator _____

Supersedes study number _____

Condition of tools _____

Observer: _____ Approved by: _____

Sketch

Element Number	Content of Operations Is as Follows (Element Description):	Time Allowed
	TOTAL TIME	
	Reason for Change:	

Work Cell Load Chart

Part number _____ **Part name** _____ **Date** _____ **Engineer** _____

Operation Number	Operation Description	Time in Minutes		Time in ____ Minutes
		Manual	Machine	Walk

352

Appendix B
Answers to Problems

CHAPTER 1

#1 page 1

#2 page 1

#3 page 2—cost and quality

#4 page 2—we can reduce the cost of any job! . . . etc.

#5 page 3

#6 page 3—eliminate, combine, change sequence and simplify; eliminate is the most important

#7 page 4—60%, 85%, 120%

#8 page 4 bottom—work hard, work long, criticize, be criticized, and become involved

CHAPTER 2

1 page 8

2 page 8

3 page 22

CHAPTER 3

CHAPTER 4

CHAPTER 5

#1 page 67, paragraph 1

#2 pages 69, 73, 76, 83

#3 page 67, paragraph 2

#4 page 68—ask 6 questions of each step in the process to seek elimination, combination, reroute, or simplification

#5 page 71

#6 page 71

#7 page 78 top

#8 on your own

#9 page 83

CHAPTER 6

#1 page 93

#2 The PTSS form is better designed

#3 $.1833 − $.1295 × 1,000,000 = $53,800

#4 pages 91, 93 activities

#5 200 pcs/hour/machine and .001 hrs/part

#6 $.10 each

#7 on your own

CHAPTER 7

#1 page 110

#2 page 109, paragraph 1—cheapest cost, anything else must be justified

#3 page 109, paragraph 1—anywhere

#4 pages a. 112, b. 112, c. 114, d. 114, e. 115, f. 117–127

#5 page 127

#6 page 127—blueprint of method and bill of material for standard.

#7
#8 } on your own project
#9

CHAPTER 8

#1 page 130, list of 9

#2 page 137, bottom right of table

#3 pages 137–142

#4 pages 144–146

#5 page 142—design change or fixture

#6 page 142, middle—eliminate

#7 pages 147–148

#8 pages 148–149, middle—3 places on minutes, 5 places on hours

#9 page 151

#10 pages 150–153

#11 on your own project

ANSWER TO EXAMPLE ON PAGE 139

R16	11	R12
M17	12	M15
R15m	9	R15m
m M21	12	m M21
M5—$\frac{50}{2}$	11	M5—$\frac{50}{2}$
M50	27	M50 (48″ is longest reach or move)
R36	21	M24

#12 PTSS form

Motion Description	Symbol	Time .001	Symbol	Motion Description	Element
1. Element #1—pack parts 1, 2, 7, and 8					time
Reach to parts 1 and 8	R36	21	R36	Same as left hand	
Grasp 1	G1	3		Same as left hand	
		3	G1	Same as left hand	
To parts 2 and 7	M9	8	M9	Same as left hand	
Grasp 2	G2	6		Same as left hand	
		4	R2	Same as left hand	
		6	G2	Same as left hand	
Move to fixture	M36	21		Same as left hand	
In fixture	AP2	10		Same as left hand	
		5	AP2	Same as left hand	
Element total		87			0.087
2. Element #2—pack parts 3 through 6					
To parts 3 and 5	R30	18	R30	Same as left hand	
Grasp 3	G3	9		Same as left hand	
		4	R2	Same as left hand	
		9	G3	Same as left hand	
To parts 4 and 5	R6	6	R6	Same as left hand	
Grasp part 4	G3	9		Same as left hand	
		9	G3	Same as left hand	
Move to fixture	M30	18	M30	Same as left hand	
	AP2	10		Same as left hand	
		5	AP2	Same as left hand	
Element total		97			0.097
				Total time	0.184
				+10% allowance	0.018
				Standard time	0.202
				Pieces per hour	297
				Hours per unit	0.00337

CHAPTER 9

#1 your classmate's project may be a good source of projects to time.

#2 page 159, paragraph 1—oldest

#3 page 160, paragraph 1

#4 pages 161–168

#5 page 169

#6 pages 172–173 list

#7 page 173

#8 page 174—Step 2- ⑬

#9 page 175—list

#10 page 175 bottom

#11 page 178- ㉑

#12 page 178 step 6, your project

#13 pages 185–187

#14 pages 195–199

#15 page 184, step 9

#16 page 185, list

#17 page 186, 4 effort

#18 page 190—136%, 109%, 87%, 97%

Standard is always .435 minutes

#19 page 191

#20 page a. 195 bottom, b. 196–198, c. 196–198, 5%, 5%, 0%

#21 pages 196–197

#22 pages 199–201

#23 pages 201–203

#24 pages 201–203

#25 Team projects are a good source of your first time studies.

FRED MEYERS & ASSOCIATES — TIME STUDY WORKSHEET

☐ SNAP BACK
☒ CONTINUOUS

OPERATION DESCRIPTION: ASSEMBLE PARTS 2 & 4, MACHINE SCREW & STAKE. INSPECT

PART NUMBER	OPERATION NO.		MACHINE NAME	MACHINE NUMBER	
4650-0950	1515		PRESS	21	
OPERATOR NAME	MONTHS ON JOB	DRAWING NO.	TOOL NUMBER	FEEDS & SPEEDS. NONE	
MEYERS	5	4650-0950	M61		
		DEPARTMENT		MACHINE CYCLE .030	
		ASSEMBLY		TIME 8:30 AM.	

QUALITY OK ? ☒
SAFETY CHECKED ? ☒
SETUP PROPER ? ☒
NOTES:

PART DESCRIPTION:
GOLF CLUB SOLE ASSEMBLY – WOOD & STEEL

MATERIAL SPECIFICATIONS:

ELEMENT #	ELEMENT DESCRIPTION		READINGS 1	2	3	4	5	6	7	8	9	10	TOTAL CYCLES	AVERAGE TIME	% R	NORMAL TIME	FREQUENCY	UNIT NORMAL TIME	RANGE	R/X	HIGHEST
1	ASSEMBLY	R	9	41	71	1.07	38	77	2.08	48	77	3.07	.76	.084	90	.076	1	.076	.03		✓
		E	09	09	09	(15)	08	08	.10	07	08	08	9								
2	DRIVE SCREW	R	15	46	79	13	43	82	14	53	82	93	.51	.057	100	.057	1	.057	.03	.53	✓
		E	06	05	08	06	05	05	06	05	05	(86)	9								
3	PRESS	R	28	59	94	27	66	95	28	66	96	4.06	1.22	.136	110	.150	1	.150	.02		
		E	13	13	15	14	(23)	13	14	13	14	13	9								
4	INSPECT	R	32	62	92	30	69	98	41	69	99	4.09	.25	.031	100	.031	1	.031	.01		
		E	04	03	-.02	03	03	03	(13)	03	03	03	8								
5	LOAD SCREWS	R										3.83	1	.76	125	.950	10	.095	—		
		E				*1			*2			.76									
		R																				
		E							*3													
		R																				
		E																				
		R																				
		E																				

FOREIGN ELEMENTS:
*1. 23 PART JAMMED.
*2. 13 TRIED TO REWORK PART.
*3. 10 RESTART FROM LOADING SCREWS.

NOTES:
LOAD SCREWS COULD BE IMPROVED
TO ELIMINATE .095 MIN.

.409
-.095 (SAVE)
.314
+.031
.345

.0750
-.00575
.00275 Hrs/Unit
X $10.00/Hr.
.0575 $/Unit
500,000/Yr.
$28,750

.00575 Hrs
174 Pieces/Hrs

R/X	.1	.2	.3	.4	.5	.6	.7	.8	.9	1.0
CYCLES	2	7	15	27	42	61	83	108	138	168
					48					

TOTAL NORMAL MIN.	.409
ALLOWANCE + ____ 10 %	.041
STANDARD MINUTES	.450
HOURS PER UNIT	0 0 7 5 0
UNITS PER HOUR	133

ON BACK
WORK STATION LAYOUT
PRODUCT SKETCH

ENGINEER: FRED MEYERS   DATE: 10/10/XX

APPROVED BY: FRED MEYERS   DATE: 10/10/XX

#26 Answer to problem in Figure 9–17 on page 185.

Element / Cycles	Load Machine	Run	Unload	Inspect	Material Handling		
		Answer to Figure 9-26 **Vertical time study form and problem**					
1	0.15	0.55	0.80				
	0.15	0.40	0.25				
2	0.94	1.34	1.61				
	0.14	0.40	0.27				
3	1.77	2.17	3.52		3.27		
	0.16	0.40	0.25		1.10		
4	3.67	4.07	5.24	5.00			
	0.15	0.40	0.24	0.93			
5	5.39	5.79	6.50*				
	0.15	0.40	eliminate 0.71				
Total time	.75		1.01	.93	1.10		
Number	5		4	10	1000		
Average time	.150	.40	.253	.093	.001		
Rating %	120	100	80	130	60		
Normal time	.180	.400	.202	.121	.001		
Allowances (add 10%)	.018	.040	.020	.012	.0001		
Standard time	.198	.400	.222	.133	.001		
					Total standard time = .994		

CHAPTER 10

#1 page 212

#2 pages 213–214

#3 page 214 (1–5)

#4 pages 217–222, $a = .00094$, $b = .02$ omit data point 104", 1.80 min

#5 page: 217: $y = a + bx$; $.00094 + .02 (100) = 2.001$ min

#6 pages 226–227

(1) $\dfrac{250 \times 12}{3.14 \times .5} = 1{,}911$ RPM   (2) $\dfrac{3}{.001} = 3{,}000$ rev.   (3) $\dfrac{3{,}000 \text{ rev.}}{1{,}911 \text{ RPM}} = 1.57$ min

#7 Drill—add .4 times the diameter (.5) to the length of cut (3.0″). Mill—add the diameter of the cutter to the length of cut and multiply the number of teeth on the cutter times the feed rate, .001.

#8 Page 223—39 bags per hour

#9 pages 230–231

#10 page 231

#11 page 231

#12 page 232

#13 page 239

#14 page 239

#15 page 239 ⑭

#16 page 239 ⑬ ⑭

#17 page 240

#18 page 240 ⑲

#19 Any improvement with a total hours per 1000 of less than 43.051 is a better layout

#20 page 244 top, try combining operation #(5 and 10), (15 and 20), (25, 30, and 35)

#21 49.36, 4.54, .45 feet per minute

#22 page 243 is an improvement of page 242

#23 page 243

CHAPTER 11

#1 page 254

#2 page 254, elemental ratio studies, performance sampling studies, and time standard development studies. b. pages 254, 255; 266; 268

#3 page 254–255

#4 a, b, c, d, e, f, g—255–261

#5 pages 259–260 list

#6 page 266

#7

JOB	NUMBER OF OBSERVATIONS	%	HOURS	NO. UNITS	HRS./UNIT NORMAL	STANDARD HRS./UNIT
1	5,000	10	425	2,900	.14655	.16121
2	10,000	20	850	8,800	.09659	.10625
3	20,000	40	1,700	25,000	.06800	.0748
Idle	15,000	30	1,275			
Total	50,000	100	4,250			
		70% + 7% = 77% efficiency				

CHAPTER 12

#1 page 275

#2 pages 275–276 ratios

#3 page 276 top list

#4 pages 278–286

#5 You are on your own. Use your knowledge of each technique.

CHAPTER 13

#1 page 287 top

#2 page 287, list

#3 page 288 middle of page

#4 85% of 85% = 72.25%

#5 page 290

#6 page 290

#7 a. 290, b. 290, c. 291

#8 pages 291–293

#9

	ACTUAL HRS.	EARNED HRS.	%
	2	2.40	120
	1.75	1.67	95
	2.75	2.67	97
	<u>1.5</u>	<u>2.00</u>	<u>133</u>
TOTALS =	8	8.74	109

#10 pages 293–300

CHAPTER 14

#1 page 302

#2 pages a. 302–303, b. 303, c. 303

#3 page 304

#4 page 305

#5 page 306

#6 individual incentive plans, group plans: page 309

#7 page 313

#8 26% page 306

#9 pages 308–309

#10 page 309

#11 page 309

#12 pages 310–313

#13 pages 313–316

#14 pages 317–318

#15 page 318

CHAPTER 15

#1 page 320, list

#2 on your own

#3 on your own

CHAPTER 16

#1 on your own

#2 pages 330–331

#3 page 330

Index